Contemporary Calculus IV

For the
students

A free, color PDF version is available online at
http://scidiv.bellevuecollege.edu/dh/Calculus_all/Calculus_all.html

Dale Hoffman
Bellevue College
Author web page: http://scidiv.bellevuecollege.edu/dh/

8/8/2013

CONTEMPORARY CALCULUS IV: Contents

11.0 MOVING BEYOND TWO DIMENSIONS

So far our study of calculus has taken place almost exclusively in the xy–plane, a 2–dimensional space. The functions we worked with typically had the form $y = f(x)$ so the graphs of these functions could be drawn in the xy–plane. And we have considered limits, derivatives, integrals, and their applications in two dimensions. However, we live in a three (or more) dimensional space, and some ideas and applications require that we move beyond two dimensions.

This chapter marks the start of our move into three dimensions and the mathematics of higher dimensions. The next several chapters extend the ideas and techniques of limits, derivatives, rates of change, maximums and minimums, and integrals beyond two dimensions. The work you have already done in two dimensions is an absolutely vital foundation for these extensions. As we work beyond two dimensions you should be alert for the the parts of the ideas and techniques that extend very easily (many of them) and those that require more extensive changes.

Section 11.1 introduces vectors and some of the vocabulary, techniques and applications of vectors in the plane. This section still takes place in two dimensions, but the ideas are important for our move into higher dimensional spaces.

Section 11.2 introduces the three–dimensional rectangular coordinate system, visualization in three dimensions, and measuring distances between points in three dimensions.

Section 11.3 extends the basic vector ideas, techniques and applications to three dimensions.

Sections 11.4 and 11.5 introduce two important types of multiplication, the dot product and the cross product, for vectors in 3--dimensional space and examines what they measure and some of their applications.

Section 11.6 considers the simplest objects, lines and planes, in 3–dimensional space.

Section 11.7 introduces surfaces described by second–degree equations and catalogs the possible shapes they can have.

Chapter 11 is the first step in our move beyond two dimensions. It contains the fundamental geometry of points and vectors in three dimensions. The concepts and techniques of this chapter are important and useful by themselves, and they are a necessary foundation for the study of calculus in three and more dimensions in Chapter 12 and beyond.

11.1 VECTORS IN THE PLANE

Measurements of some quantities such as mass, speed, temperature, and height can be given by a single number, but a single number is not enough to describe measurements of some quantities in the plane such as displacement or velocity. Displacement or velocity not only tell us how much or how fast something has moved but also tell the direction of that movement. For quantities that have both length (magnitude) and direction, we use vectors.

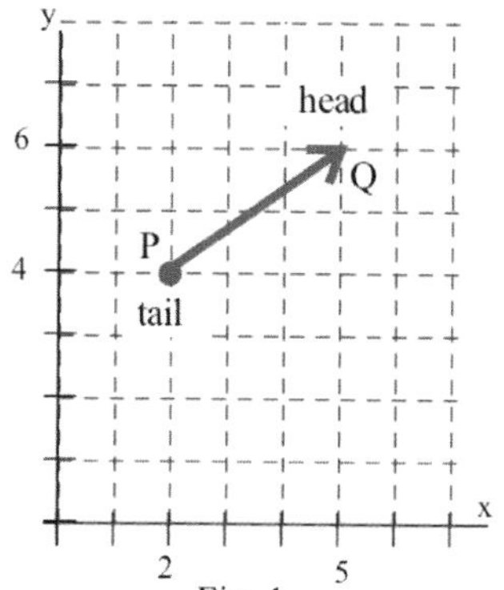

Fig. 1

A **vector** is a quantity that has both a magnitude and a direction, and vectors are represented geometrically as directed line segments (arrows). The vector **V** given by the directed line segment from the starting point P = (2,4) to the point Q = (5,6) is shown in Fig. 1. The starting point is called the **tail** of the vector and the ending point is called the **head** of the vector. Geometrically, two vectors are equal if they have the same length and point in the same direction. Fig. 2 shows a number of vectors that are equal to vector **V** in Fig. 1. The equality of vectors in the plane depends only on their lengths and directions. Equality of vectors does not depend on their locations in the plane.

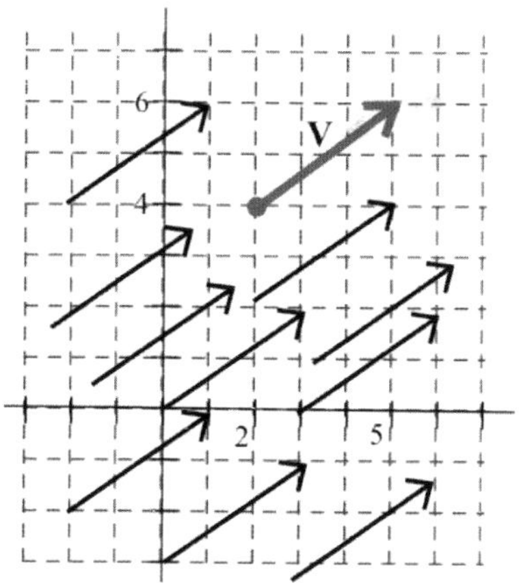

All of these directed line segments represent the vector V
Fig. 2

A vector in the plane can be represented algebraically as an ordered pair of numbers measuring the horizontal and vertical displacement of the endpoint of the vector from the beginning point of the vector. The numbers in the ordered pair are called the **components** of the vector. Vector **V** in Fig. 1 can be represented as $\mathbf{V} = \langle 5-2, 6-4 \rangle = \langle 3, 2 \rangle$, an ordered pair of numbers enclosed by "bent" brackets. All of the vectors in Fig. 2 are also represented algebraically by $\langle 3, 2 \rangle$.

Notation: In our work with vectors, it is important to recognize when we are describing a number or a point or a vector. To help keep those distinctions clear, we use different notations for numbers, points and vectors:

a number: regular lower case letter: $a, b, x, y, x_1, y_1, \ldots$
A number is called a scalar quantity or simply a **scalar**.

a point: regular upper case letter: A, B, ...
ordered pair of numbers enclosed by ( ): $(2, 3), (a, b), (x_1, y_1), \ldots$

a vector: **bold** upper case letter: **A**, **B**, **U**, **V**, ...
ordered pair of numbers enclosed by $\langle\ \rangle$: $\langle 2, 3 \rangle, \langle a, b \rangle, \langle x_1, y_1 \rangle, \ldots$
a letter with an arrow over it: $\vec{A}, \vec{B}, \vec{U}, \vec{V}, \ldots$

Definition: Equality of Vectors

Geometrically, two vectors are equal if their lengths are equal and their directions are the same.

Algebraically, two vectors are equal if their respective components are equal:

if $\mathbf{U} = \langle a, b \rangle$ and $\mathbf{V} = \langle x, y \rangle$, then $\mathbf{U} = \mathbf{V}$ if and only if $a = x$ and $b = y$.

Example 1: Vectors **U** and **V** are given in Fig. 3.

(a) Represent **U** and **V** using the $\langle\ \rangle$ notation.

(b) Sketch **U** and **V** as line segments starting at the point (0,0).

(c) Sketch **U** and **V** as line segments starting at the point (–1,5).

(d) If (x,y) is the starting point, what is the ending point of the line segment representing the vector **U** ?

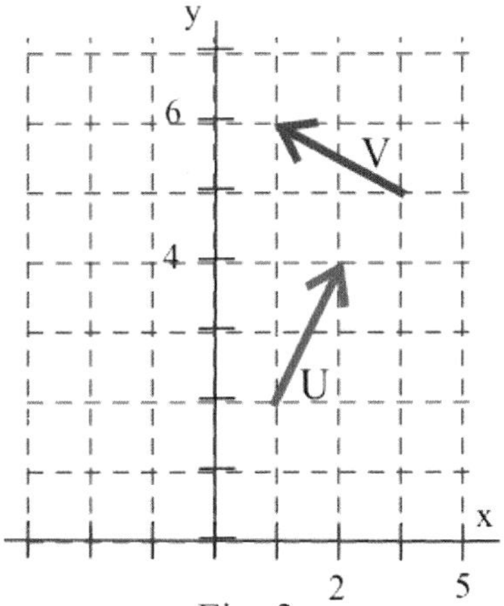

Fig. 3

Solution:

(a) The components of a vector are the displacements from the starting to the ending points so $\mathbf{U} = \langle 2{-}1, 4{-}2 \rangle = \langle 1, 2 \rangle$ and $\mathbf{V} = \langle 1{-}3, 6{-}5 \rangle = \langle -2, 1 \rangle$.

(b) If **U** starts at (0,0), then the ending point of the line segment is (0+1, 0+2) = (1,2).
The ending point of **V** is (0–2, 0+1) = (–2, 1). See Fig. 4.

(c) The ending point of the line segment for **U** is (–1+1, 5+2) = (0,7). For **V**, the ending point is (–1–2, 5+1) = (–3, 6). See Fig. 4.

(d) If (x,y) is the starting point for $\mathbf{U} = \langle 1, 2 \rangle$ then the ending point is (x+1, y+2).

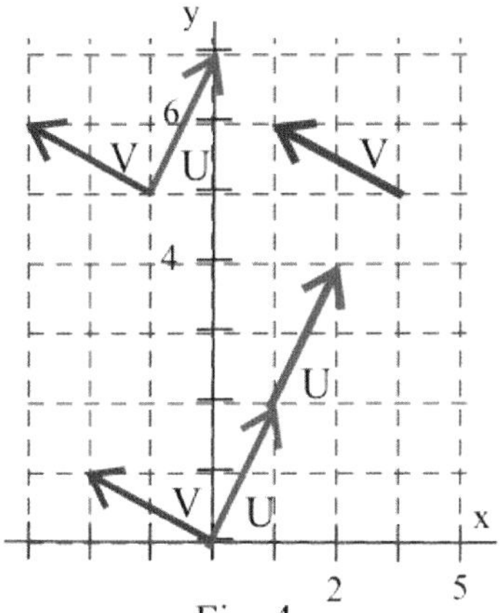

Fig. 4

Practice 1: $\mathbf{A} = \langle 3, 4 \rangle$ and $\mathbf{W} = \langle -2, 3 \rangle$.

(a) Represent **A** and **W** as line segments beginning at the point (0, 0).

(b) Represent **A** and **W** as line segments beginning at the point (2, –4).

(c) If **A** and **W** begin at the point (p, q), at which points do they end?

(d) How long is a line segment representing vector **A**? **W**?

(e) What is the slope of a line segment representing vector **A**? **W**?

(f) Find a vector whose line segment representation is perpendicular to **A**. To **W**.

The magnitude of a vector **V**, written $|\mathbf{V}|$, is the length of the line segment representing the vector. That length is the distance between the starting point and the ending point of the line segment. The magnitude can be calculated by using the distance formula.

> The **magnitude** or **length** of a vector $\mathbf{V} = \langle a, b \rangle$ is $|\mathbf{V}| = \sqrt{a^2 + b^2}$.

The only vector in the plane with magnitude 0 is $\langle 0, 0 \rangle$, called the **zero vector** and written **0** or $\vec{0}$. The zero vector is a line segment of length 0 (a point), and it has no specific direction or slope.

Adding Vectors

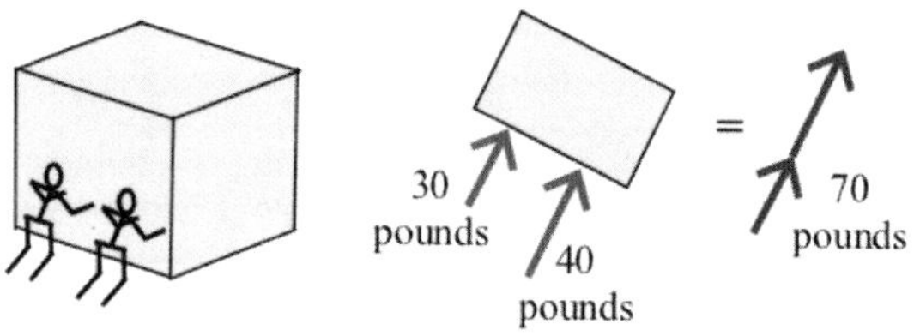

Fig. 5

If two people are pushing a box in the same direction along a line, one with a force of 30 pounds and the other with a force of 40 pounds (Fig. 5), then the result of their efforts is equivalent to a single force of 70 pounds along the same line. However, if the people are pushing in different directions (Fig. 6), the problem of finding the result of their combined effort is slightly more difficult. Vector addition provides a simple solution.

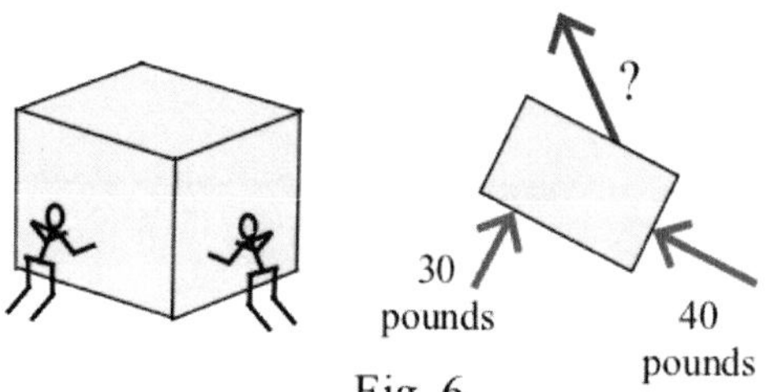

Fig. 6

> **Definition:** Vector Addition
>
> If $\mathbf{A} = \langle a_1, a_2 \rangle$ and $\mathbf{B} = \langle b_1, b_2 \rangle$, then $\mathbf{A} + \mathbf{B} = \langle a_1 + b_1, a_2 + b_2 \rangle$.

The result of applying two forces, represented by the vectors **A** and **B**, is equivalent to the single force represented by the vector **A + B**. In Fig. 6, the effort of person A can be represented by the vector $\mathbf{A} = \langle 30, 0 \rangle$ and the effort of person B by the vector $\mathbf{B} = \langle 0, 40 \rangle$. Their combined effort is equivalent to a single force vector $\mathbf{C} = \mathbf{A} + \mathbf{B} = \langle 30, 40 \rangle$. (Since $|\mathbf{C}| = 50$ pounds, if the two people cooperated and pushed in the direction of **C**, they could achieve the same result by each exerting 10 pounds less force.)

Example 2: Let $\mathbf{A} = \langle 3, 5 \rangle$ and $\mathbf{B} = \langle -1, 4 \rangle$. (a) Graph **A** and **B** each starting at the origin.

(b) Calculate $\mathbf{C} = \mathbf{A} + \mathbf{B}$ and graph it, starting at the origin.

(c) Calculate the magnitudes of **A**, **B**, and **C**.

(d) Find a vector **V** so $\mathbf{A} + \mathbf{V} = \langle 4, 2 \rangle$.

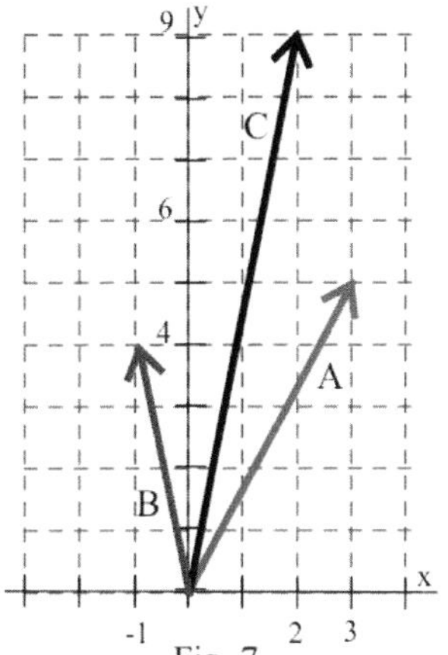

Fig. 7

Solution: (a) The graphs of **A** and **B** are shown in Fig. 7.

(b) $\mathbf{C} = \langle 3, 5 \rangle + \langle -1, 4 \rangle = \langle 2, 9 \rangle$. The graph of **C** is shown in Fig. 7.

(c) $|\mathbf{A}| = \sqrt{3^2 + 5^2} = \sqrt{34} \approx 5.8$, $|\mathbf{B}| = \sqrt{17} \approx 4.1$, and $|\mathbf{C}| = \sqrt{85} \approx 9.2$.

(d) Let $\mathbf{V} = \langle x, y \rangle$. Then $\mathbf{A} + \mathbf{V} = \langle 3+x, 5+y \rangle = \langle 4, 2 \rangle$ so $3+x = 4$ and $x = 1$. Also, $5+y = 2$ so $y = -3$ and $\mathbf{V} = \langle 1, -3 \rangle$

Practice 2: Let $\mathbf{A} = \langle -2, 5 \rangle$ and $\mathbf{B} = \langle 7, -4 \rangle$. (a) Graph **A** and **B** each starting at the origin.

(b) Calculate $\mathbf{C} = \mathbf{A} + \mathbf{B}$ and graph it, starting at the origin.

(c) Find and graph a vector **V** so $\mathbf{V} + \mathbf{C} = 0$.

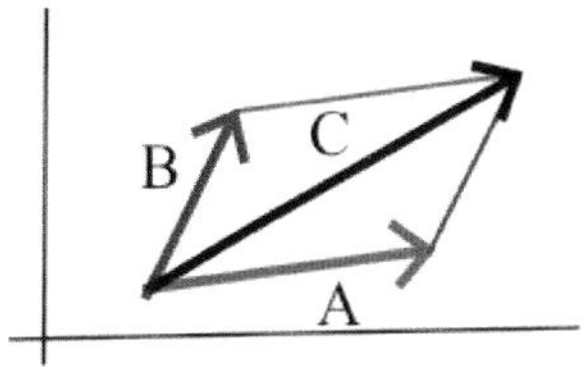

Fig. 8

The parallelogram method and the head–to–tail method are two commonly used methods for adding vectors graphically.

The parallelogram method (Fig. 8):

i) arrange the vectors **A** and **B** to have a common starting point

ii) use the two given vectors to complete a parallelogram

iii) draw a vector **C** from the common starting point of the two original vectors to the opposite corner of the parallelogram: $\mathbf{C} = \mathbf{A} + \mathbf{B}$

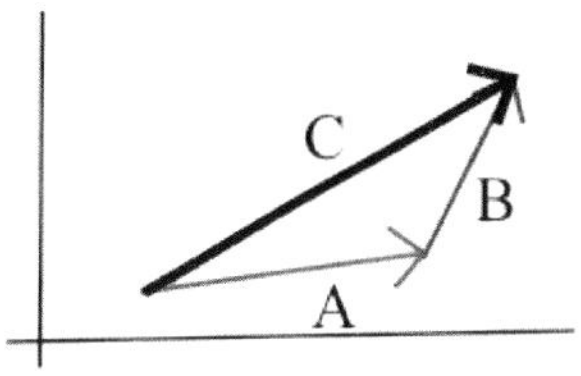

Fig. 9

The head–to–tail method (Fig. 9):

i) position the tail of vector **B** at the head of the vector **A**

ii) draw a vector **C** from the tail of **A** to the head of **B**: $\mathbf{C} = \mathbf{A} + \mathbf{B}$

The head–to–tail method is particularly useful when we need to add several vectors together. We can simply string them along head–to–tail, head–to–tail, ... and finally draw their sum as a directed line segment from the tail of the first vector to the head of the last vector (Fig. 10).

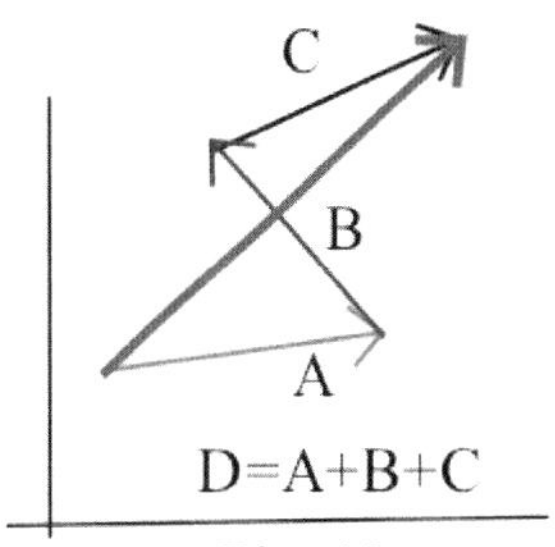

Fig. 10

Practice 3: Draw the vectors $\mathbf{U} = \mathbf{A} + \mathbf{C}$ and $\mathbf{V} = \mathbf{A} + \mathbf{B} + \mathbf{C}$ for $\mathbf{A}, \mathbf{B}$, and $\mathbf{C}$ given in Fig. 11.

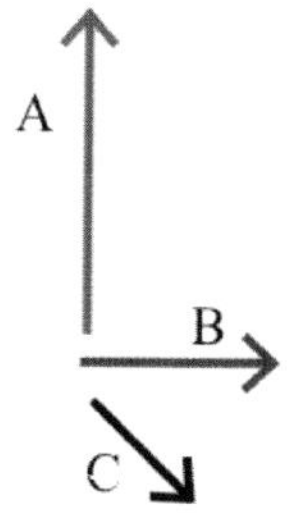

Fig. 11

Scalar Multiplication

We can multiply a vector by a number (a scalar) by multiplying each component by that number.

> **Definition**: Scalar Multiplication
>
> If $\quad$ k is a scalar and $\mathbf{A} = \langle a_1, a_2 \rangle$ is a vector,
> then $\quad k\mathbf{A} = \langle ka_1, ka_2 \rangle$, a vector.

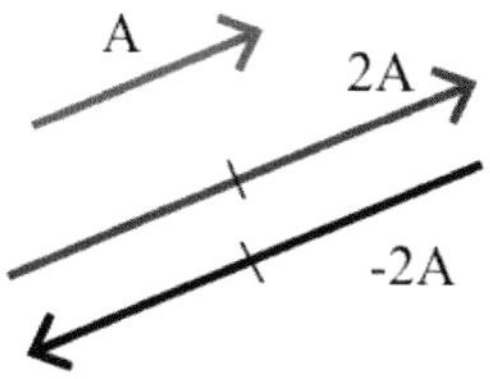

Fig. 12

Multiplying by a scalar k gives a vector that is $|k|$ times as long as the original vector. If k is positive, then $\mathbf{A}$ and $k\mathbf{A}$ have the same direction. If k is negative, then $\mathbf{A}$ and $k\mathbf{A}$ point in opposite directions (Fig. 12).

Example 3: Vectors $\mathbf{A}$ and $\mathbf{B}$ are shown in Fig. 13.

Graph and label $\mathbf{C} = 2\mathbf{A}, \mathbf{D} = \frac{1}{2}\mathbf{B}, \mathbf{E} = -2\mathbf{B}, \mathbf{F} = \mathbf{B} + 2\mathbf{A}$, and $\mathbf{G} = \mathbf{A} + (-1)\mathbf{A}$.

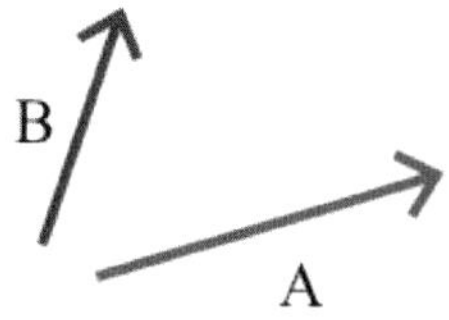

Fig. 13

Solution: The vectors are shown in Fig. 14.

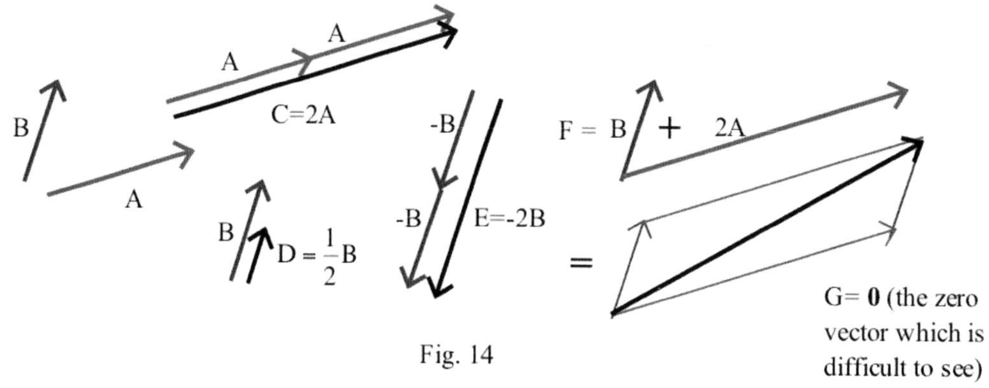

Fig. 14

Practice 4: Vectors $\mathbf{U}$ and $\mathbf{V}$ are shown in Fig. 15.

Graph and label $\mathbf{A} = 2\mathbf{U}, \mathbf{B} = -\frac{1}{2}\mathbf{U}, \mathbf{C} = (-1)\mathbf{V}$, and $\mathbf{D} = \mathbf{U} + (-1)\mathbf{V}$.

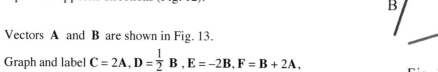

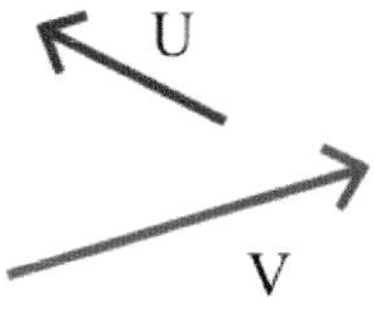

Fig. 15

Subtracting Vectors

We can also subtract vectors algebraically and graphically.

Definition: Vector Subtraction

If $\mathbf{A} = \langle a_1, a_2 \rangle$ and $\mathbf{B} = \langle b_1, b_2 \rangle$, then $\mathbf{A} - \mathbf{B} = \mathbf{A} + (-1)\mathbf{B} = \langle a_1 - b_1, a_2 - b_2 \rangle$, a vector.

Graphically, we can construct the line segment representing the vector $\mathbf{A} - \mathbf{B}$ either using the head–to–tail method (Fig. 16) to add $\mathbf{A}$ and $-\mathbf{B}$, or by moving $\mathbf{A}$ and $\mathbf{B}$ so they have a common starting point (Fig. 17) and then drawing the line segment from the head of $\mathbf{B}$ to the head of $\mathbf{A}$.

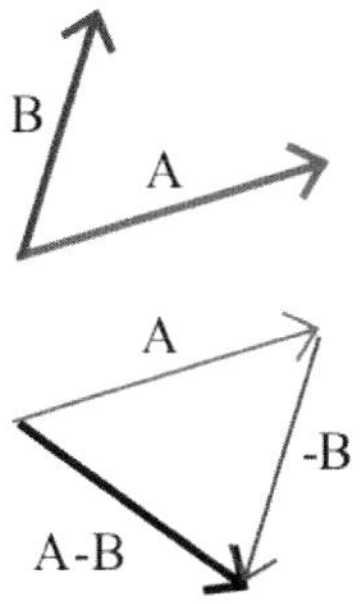

Fig. 16

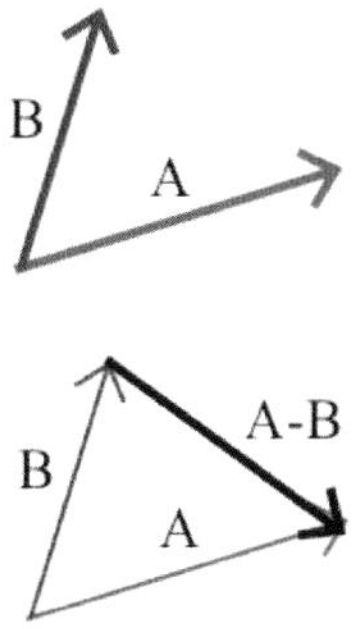

Fig. 17

Practice 5: Vectors $\mathbf{A}, \mathbf{B}$ and $\mathbf{C}$ are shown in Fig. 18. Graph $\mathbf{U} = \mathbf{A} - \mathbf{B}$ and $\mathbf{V} = \mathbf{C} - \mathbf{A}$.

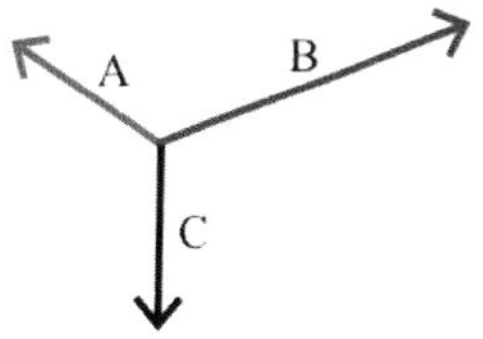

Fig. 18

Algebraic Properties of Vector Operations

Some of the properties of vectors are given below.

If $\mathbf{A}, \mathbf{B}$, and $\mathbf{C}$ are vectors in the plane and x and y are scalars, then

1. $\mathbf{A} + \mathbf{B} = \mathbf{B} + \mathbf{A}$
2. $(\mathbf{A} + \mathbf{B}) + \mathbf{C} = \mathbf{A} + (\mathbf{B} + \mathbf{C})$
3. $\mathbf{A} + \mathbf{0} = \mathbf{A}$ ($\mathbf{0} = \langle 0,0 \rangle$ is called the additive identity vector)
4. $\mathbf{A} + (-1)\mathbf{A} = \mathbf{0}$ ($-\mathbf{A} = (-1)\mathbf{A}$ is called the additive inverse vector of A)
5. $x(\mathbf{A} + \mathbf{B}) = x\mathbf{A} + x\mathbf{B}$
6. $(x + y)\mathbf{A} = x\mathbf{A} + y\mathbf{A}$

These and additional properties of vectors are easily verified using the definitions of vector equality, vector addition and scalar multiplication.

Unit Vector, Direction of a Vector, Standard Basis Vectors

Definitions:

A **unit vector** is a vector whose length is 1.

The **direction** of a nonzero vector **A** is the unit vector $\frac{1}{|\mathbf{A}|}\mathbf{A} = \frac{\mathbf{A}}{|\mathbf{A}|}$.

The **standard basis vectors** in the plane are $\mathbf{i} = \langle 1, 0 \rangle$ and $\mathbf{j} = \langle 0, 1 \rangle$. (Fig. 19)

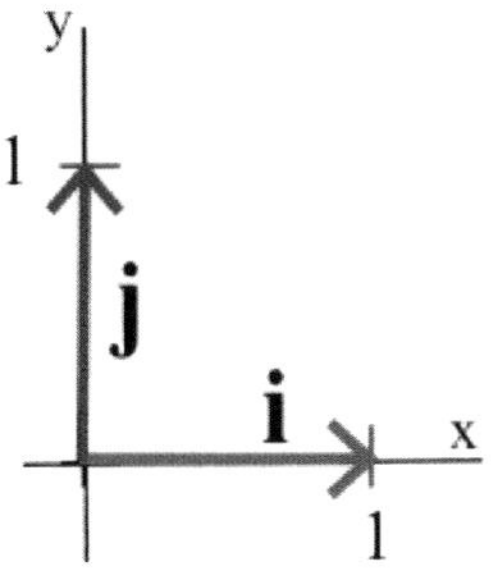

Fig. 19

Every nonzero vector **A** is the product of its magnitude and its direction: $\mathbf{A} = |\mathbf{A}| \frac{\mathbf{A}}{|\mathbf{A}|}$.

The standard basis vectors are unit vectors.

Every vector in the plane can be written as a sum of scalar multiples of these two basis vectors:
if $\mathbf{A} = \langle a_1, a_2 \rangle$, then $\mathbf{A} = \langle a_1, 0 \rangle + \langle 0, a_2 \rangle = a_1\mathbf{i} + a_2\mathbf{j}$.

Example 4: Let $\mathbf{A} = 5\mathbf{i} + 12\mathbf{j}$ and $\mathbf{B} = 3\mathbf{i} - 4\mathbf{j}$.

(a) Determine the directions of **A**, **B**, and $\mathbf{C} = \mathbf{A} - 3\mathbf{j}$.

(b) Write $3\mathbf{A} + 2\mathbf{B}$ and $4\mathbf{A} - 5\mathbf{B}$ in terms of the standard basis vectors.

Solution: (a) $|\mathbf{A}| = \sqrt{25 + 144} = \sqrt{169} = 13$ so the direction of **A** is

$\frac{1}{13}\langle 5, 12 \rangle = \langle \frac{5}{13}, \frac{12}{13} \rangle = \frac{5}{13}\mathbf{i} + \frac{12}{13}\mathbf{j}$. The direction of B is $\frac{3}{5}\mathbf{i} - \frac{4}{5}\mathbf{j}$.

$|\mathbf{C}| = |\,5\mathbf{i} + 9\mathbf{j}\,| = \sqrt{106}$ so the direction of **C** is $\frac{5}{\sqrt{106}}\mathbf{i} + \frac{9}{\sqrt{106}}\mathbf{j}$.

(b) $3\mathbf{A} + 2\mathbf{B} = 3(5\mathbf{i} + 12\mathbf{j}) + 2(3\mathbf{i} - 4\mathbf{j}) = 15\mathbf{i} + 36\mathbf{j} + 6\mathbf{i} - 8\mathbf{j} = 21\mathbf{i} + 28\mathbf{j}$.

$4\mathbf{A} - 5\mathbf{B} = 4(5\mathbf{i} + 12\mathbf{j}) - 5(3\mathbf{i} - 4\mathbf{j}) = 20\mathbf{i} + 48\mathbf{j} - 15\mathbf{i} + 20\mathbf{j} = 5\mathbf{i} + 68\mathbf{j}$.

Practice 6: Let $\mathbf{U} = 7\mathbf{i} + 24\mathbf{j}$ and $\mathbf{V} = 15\mathbf{i} - 8\mathbf{j}$.
Determine the directions of **U**, **V**, and $\mathbf{W} = \mathbf{U} + 3\mathbf{i}$.

Often in applications a vector is described in terms of a magnitude at an angle to some line. In those situations we typically need to use trigonometry to determine the components of the vector.

Example 5: Suppose you are pulling on the rope with a force of 50 pounds at an angle of 25° to the horizontal ground (Fig. 20). What are the components of this force vector parallel and perpendicular to the ground?

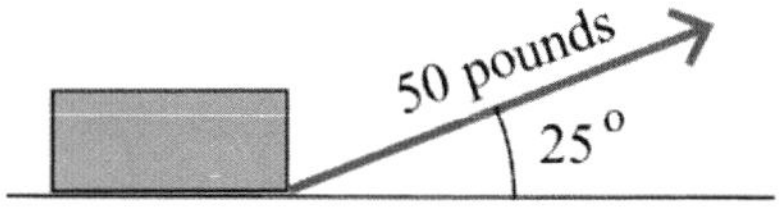

Fig. 20

Solution: The horizontal component (parallel to the ground) is 50 cos(25°) ≈ 45.3 pounds.

The vertical component (perpendicular to the ground) is 50 sin(25°) ≈ 21.1 pounds.

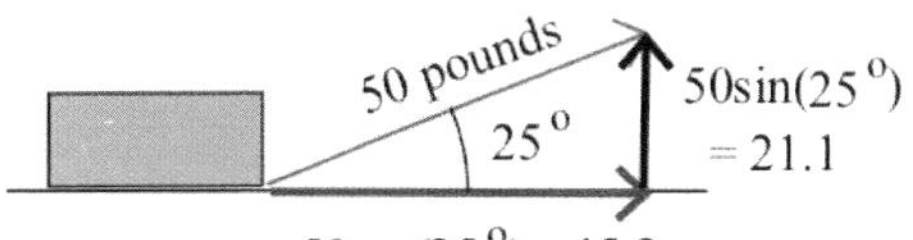

Fig. 21

A force of approximately 45.3 pounds operates to pull the box along the ground (Fig. 21), and a force of approximately 21.1 pounds is operating to lift the box.

The following result from trigonometry is used to find the components of vectors in the plane.

> If a vector **V** with magnitude |**V**| makes an angle of θ with a horizontal line,
>
> then $\mathbf{V} = |\mathbf{V}|\cos(\theta)\,\mathbf{i} + |\mathbf{V}|\sin(\theta)\,\mathbf{j} = \langle\, |\mathbf{V}|\cos(\theta), |\mathbf{V}|\sin(\theta)\,\rangle$.

Practice 7: A horizontal force of 50 pounds is required to move the box in Fig. 22. If you can pull on the rope with a total force of 70 pounds, what is the largest angle that the rope can make with the ground and still move the box?

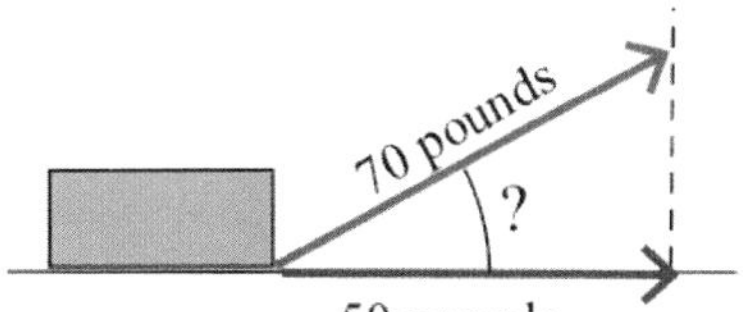

Fig. 22

Additional Applications of Vectors in the Plane

The following applications are more complicated than the previous ones, but they begin to illustrate the range and power of vector methods for solving applied problems. In general, vector methods allow us to work separately with the horizontal and vertical components of a problem, and then put the results together into a complete answer.

Example 6: In water with no current, your boat can travel at 20 knots (nautical miles per hour). Suppose your boat is on the ocean and you want to follow a course to travel due north, but the water current is $\mathbf{W} = -6\mathbf{i} - 8\mathbf{j}$ (Fig. 23). At what angle θ, east of due north, should you steer your boat so the resulting course **R** is due north?

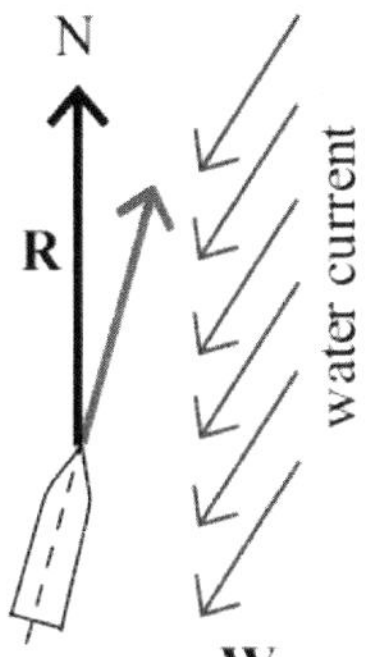

Fig. 23

Solution: First, we should notice that since **R** points due north, the **i** component **R** is 0: $\mathbf{R} = \langle\, 0, s\,\rangle$. We also need to recognize that **V** makes an angle of $90° - \theta$ with the horizontal so $\mathbf{V} = \langle\, 20\cos(90° - \theta), 20\sin(90° - \theta)\,\rangle$. Finally $\mathbf{V} + \mathbf{W} = \mathbf{R}$ so $\langle\, 20\cos(90° - \theta) - 6, 20\sin(90° - \theta) - 8\,\rangle = \langle\, 0, s\,\rangle$. Equating the first components of this vector equation, we have $20\cos(90° - \theta) - 6 = 0$ so $\cos(90° - \theta) = 6/20 = 0.3$, $90° - \theta \approx 72.5°$, and $\theta \approx 17.5°$. You should steer your boat approximately 17.5° east of due north in order to maintain a course taking you due north. Your speed along this course is $|\mathbf{R}| = s = 20\sin(90° - \theta) - 8 \approx 20\sin(72.5°) - 8 \approx 11.1$ knots.

Practice 8: With the same boat and water current as in Example 6, at what angle θ, east of due north, should you steer your boat so the resulting course **R** is due east? What is your resulting speed due east?

Example 7: A video camera weighing 15 pounds is going to be suspended by two wires from the ceiling of a room as shown in Fig. 24. What is the resulting tension in each wire? (The tension in a wire is the magnitude of the force vector.)

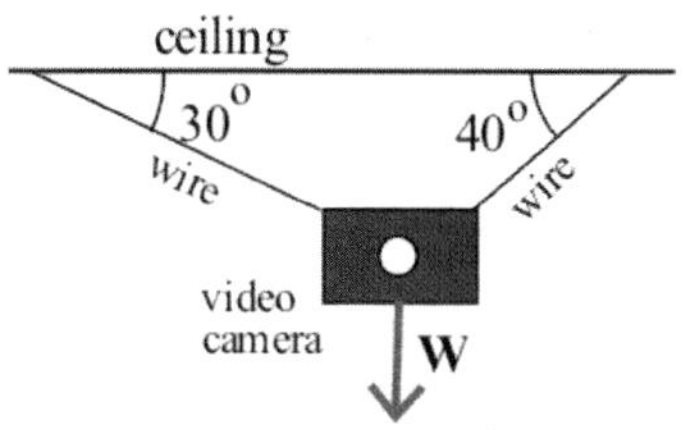

Fig. 24

Solution: The force vector of the camera is straight down so $\mathbf{W} = \langle 0, -15 \rangle$. Let vector **A** be the force vector for the left (30°) wire, and vector **B** be the force vector for the right (40°) vector. Vector **A** has magnitude $|\mathbf{A}|$ and can be represented as $\langle -|\mathbf{A}|\cos(30°), |\mathbf{A}|\sin(30°) \rangle$. Similarly, $\mathbf{B} = \langle |\mathbf{B}|\cos(40°), |\mathbf{B}|\sin(40°) \rangle$. Since the system is in equilibrium, the sum of the force vectors is 0 so

$$\mathbf{0} = \mathbf{A} + \mathbf{B} + \mathbf{W} = \langle -|\mathbf{A}|\cos(30°) + |\mathbf{B}|\cos(40°) + 0, |\mathbf{A}|\sin(30°) + |\mathbf{B}|\sin(40°) - 15 \rangle .$$

From the components of the vector equation we have two equations,

$$0 = -|\mathbf{A}|\cos(30°) + |\mathbf{B}|\cos(40°) + 0 \text{ and } 0 = |\mathbf{A}|\sin(30°) + |\mathbf{B}|\sin(40°) - 15 ,$$

that we want to solve for the tensions $|\mathbf{A}|$ and $|\mathbf{B}|$.

From the first, we get $|\mathbf{A}|\cos(30°) = |\mathbf{B}|\cos(40°)$ so $|\mathbf{B}| = |\mathbf{A}| \dfrac{\cos(30°)}{\cos(40°)}$.

Substituting this value for $|\mathbf{B}|$ into the second equation we have

$$0 = |\mathbf{A}|\sin(30°) + |\mathbf{A}|\frac{\cos(30°)}{\cos(40°)}\sin(40°) - 15 = |\mathbf{A}|\{\sin(30°) + \cos(30°)\tan(40°)\} - 15$$

so $|\mathbf{A}| = \dfrac{15}{\sin(30°) + \cos(30°)\tan(40°)} \approx 12.2$ pounds. Putting this value back into

$|\mathbf{B}| = |\mathbf{A}|\dfrac{\cos(30°)}{\cos(40°)}$, we get $|\mathbf{B}| = (12.2)\dfrac{\cos(30°)}{\cos(40°)} \approx 13.9$ pounds .

Practice 9: What are the tensions in the wires if the angles are changed to 35° and 50° ?

PROBLEMS

In problems 1 – 4, vectors **U** and **V** are given graphically. Sketch the vectors 3**U**, –2**V** , **U** + **V** , and **U** – **V** .

1. **U** and **V** are given in Fig. 25.
2. **U** and **V** are given in Fig. 26.
3. **U** and **V** are given in Fig. 27.
4. **U** and **V** are given in Fig. 28.

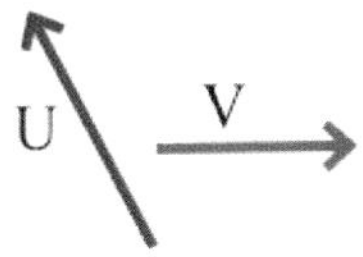

Fig. 25

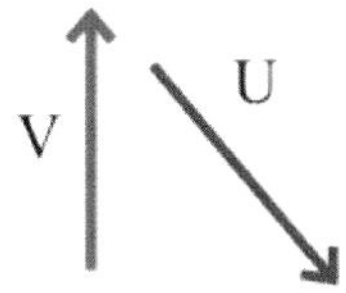

Fig. 26

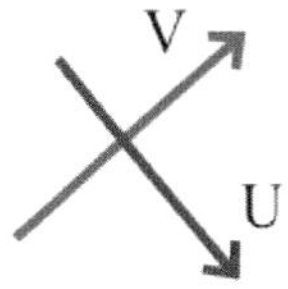

Fig. 27

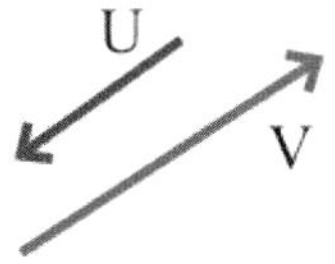

Fig. 28

In problems 5 – 10, vectors **U** and **V** are given.

(a) Sketch the vectors **U**, **V**, 3**U**, –2**V** , **U** + **V** , and **U** – **V** .

(b) Calculate | **U** | and | **V** | and find the directions of **U** and **V**.

(c) Find the slopes of the line segments representing **U** and **V** and their angles with the x–axis.

5. $\mathbf{U} = \langle 1, 4 \rangle$ and $\mathbf{V} = \langle 3, 2 \rangle$
6. $\mathbf{U} = \langle 2, 5 \rangle$ and $\mathbf{V} = \langle 6, 1 \rangle$
7. $\mathbf{U} = \langle -2, 5 \rangle$ and $\mathbf{V} = \langle 3, -7 \rangle$
8. $\mathbf{U} = \langle -3, 4 \rangle$ and $\mathbf{V} = \langle 1, -5 \rangle$
9. $\mathbf{U} = \langle -4, -3 \rangle$ and $\mathbf{V} = \langle 3, -4 \rangle$
10. $\mathbf{U} = \langle -5, 2 \rangle$ and $\mathbf{V} = \langle -2, -5 \rangle$

In problems 11 – 14, vectors **A**, **B**, and **C** are given. Calculate **U** = **A** + **B** – **C** and **V** = **A** – **B** + **C**.

11. **A**, **B**, and **C** in Fig. 29.
12. **A**, **B**, and **C** in Fig. 30.
13. $\mathbf{A} = \langle 1, 4 \rangle, \mathbf{B} = \langle 3, 1 \rangle, \mathbf{C} = \langle 5, 2 \rangle$
14. $\mathbf{A} = \langle -1, 5 \rangle, \mathbf{B} = \langle 2, 0 \rangle, \mathbf{C} = \langle 6, -2 \rangle$

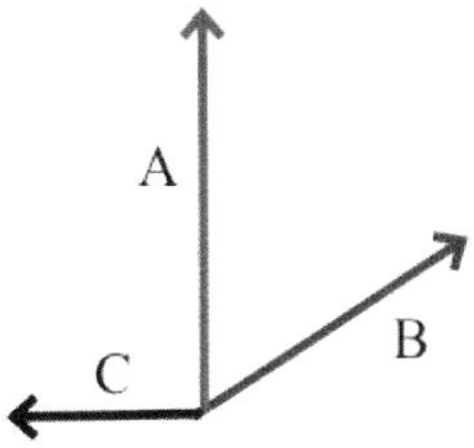

Fig. 29

In problems 15 – 20, find a vector **V** with the given properties.

15. | **V** | = 3 and direction $0.6\mathbf{i} + 0.8\mathbf{j}$
16. | **V** | = 2 and direction $0.6\mathbf{i} - 0.8\mathbf{j}$
17. | **V** | = 5 and **V** makes an angle of 35° with the positive x–axis.
18. | **V** | = 2 and **V** makes an angle of 150° with the positive x–axis.
19. | **V** | = 7 and **V** has slope 3.
20. | **V** | = 7 and **V** has slope –2.

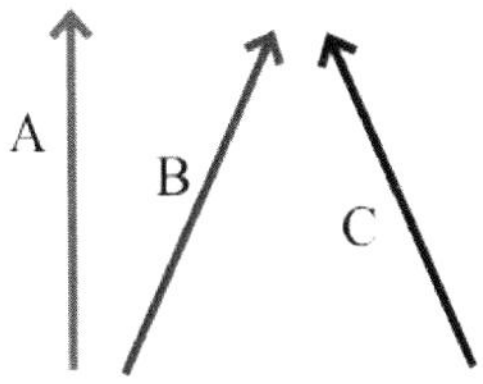

Fig. 30

In problems 21 – 26, find the direction of each function at the given point. That is, find a unit vector, there are two, parallel to the tangent line to the curve at the given point).

21. $f(x) = x^2 + 3x - 2$ at $(1, 2)$ 22. $f(x) = \sin(3x) + e^x$ at $(0, 1)$

23. $f(x) = \ln(x^2 + 1)$ at $(0, 0)$ 24. $f(x) = \dfrac{2 + \sin(x)}{1 + x}$ at $(0, 2)$

25. $f(x) = \cos^2(3x)$ at $(\pi, 1)$ 26. $f(x) = \arctan(x)$ at $(0, 0)$

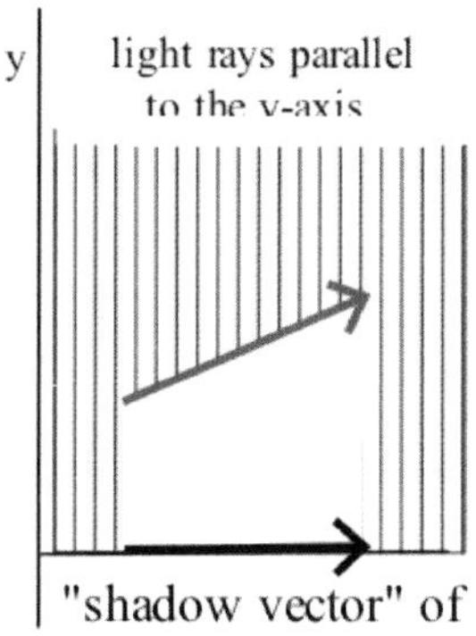

"shadow vector" of U on the x-axis
Fig. 31

In problems 27 – 34, a vector **U** is shown or given as $\mathbf{U} = \langle a, b \rangle$.

(a) Sketch the "shadow vector" (Fig. 31) of **U** on the x–axis. (This is the "projection of **U**" onto the x–axis.)

(b) Sketch the "shadow vector" of **U** on the y–axis. (This is the "projection of **U**" onto the y–axis.)

(c) Represent each of these "shadow vectors" in the form $a\mathbf{i} + b\mathbf{j}$.

27. **U** is given in Fig. 32. 28. **U** is given in Fig. 33.

29. **U** is given in Fig. 34. 30. **U** is given in Fig. 35.

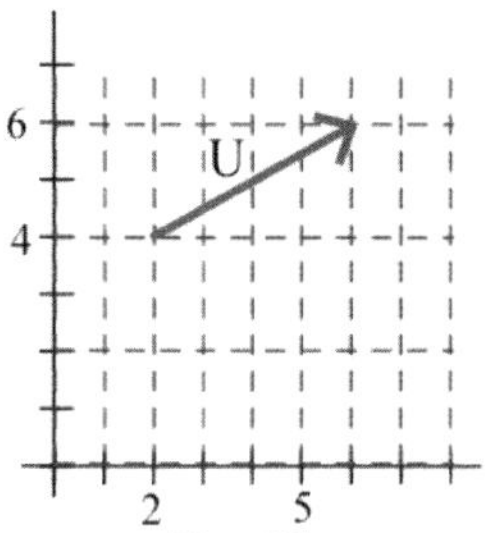

Fig. 32

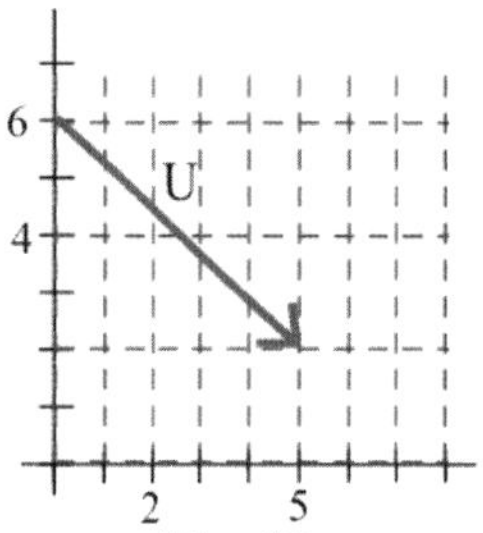

Fig. 33

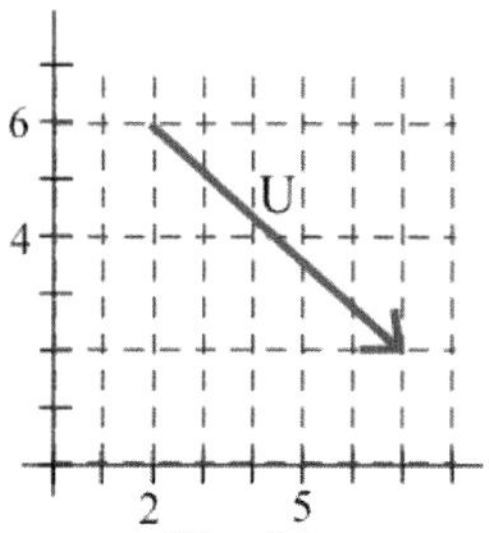

Fig. 34

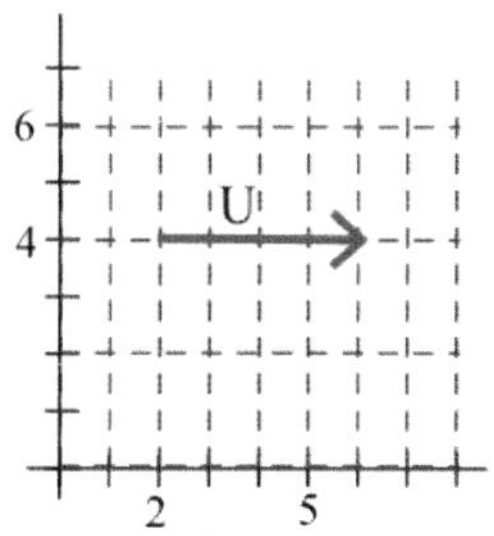

Fig. 35

31. $\mathbf{U} = \langle 1, 4 \rangle$ 32. $\mathbf{U} = \langle -2, 3 \rangle$ 33. $\mathbf{U} = \langle 5, -2 \rangle$ 34. $\mathbf{U} = \langle -1, -3 \rangle$

In problems 35 – 38, vectors **A** and **B** are given. Find a vector **C** so that $\mathbf{A} + \mathbf{B} + \mathbf{C} = \mathbf{0}$.

35. **A** and **B** are shown in Fig. 36. 36. **A** and **B** are shown in Fig. 37.

37. $\mathbf{A} = \langle 7, 4 \rangle$ and $\mathbf{B} = \langle -3, 2 \rangle$ 38. $\mathbf{A} = \langle -5, -3 \rangle$ and $\mathbf{B} = \langle 2, -4 \rangle$

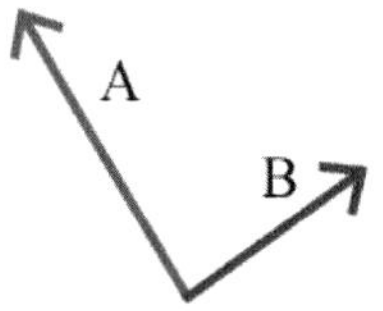

Fig. 36

39. Suppose you are pulling on the rope with a force of 60 pounds at an angle of 65° to the horizontal ground. What are the components of this force vector parallel and perpendicular to the ground?

40. Suppose you are pulling on the rope with a force of 100 pounds at an angle of 15° to the horizontal ground. What are the components of this force vector parallel and perpendicular to the ground?

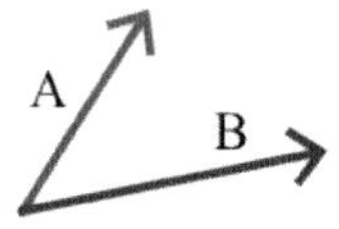

Fig. 37

41. Two ropes are attached to the bumper of a car. Rope A is pulled with a force of 50 pounds at an angle of 45° to the horizontal ground, and rope B is pulled with a force of 70 pounds at an angle of 35° to the horizontal ground. The same effect can be produced by a single rope pulling with what force and at what angle to the ground?

42. Two ropes are attached to the bumper of a car. Rope A is pulled with a force of 60 pounds at an angle of 30° to the horizontal ground, and rope B is pulled with a force of 80 pounds at an angle of 15° to the horizontal ground. The same effect can be produced by a single rope pulling with what force and at what angle to the ground?

43. Rope A is pulled with a force of 100 pounds at an angle of 30° to the horizontal ground, rope B is pulled with a force of 90 pounds at an angle of 25° to the horizontal, and rope C is pulled with a force of 80 pounds at an angle of 15° to the horizontal. The same effect can be produced pulling on a single rope with what force and at what angle to the ground?

44. A plane is flying due east at 200 miles per hour when it encounters a wind $\mathbf{W} = 30\mathbf{i} + 40\mathbf{j}$.

(a) What is the path of the plane in this wind if the pilot keeps it pointed due east?

(b) What direction should the pilot point the plane in order to fly due east?

45. A boat is moving due north at 18 miles per hour when it encounters a current $\mathbf{C} = -3\mathbf{i} + 4\mathbf{j}$.

(a) What is the path of the boat in this current if the boater keeps it pointed due north?

(b) What direction should the boater steer in order to go due north?

46. Suppose you leave home and hike 10 miles due north, then 8 miles in the direction 40° east of north, and then 6 miles due east. (a) How far are you from home? (b) What direction should you hike in order to return home?

47. A 60 foot rope is attached to the tops of two poles that are 50 feet apart, and a 100 pound person is going hand–over–hand from one end of the rope to the other (Fig. 38).

(a) What is the tension in each part of the rope when the person is half way?

(b) What is the tension in the rope when the person is 10 feet (horizontally) from the start?

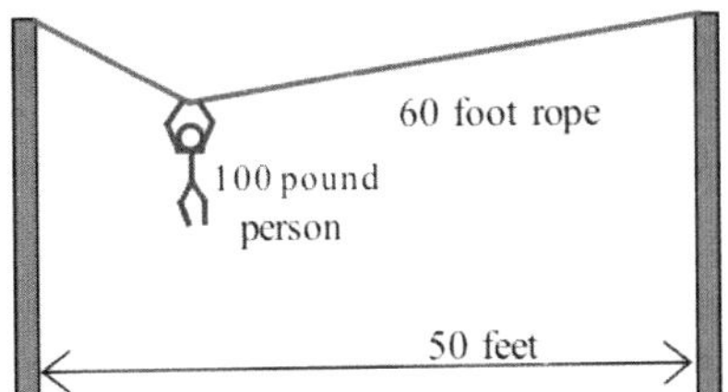

Fig. 38

Practice Answers

Practice 1: (a) and (b) See Fig. 39.

(c) **A** ends at the point (p+3, q+4). **W** ends at the point (p–2, q+3).

(d) We can use the Pythagorean distance formula to determine the length of the hypotenuse.

For **A**, length = $\sqrt{3^2+4^2} = \sqrt{25} = 5$.

For **W**, length = $\sqrt{(-2)^2+3^2} = \sqrt{13} \approx 3.6$.

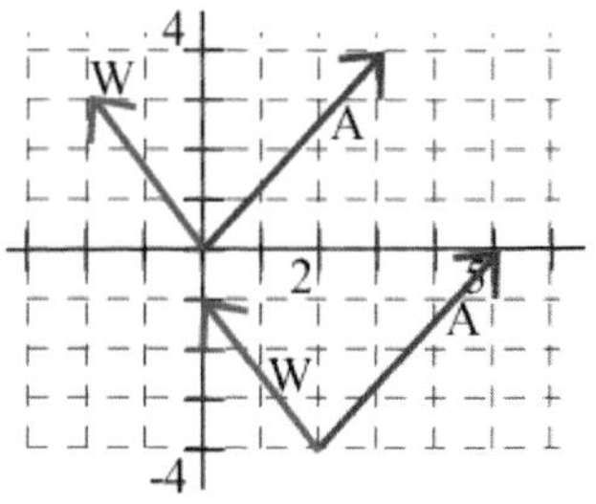

Fig. 39

(e) Slope = rise/run. For **A**, slope = 4/3. For **W**, slope = 3/(–2).

(f) Two lines are perpendicular if one slope is the negative reciprocal of the other. If **B** is perpendicular to **A**, then the slope of **B** is –1/(slope of **A**) = –3/4. The vector **B** = $\langle 4, -3 \rangle$ is one vector that has the slope we want. **B** = $\langle -4, 3 \rangle$, and **B** = $\langle 2, -3/2 \rangle$ are two other vectors with the same slope, and there are lots of others. **U** = $\langle 3, 2 \rangle$ has slope 2/3 and is perpendicular to **W**.

Practice 2: (a) **A** and **B** are shown in Fig. 40.

(b) **C** = $\langle -2, 5 \rangle + \langle 7, -4 \rangle = \langle 5, 1 \rangle$. The graph of **C** is shown in Fig. 40.

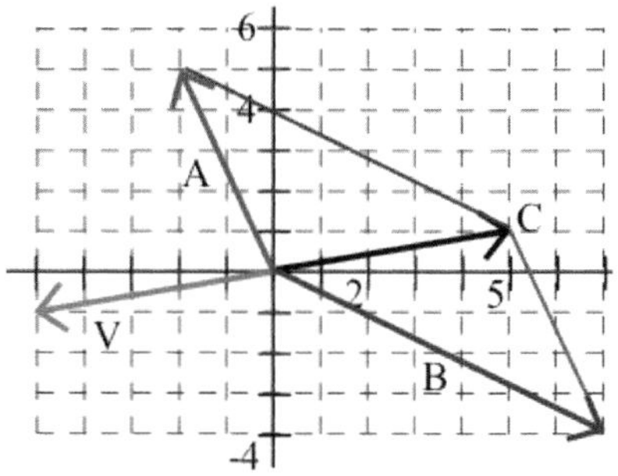

Fig. 40

(c) **V** = $\langle -5, -1 \rangle$. The graph of **V** is shown in Fig. 40.

Practice 3: See Fig. 41.

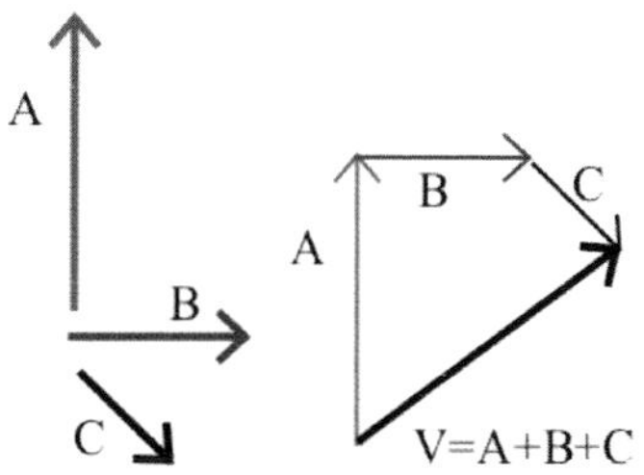

Fig. 41

Practice 4: See Fig. 42.

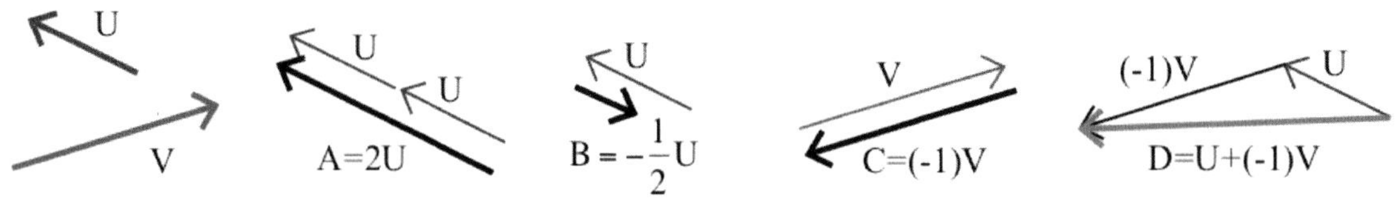

Fig. 42

Practice 5: See Fig. 43.

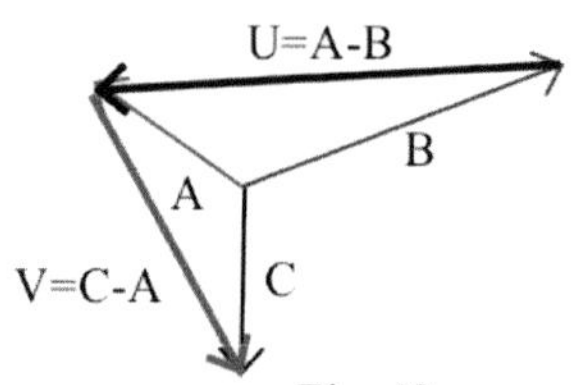

Fig. 43

Practice 6: $|\mathbf{U}| = \sqrt{(7)^2 + (24)^2} = \sqrt{625} = 25$, so the direction of $\mathbf{U}$ is

$$\frac{\mathbf{U}}{|\mathbf{U}|} = \frac{7\mathbf{i} + 24\mathbf{j}}{25} = \frac{7}{25}\mathbf{i} + \frac{24}{25}\mathbf{j} .$$

$|\mathbf{V}| = \sqrt{(15)^2 + (-8)^2} = \sqrt{289} = 17$, so the direction of $\mathbf{V}$ is $\frac{\mathbf{V}}{|\mathbf{V}|} = \frac{15\mathbf{i} - 8\mathbf{j}}{17} = \frac{15}{17}\mathbf{i} - \frac{8}{17}\mathbf{j}$.

$\mathbf{W} = \mathbf{U} + 3\mathbf{i} = (7\mathbf{i} + 24\mathbf{j}) + 3\mathbf{i} = 10\mathbf{i} + 24\mathbf{j}$, so $|\mathbf{W}| \sqrt{(10)^2 + (24)^2} = \sqrt{676} = 26$.

The direction of $\mathbf{W}$ is $\frac{10\mathbf{i} + 24\mathbf{j}}{26} = \frac{5}{13}\mathbf{i} + \frac{12}{13}\mathbf{j}$.

Practice 7: $|\mathbf{V}| = 70$ pounds, and you want the horizontal component, $|\mathbf{V}|\cos(\theta)$, to be 50 pounds, so $70\cos(\theta) = 50$ and $\theta = \arccos(5/7) \approx 44.4°$.

Practice 8: $\mathbf{R} = \langle s, 0 \rangle$. $\mathbf{V}$ makes an angle of $90° - \theta$ with the horizontal so

$\mathbf{V} = \langle 20\cos(90° - \theta), 20\sin(90° - \theta) \rangle$.

Also, $\mathbf{V} + \mathbf{W} = \mathbf{R}$ so $\langle 20\cos(90° - \theta) - 6, 20\sin(90° - \theta) - 8 \rangle = \langle s, 0 \rangle$.

Equating the second components of this vector equation, we have

$20\sin(90° - \theta) - 8 = 0$ so $\sin(90° - \theta) = 8/20 = 0.8$, $90° - \theta \approx 23.6°$, and $\theta \approx 66.4°$.

You should steer your boat approximately $66.4°$ east of due north in order to maintain a course taking you due north. Your speed due east is

$|\mathbf{R}| = s = 20\cos(90° - \theta) - 6 \approx 20\cos(90° - 66.4°) - 6 = 12.3$ knots.

Practice 9: The method of solution is the same as Example 7.

$\mathbf{W} = \langle 0, -15 \rangle$, $\mathbf{A} = \langle -|\mathbf{A}|\cos(35°), |\mathbf{A}|\sin(35°) \rangle$, $\mathbf{B} = \langle |\mathbf{B}|\cos(50°), |\mathbf{B}|\sin(50°) \rangle$, and

$\mathbf{0} = \mathbf{A} + \mathbf{B} + \mathbf{W} = \langle -|\mathbf{A}|\cos(35°) + |\mathbf{B}|\cos(50°) + 0, |\mathbf{A}|\sin(35°) + |\mathbf{B}|\sin(50°) - 15 \rangle$.

Imitating the algebraic steps of Example 7, we get

$|\mathbf{A}| = \frac{15}{\sin(35°) + \cos(35°)\tan(50°)} \approx 9.68$ pounds and

$|\mathbf{B}| = |\mathbf{A}| \frac{\cos(35°)}{\cos(50°)} \approx 12.33$ pounds.

11.2 RECTANGULAR COORDINATES IN THREE DIMENSIONS

In this section we move into 3–dimensional space. First we examine the 3–dimensional rectangular coordinate system, how to locate points in three dimensions, distance between points in three dimensions, and the graphs of some simple 3–dimensional objects. Then, as we did in two dimensions, we discuss vectors in three dimensions, the basic properties and techniques with 3–dimensional vectors, and some of their applications. The extension of the algebraic representations and techniques from 2 to three dimensions is straightforward, but it usually takes practice to visualize 3–dimensional objects and to sketch them on a 2–dimensional piece of paper.

3–Dimensional Rectangular Coordinate System

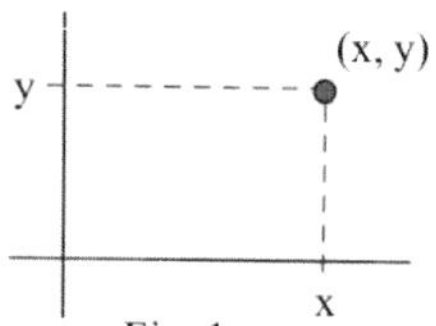

Fig. 1

In the 2–dimensional rectangular coordinate system we have two coordinate axes that meet at right angles at the origin (Fig. 1), and it takes two numbers, an ordered pair (x, y), to specify the rectangular coordinate location of a point in the plane (2 dimensions). Each ordered pair (x, y) specifies the location of exactly one point, and the location of each point is given by exactly one ordered pair (x, y). The x and y values are the coordinates of the point (x, y).

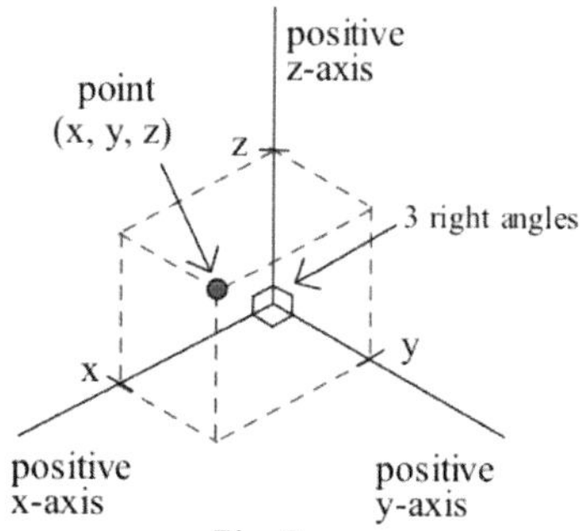

Fig. 2

The situation in three dimensions is very similar. In the 3–dimensional rectangular coordinate system we have three coordinate axes that meet at right angles (Fig. 2), and three numbers, an ordered triple (x, y, z), are needed to specify the location of a point. Each ordered triple (x, y, z) specifies the location of exactly one point, and the location of each point is given by exactly one ordered triple (x, y, z). The x, y and z values are the coordinates of the point (x, y, z). Fig. 3 shows the location of the point (4, 2, 3).

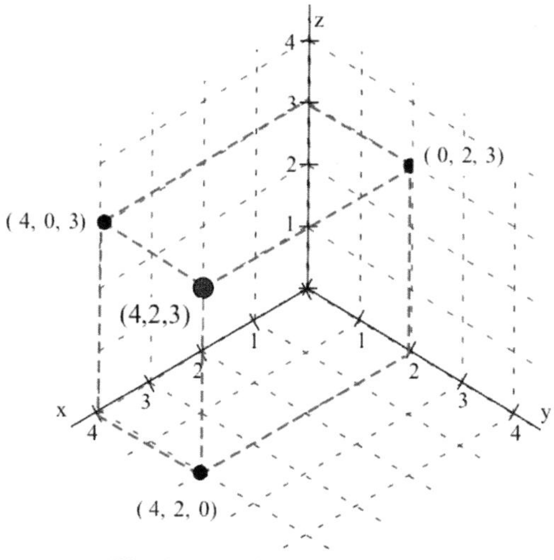

Fig. 3: Locating the point (4, 2, 3)

Right–hand orientation of the coordinate axes (Fig. 4): Imagine your right hand in front hand in front of you with the palm toward your face, your thumb pointing up, you index finger straight out, and your next finger toward your face (and the two bottom fingers bent into the palm (Fig. 4). Then, in the right hand coordinate system, your thumb points along the positive z-axis, your index finger along the positive x-axis, and the other finger along the positive y-axis. Other orientations of the axes are possible and valid (with appropriate labeling), but the right–hand system is the most common orientation and is the one we will generally use.

The three coordinate axes determine three planes (Fig.5): the xy–plane consisting of all points with z–coordinate 0, the xz–plane consisting of all points with y–coordinate 0, and the yz–plane with x–coordinate 0. These three planes then divide the 3–dimensional space into 8 pieces called **octants**. The only octant we shall refer to by name is the **first octant** which is the octant determined by the positive x, y, and z–axes (Fig. 6).

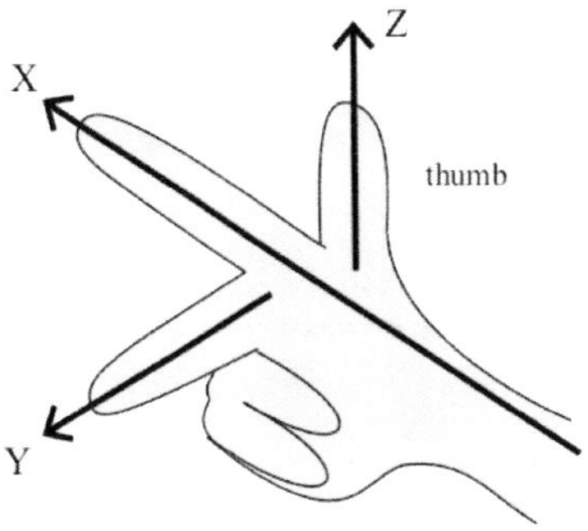

Fig. 4: Right-hand coordinate system

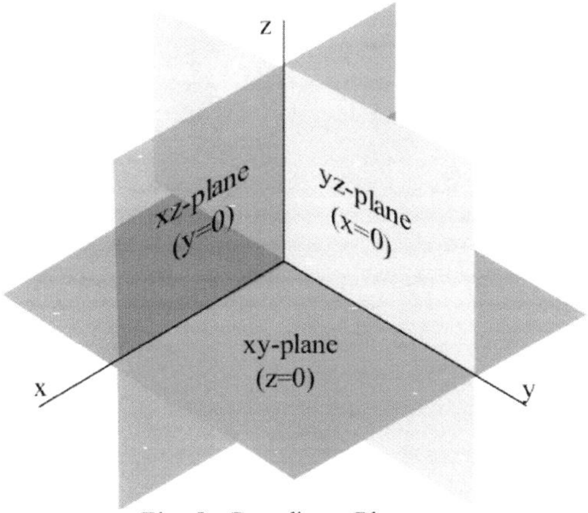

Fig. 5: Coordinate Planes

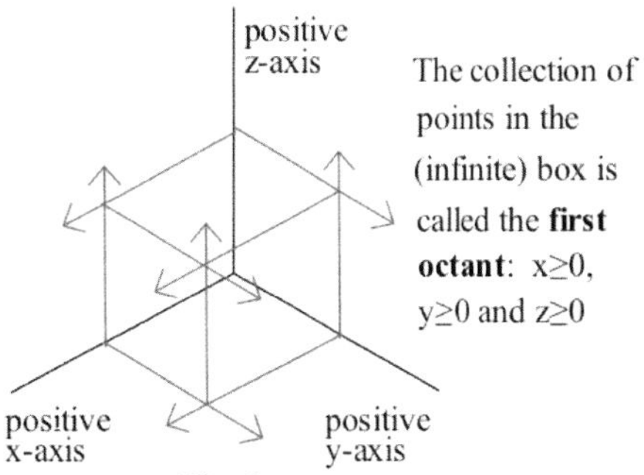

Fig. 6

Visualization in three dimensions: Some people have difficulty visualizing points and other objects in three dimensions, and it may be useful for you to spend a few minutes to create a small model of the 3–dimensional axis system for your desk. One model consists of a corner of a box (or room) as in Fig. 7: the floor is the xy–plane; the wall with the window is the xz–plane; and the wall with the door is the yz–plane. Another simple model uses a small Styrofoam ball and three pencils (Fig. 8): just stick the pencils into the ball as in Fig. 8, label each pencil as the appropriate axis, and mark a few units along each axis (pencil). By referring to such a model for your early work in three dimensions, it becomes easier to visualize the locations of points and the shapes of other objects later.

This visualization can be very helpful.

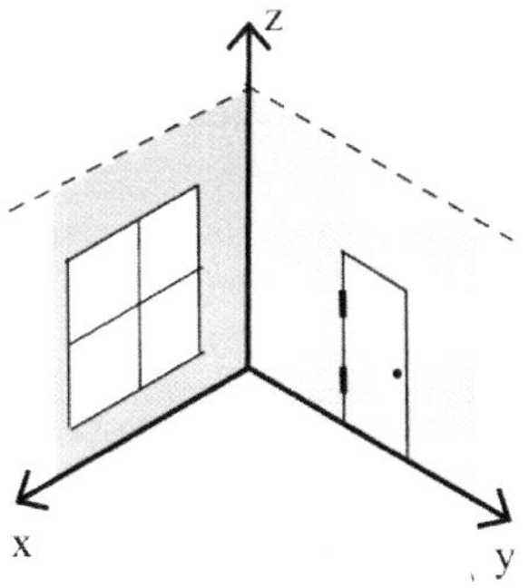

Fig. 7

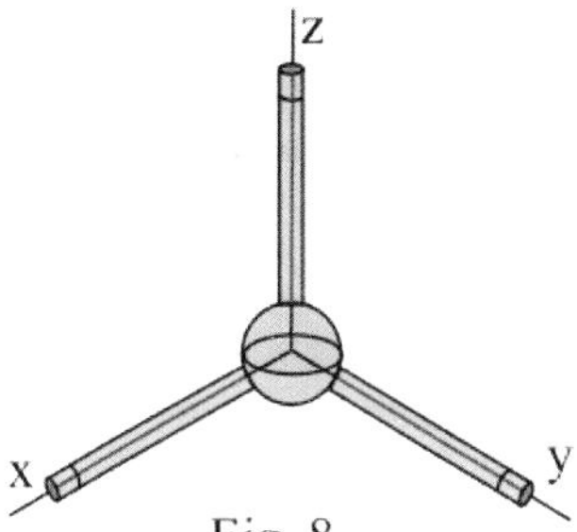

Fig. 8

Each ordered triple (x, y, z) specifies the location of a single point, and this location point can be plotted by locating the point $(x, y, 0)$ on the xy–plane and then going up z units (Fig. 9). (We could also get to the same (x, y, z) point by finding the point $(x, 0, z)$ on the xz–plane and then going y units parallel to the y–axis, or by finding $(0, y, z)$ on the yz–plane and then going x units parallel to the x–axis.)

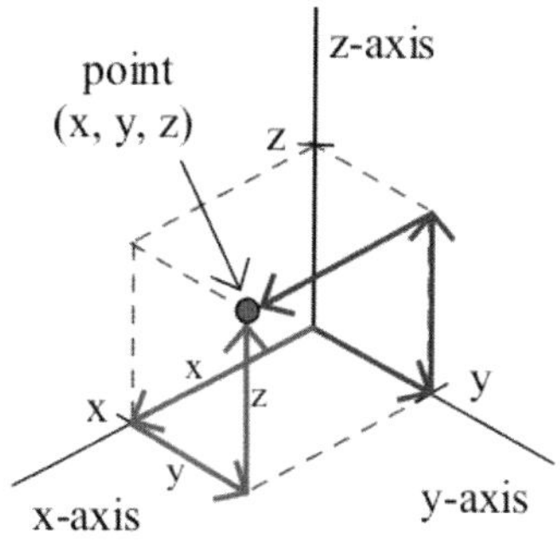

Fig. 9

Example 1: Plot the locations of the points $P = (0, 3, 4), Q = (2, 0, 4)$,

$R = (1, 4, 0)$, $S = (3, 2, 1)$, and $T(-1, 2, 1)$.

Solution: The points are shown in Fig. 10.

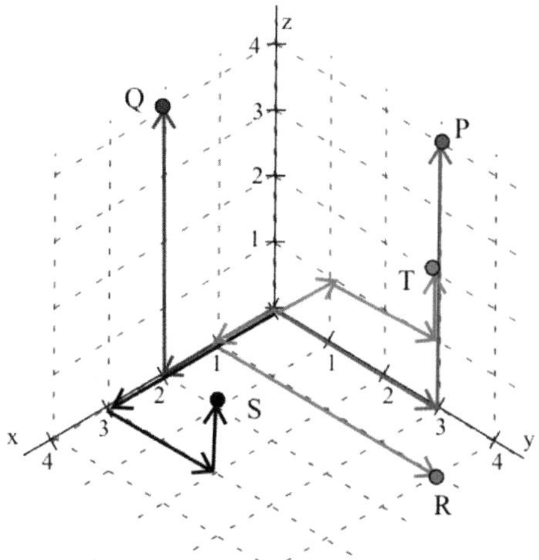

Fig. 10

Practice 1: Plot and label the locations of the points $A = (0, -2, 3), B = (1, 0, -5), C = (-1, 3, 0)$, and $D = (1, -2, 3)$ on the coordinate system in Fig. 11.

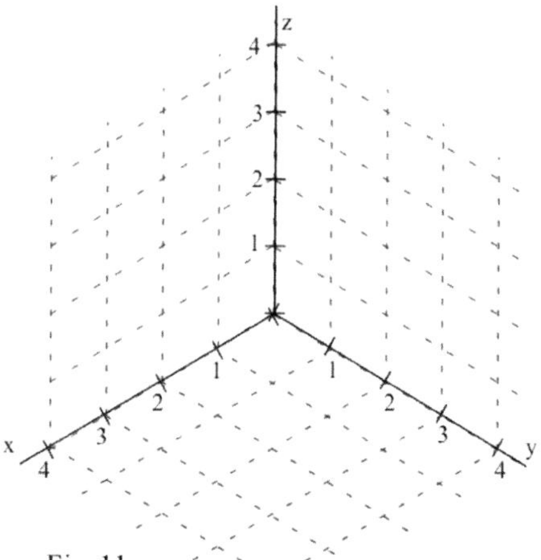
Fig. 11

Practice 2: The opposite corners of a rectangular box are at $(0, 1, 2)$ and $(2, 4, 3)$. Sketch the box and find its volume.

Once we can locate points, we can begin to consider the graphs of various collections of points. By the graph of "$z = 2$" we mean the collection of all points (x, y, z) which have the form "$(x, y, 2)$". Since no condition is imposed on the x and y variables, they take all possible values. The graph of $z = 2$ (Fig. 12) is a plane parallel to the xy–plane and 2 units above the xy–plane. Similarly, the graph of $y = 3$ is a plane parallel to the xz–plane (Fig. 13a), and $x = 4$ is a plane parallel to the yz–plane (Fig. 13b). (Note: The planes have been drawn as rectangles, but they actually extend infinitely far.)

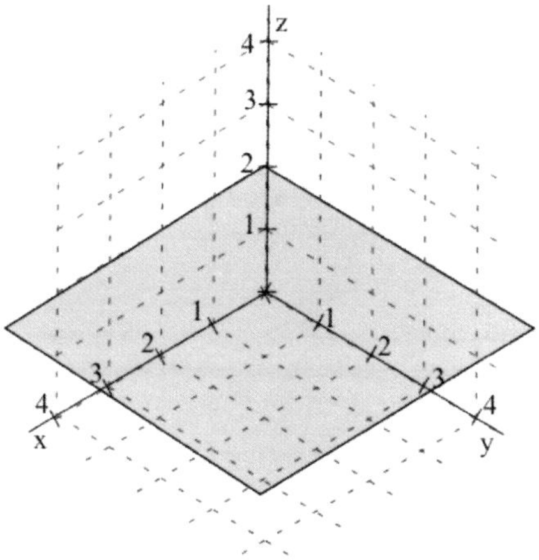
Fig. 12: Plane z=2

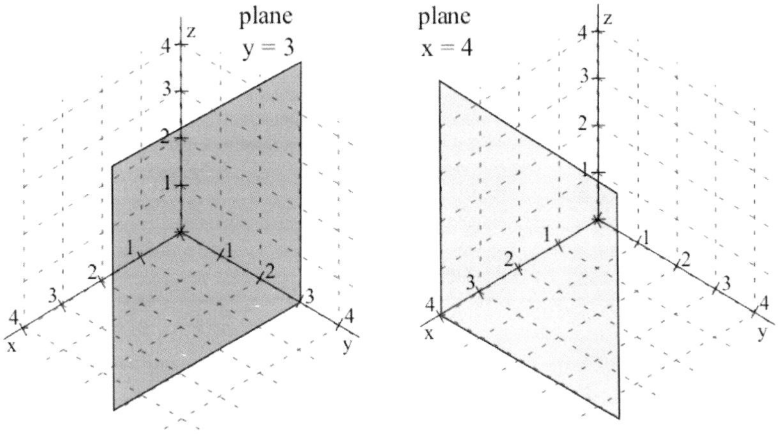

Fig. 13: Planes $y = 3$ and $x = 4$

Practice 3: Graph the planes (a) $x = 2$, (b) $y = -1$, and (c) $z = 3$ in Fig. 14. Give the coordinates of the point that lies on all three planes.

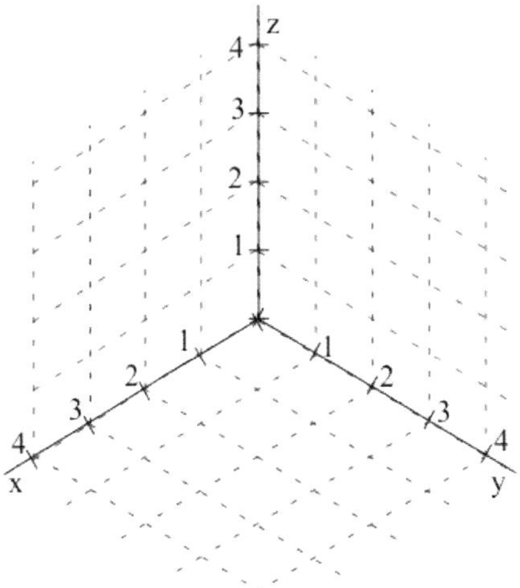

Fig. 14

Example 2: Graph the set of points (x, y, z) such that $x = 2$ and $y = 3$.

Solution: The points that satisfy the conditions all have the form $(2, 3, z)$, and, since no restriction has been placed on the z–variable, z takes all values. The result is the line (Fig. 15) through the point $(2, 3, 0)$ and parallel to the z–axis.

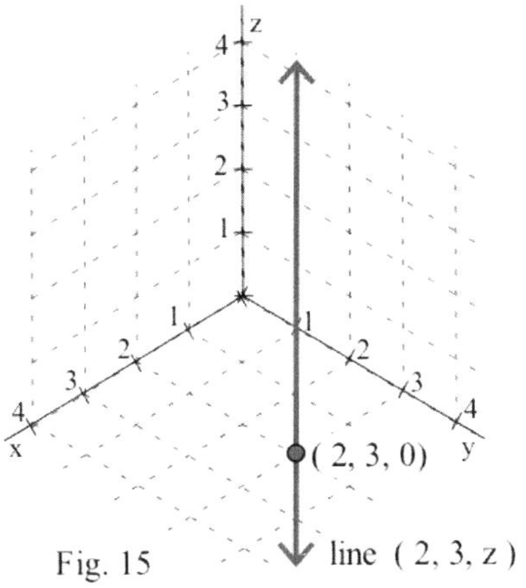

Fig. 15

Practice 4: On Fig. 15, graph the points that have the form (a) $(x, 1, 4)$ and (b) $(2, y, -1)$.

In Section 11.5 we will examine planes and lines that are not parallel to any of the coordinate planes or axes.

Example 3: Graph the set of points (x, y, z) such that $x^2 + z^2 = 1$.

Solution: In the xz–plane (y = 0), the graph of $x^2 + z^2 = 1$ is a circle centered at the origin and with radius 1 (Fig. 16a). Since no restriction has been placed on the y–variable, y takes all values. The result is the cylinder in Figs. 16b and 16c, a circle moved parallel to the y–axis.

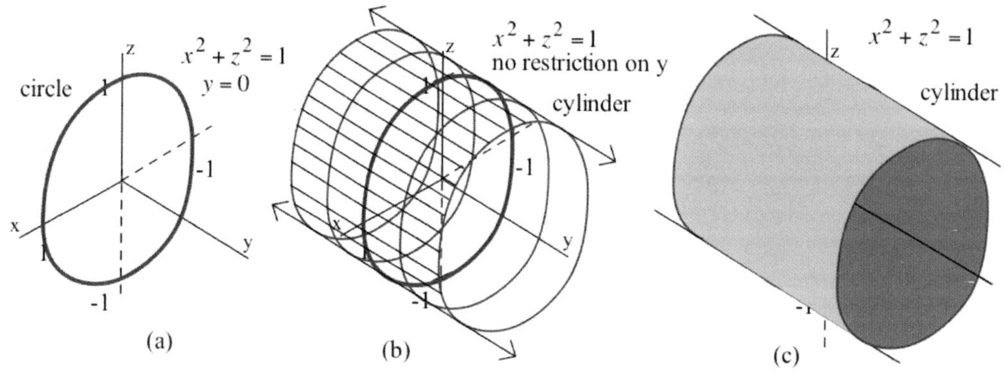

Fig. 16

Practice 5: Graph the set of points (x, y, z) such that $y^2 + z^2 = 4$. (Suggestion: First graph $y^2 + z^2 = 4$ in the yz–plane (x = 0) and then extend the result as x takes on all values.)

Distance Between Points

In two dimensions we can think of the distance between points as the length of the hypothenuse of a right triangle (Fig. 17), and that leads to the Pythagorean formula: distance $= \sqrt{\Delta x^2 + \Delta y^2}$. In three dimensions we can also think of the distance between points as the length of the hypothenuse of a right triangle (Fig. 18), but in this situation the calculations appear more complicated. Fortunately, they are straightforward:

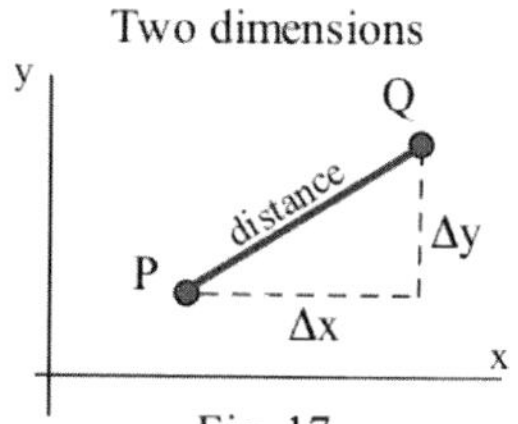

Fig. 17

$$\text{distance}^2 = \text{base}^2 + \text{height}^2 = \left(\sqrt{\Delta x^2 + \Delta y^2}\right)^2 + \Delta z^2$$

$$= \Delta x^2 + \Delta y^2 + \Delta z^2 \quad \text{so distance} = \sqrt{\Delta x^2 + \Delta y^2 + \Delta z^2} \; .$$

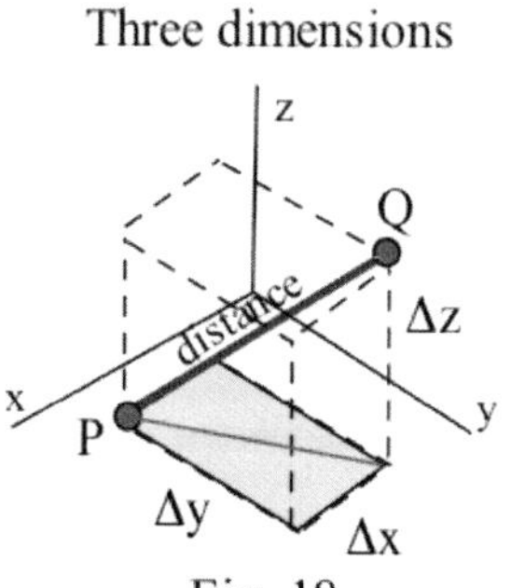

Fig. 18

If $P = (x_1, y_1, z_1)$ and $Q = (x_2, y_2, z_2)$ are points in space,

then the distance between P and Q is

$$\text{distance} = \sqrt{\Delta x^2 + \Delta y^2 + \Delta z^2}$$

$$= \sqrt{(x_2 - x_1)^2 + (y_2 - y_1)^2 + (z_2 - z_1)^2} \; .$$

The 3–dimensional pattern is very similar to the 2–dimensional pattern with the additional piece Δz^2 .

Example 4: Find the distances between all of the pairs of the given points. Do any three of these points form a right triangle? Do any three of these points lie on a straight line?

Points: $A = (1, 2, 3), B = (7, 5, -3), C = (8, 7, -1), D = (11, 13, 5)$.

Solution: $\text{Dist}(A, B) = \sqrt{6^2 + 3^2 + (-6)^2} = \sqrt{36 + 9 + 36} = \sqrt{81} = 9$. Similarly,

$\text{Dist}(A,C) = \sqrt{90}$, $\text{Dist}(A, D) = 15$, $\text{Dist}(B, C) = 3$, $\text{Dist}(B, D) = 12$, and $\text{Dist}(C, D) = 9$.

$\{ \text{Dist}(A, B) \}^2 + \{ \text{Dist}(B, D) \}^2 = \{ \text{Dist}(A, D) \}^2$ so the points A, B, and D form a right triangle with the right angle at point B. Also, the points A, B, and C form a right triangle with the right angle at point B since $9^2 + 3^2 = (\sqrt{90})^2$

$\text{Dist}(B, C) + \text{Dist}(C, D) = \text{Dist}(B, D)$ so the points B, C, and D line on a straight line. The points are shown in Fig. 19. In three dimensions it is often difficult to determine the size of an angle from a graph or to determine whether points are collinear.

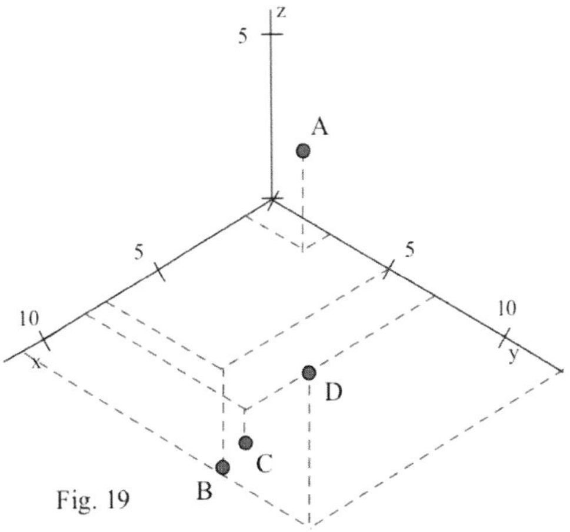

Fig. 19

Practice 6: Find the distances between all of the pairs of the points $A = (3, 1, 2), B = (9, 7, 5), C = (9, 7, 9)$. Which two of these points are closest together? Which two are farthest apart? Do these three points form a right triangle?

In two dimensions, the set of points at a fixed distance from a given point is a circle, and we used the distance formula to determine equations describing circles: the circle with center (2, 3) and radius 5 (Fig. 20) is given by $(x-2)^2 + (y-3)^2 = 5^2$ or $x^2 + y^2 - 4x - 6y = 12$. The same ideas work for spheres in three dimensions.

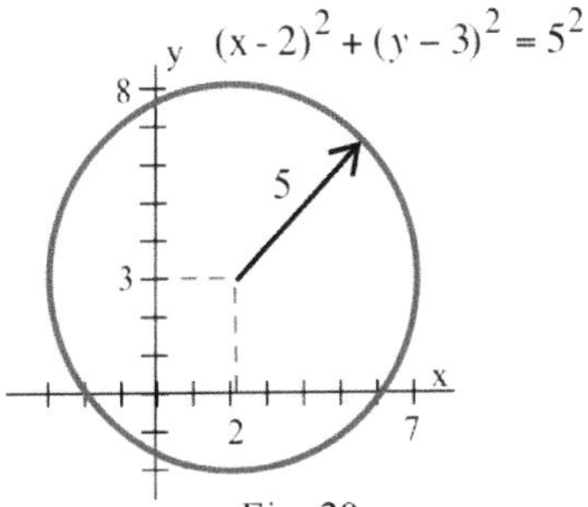

Fig. 20

Spheres: The set of points (x, y, z) at a fixed distance r from a point (a, b, c) is a sphere (Fig. 21) with center (a, b, c) and radius r.

The sphere is given by the equation $(x-a)^2 + (y-b)^2 + (z-c)^2 = r^2$.

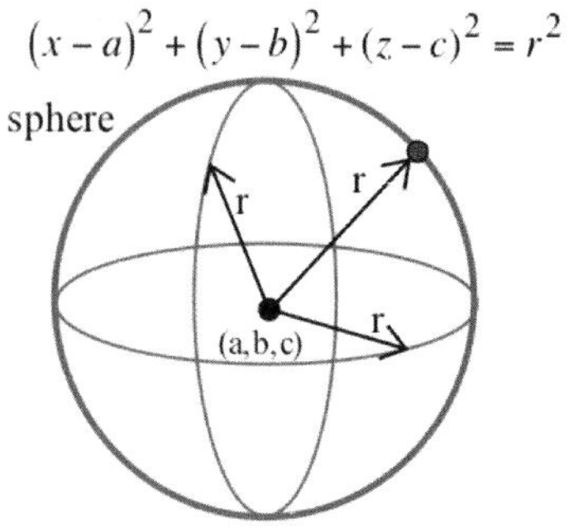

Fig. 21

Example 5: Write the equations of the following two spheres: (A) center $(2,-3,4)$ and radius 3, and (B) center $(4,3,-5)$ and radius 4. What is the minimum distance between a point on A and a point on B? What is the maximum distance between a point on A and a point on B?

Solution: (A) $(x-2)^2+(y+3)^2+(z-4)^2=3^2$.

(B) $(x-4)^2+(y-3)^2+(z+5)^2=4^2$.

The distance between the centers is

$\sqrt{(4-2)^2+(3+3)^2+(-5-4)^2} = \sqrt{121} = 11$, so

the minimum distance between points on the spheres (Fig. 22) is

$$11 - (\text{one radius}) - (\text{other radius}) = 11 - 3 - 4 = 4.$$

The maximum distance between points on the spheres is $11 + 3 + 4 = 18$.

Practice 7: Write the equations of the following two spheres:

(A) center $(1,-5,3)$ and radius 10, and (B) center $(7,-7,0)$ and radius 2.

What is the minimum distance between a point on A and a point on B?

What is the maximum distance between a point on A and a point on B?

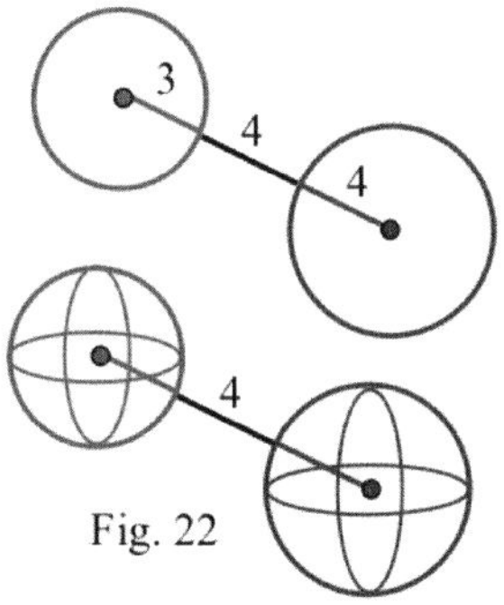

Fig. 22

Beyond Three Dimensions

At first it may seem strange that there is anything beyond three dimensions, but fields as different as physics, statistics, psychology, economics and genetics routinely work in higher dimensional spaces. In three dimensions we use an ordered 3–tuple (x, y, z) to represent and locate a point, but there is no logical or mathematical reason to stop at three. Physicists talk about "space–time space," a 4–dimensional space where a point is represented by a 4–tuple (x, y, z, t) with x, y, and z representing a location and t represents time. This is very handy for describing complex motions, and the "distance" between two "space–time" points tells how far apart they are in (3–dimensional) distance and time. "String theorists," trying to model the early behavior and development of the universe, work in 10–dimensional space and use 10–tuples to represent points in that space.

On a more down to earth scale, any object described by 5 separate measures (numbers) can be thought of as a point in 5–dimensional space. If a pollster asks students 5 questions, then one student's responses can be represented as an ordered 5–tuple (a, b, c, d, e) and can be thought of as a point in five dimensional space. The collection of responses from an entire class of students is a cloud of points in 5–space, and the center of mass of that cloud (the point formed as the mean of all of the individual points) is often used as a group response. Psychologists and counselors sometimes use a personality profile that rates people on four independent scales (IE, SN, TF, JP). The "personality type" of each person can be represented as an ordered 4–tuple, a point in 4–dimensional "personality–type space." If the distance between two people is small in "personality–type space," then they have similar "personality types." Some matchmaking services ask clients a number of questions (each question is a dimension in this "matching space") and then try to find a match a small distance away.

Many biologists must deal with huge amounts of data, and often this data is represented as ordered n–tuples, points in n–dimensional space. In the book The History and Geography of Human Genes (1994), the authors summarize more than 75,000 allele frequencies in nearly 7,000 human populations in the form of maps. "To construct one of the maps, eighty–two genes were examined in many populations throughout the world. Each population was represented on a computer grid as a point in eighty–two dimensional space, with its position along each dimensional axis representing the frequency of one of the alleles in question." (Natural History, 6/94, p. 84)

Geometrically, it is difficult to work in more than three dimensions, but length/distance calculations are still easy.

> Definitions for n dimensions:
>
> A point in n–dimensional space is an ordered n–tuple $(a_1, a_2, a_3, \ldots, a_n)$.
>
> If $A = (a_1, a_2, a_3, \ldots, a_n)$ and $B = (b_1, b_2, b_3, \ldots, b_n)$ are points in n–dimensional space,
>
> then the distance between A and B is
>
> $$\text{distance} = \sqrt{(b_1-a_1)^2 + (b_2-a_2)^2 + (b_3-a_3)^2 + \ldots + (b_n-a_n)^2}\ .$$

Example 6: Find the distance between the points $P = (1, 2, -3, 5, 6)$ and $Q = (5, -1, 4, 0, 7)$.

Solution: $\text{Distance} = \sqrt{(5-1)^2+(-1-2)^2+(4--3)^2+(0-5)^2+(7-6)^2} = \sqrt{16+9+49+25+1} = \sqrt{100} = 10.$

Practice 8: Write an equation for the 5–dimensional sphere with radius 8 and center $(3, 5, 0, -2, 4)$.

PROBLEMS

In problems 1 – 4, plot the given points.

1. A = (0,3,4), B = (1,4,0), C = (1,3,4), D = (1,4,2)

2. E = (4,3,0), F = (3,0,1), G = (0,4,1), H = (3,3,1)

3. P = (2,3,–4), Q = (1,–2,3), R = (4,–1,–2), S = (–2,1,3)

4. T = (–2,3,–4), U = (2,0,–3), V = (–2,0,0), W = (–3,–1,–2)

In problems 5 – 8, plot the lines.

5. (3, y, 2) and (1, 4, z)

6. (x, 3, 1) and (2, 4, z)

7. (x, –2, 3) and (–1, y, 4)

8. (3, y, –2) and (–2, 4, z)

In problems 9 – 12, three collinear points are given. Plot the points and the draw a line through them.

9. (4,0,0), (5,2,1), and (6,4,2)

10. (1,2,3), (3,4,4), and (5,6,5)

11. (3,0,2), (3,2,3), and (3,6,5)

12. (–1,3,4), (2,3,2), and (5,3,0)

In problems 13 – 16, calculate the distances between the given points and determine if any three of them are collinear. (Note: P, Q, R are collinear if dist(P,Q) + dist(Q,R) = dist(P,R))

13. A = (5,3,4), B = (3,4,4), C = (2,2,3), D = (1,6,4)

14. A = (6,2,1), B = (3,2,1), C = (3,2,5), D = (1,–4,2)

15. A = (3,4,2), B = (–1,6,–2), C = (5,3,4), D = (2,2,3)

16. A = (–1,5,0), B = (1,3,2), C = (5,–1,3), D = (3,1,2)

In problems 17 – 20 you are given three corners of a box whose sides are parallel to the xy, xz, and yz planes. Find the other five corners and calculate the volume of the box.

17. (1,2,1), (4,2,1), and (1,4,3)

18. (5,0,2), (1,0,5), and (1,5,2)

19. (4,5,0), (1,4,3), and (1,5,3)

20. (4,0,1), (0,3,1), and (0,0,5)

In problems 21 – 24, graph the given planes.

21. y = 1 and z = 2

22. x = 4 and y = 2

23. x = 1 and y = 0

24. x = 2 and z = 0

In problems 25 – 28, the center and radius of a sphere are given. Find an equation for the sphere.

25. Center = (4, 3, 5), radius = 3

26. Center = (0, 3, 6), radius = 2

27. Center = (5, 1, 0), radius = 5

28. Center = (1, 2, 3), radius = 4

In problems 29 – 32, the equation of a sphere is given. Find the center and radius of the sphere.

29. $(x-3)^2 + (y+4)^2 + (z-1)^2 = 16$

30. $(x+2)^2 + y^2 + (z-4)^2 = 25$

31. $x^2 + y^2 + z^2 - 4x - 6y - 8z = 71$

32. $x^2 + y^2 + z^2 + 6x - 4y = 12$

Problems 33 – 36 name all of the shapes that are possible for the intersection of the two given shapes in three dimensions.

33. A line and a plane

34. Two planes

35. A plane and a sphere

36. Two spheres

In problems 37 – 44, sketch the graphs of each collection of points. Name the shape of each graph.

37. All (x, y, z) such that (a) $x^2 + y^2 = 4$ and $z = 0$, (b) $x^2 + y^2 = 4$ and $z = 2$.

38. All (x, y, z) such that (a) $x^2 + z^2 = 4$ and $y = 0$, (b) $x^2 + z^2 = 4$ and $y = 1$.

39. All (x, y, z) such that $x^2 + y^2 = 4$ and no restriction on z.

40. All (x, y, z) such that $x^2 + z^2 = 4$ and no restriction on y.

41. All (x,y,z) such that (a) $y = \sin(x)$ and $z = 0$, (b) $y = \sin(x)$ and $z = 1$,
(c) $y = \sin(x)$ and no restriction on z.

42. All (x,y,z) such that (a) $z = x^2$ and $y = 0$, (b) $z = x^2$ and $y = 2$,
(c) $z = x^2$ and no restriction on y.

43. All (x,y,z) such that (a) $z = 3 - y$ and $x = 0$, (b) $z = 3 - y$ and $x = 2$,
(c) $z = 3 - y$ and no restriction on x.

44. All (x,y,z) such that (a) $z = 3 - x$ and $y = 0$, (b) $z = 3 - x$ and $y = 2$,
(c) $z = 3 - x$ and no restriction on y.

The volume of a sphere with radius r is $\frac{4}{3}\pi r^3$. Use that formula to help determine the volumes of the following parts of spheres in problems 45 and 46.

45. All (x,y,z) such that (a) $x^2 + y^2 + z^2 \le 4$ and $z \ge 0$, (b) $x^2 + y^2 + z^2 \le 4$, $z \ge 0$, and $y \ge 0$,
(c) $x^2 + y^2 + z^2 \le 4$, $z \ge 0$, $y \ge 0$, and $x \ge 0$.

46. All (x,y,z) such that (a) $x^2 + y^2 + z^2 \le 9$ and $x \ge 0$, (b) $x^2 + y^2 + z^2 \le 9$, $x \ge 0$, and $z \ge 0$,
(c) $x^2 + y^2 + z^2 \le 9$, $x \ge 0$, $z \ge 0$, and $y \ge 0$.

"Shadow" Problems

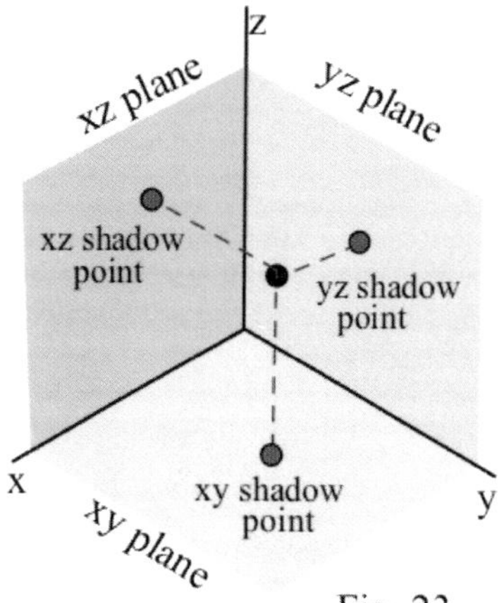

Fig. 23

The following "shadow" problems assume that we have an object in the first octant. Then light rays parallel to the x–axis cast a shadow of the object on the yz–plane (Fig. 23). Similarly, light rays parallel to the y–axis cast a shadow of the object on the xz–plane, and rays parallel to the z–axis cast a shadow on the xy–plane. (The point of these and many of the previous problems is to get you thinking and visualizing in three dimensions.)

S1. Give the coordinates of the shadow points of the point (1,2,3) on each of the coordinate planes.

S2. Give the coordinates of the shadow points of the point (4,1,2) on each of the coordinate planes.

S3. Give the coordinates of the shadow points of the point (a,b,c) on each of the coordinate planes.

S4. A line segment in the first octant begins at the point (4,2,1) and ends at (1,3,3). Where do the shadows of the line segment begin and end on each of the coordinate planes? Are the shadows of the line segment also line segments?

S5. A line segment begins at the point (1,2,4) and ends at (1,4,3). Where do the shadows of the line segment begin and end on each of the coordinate planes? Are the shadows of the line segment also line segments?

S6. A line segment begins at the point (a,b,c) and ends at (p,q,r). Where do the shadows of the line segment begin and end on each of the coordinate planes? Are the shadows of the line segment also line segments?

S7. The three points (0,0,0), (4,0,3), and (4,0,2) are the vertices of a triangle in the first octant. Describe the shadow of this triangle on each of the coordinate planes. Are the shadows always triangles?

S8. The three points (1,2,3), (4,3,1), and (2,3,4) are the vertices of a triangle in the first octant. Describe the shadow of this triangle on each of the coordinate planes. Are the shadows always triangles?

S9. The three points (a,b,c), (p,q,r), and (x,y,z) are the vertices of a triangle in the first octant. Describe the shadow of this triangle on each of the coordinate planes. Are the shadows always triangles?

S10. A line segment in the first octant is 10 inches long. (a) What is the shortest shadow it can have on a coordinate plane? (b) What is the longest shadow it can have on a coordinate plane?

S11. A triangle in the first octant has an area of 12 square inches. (a) What is the smallest area its shadow can have on a coordinate plane? (b) What is the largest area?

S12. Design a solid 3–dimensional object whose shadow on one coordinate plane is a square, on another coordinate plane a circle, and on the third coordinate plane a triangle.

Practice Answers

Practice 1: The points are plotted in Fig. 24 .

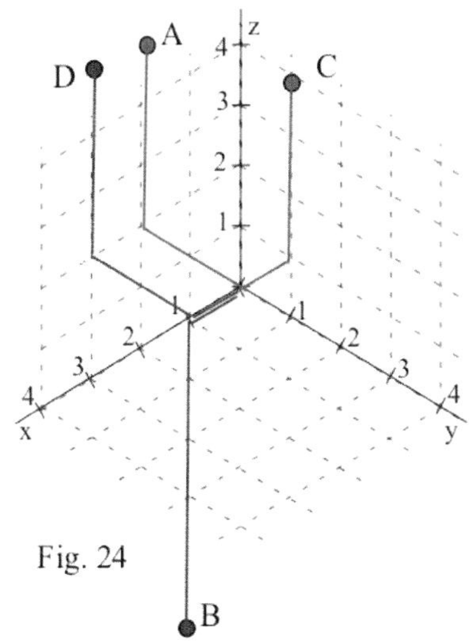

Fig. 24

Practice 2: The box is shown in Fig. 25.

$\Delta x = 2 =$ width, $\Delta y = 3 =$ length,

and $\Delta z = 1 =$ height, so

volume $= (2)(3)(1) = 6$ cubic units.

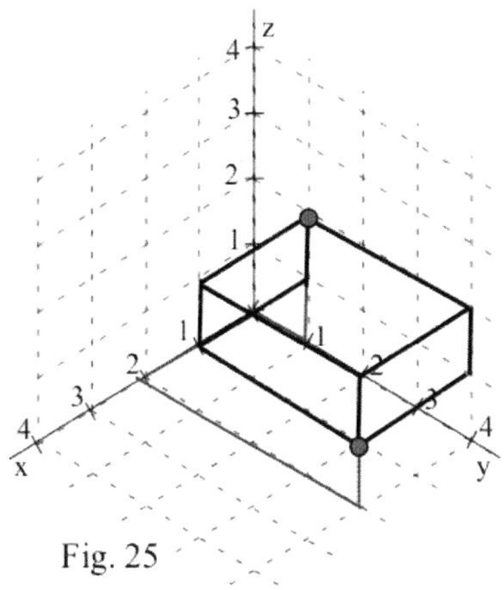
Fig. 25

Practice 3: The planes are shown in Fig. 26(a).

Each pair of planes intersects along a line, shown as a dark lines in Fig. 26(b), and the three lines intersect at the point $(2, -1, 3)$. This is the only point that lies on all three of the planes.

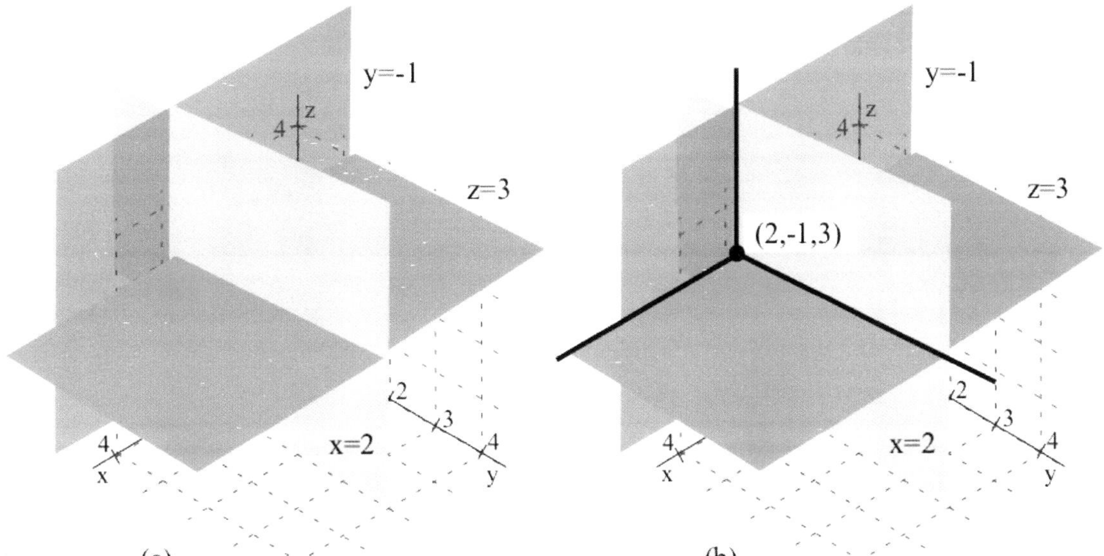

Fig. 26

Practice 4: The points that satisfy $(x, 1, 4)$ are shown in Fig. 27. The collection of these points form a line. One way to sketch the graph of the line is to first plot the point where the line crosses one of the coordinate planes, $(0, 1, 4)$ in this case, and then sketch a line through that point and parallel to the appropriate axis.

The line of points that satisfy $(2, y, -1)$ are also shown in Fig. 27, as well as the point $(2, 0, -1)$ where the line intersects the xz–plane.

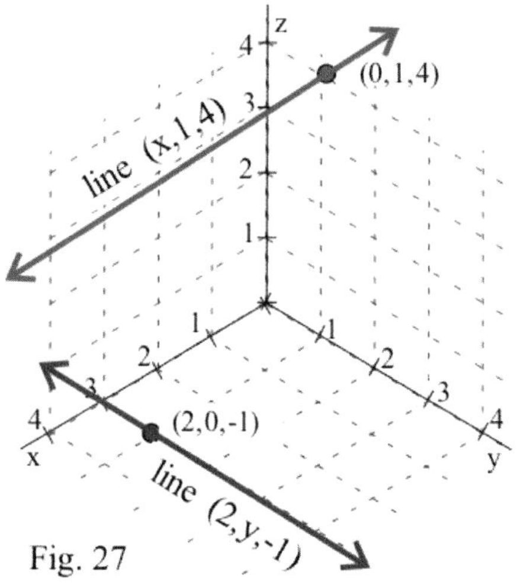

Fig. 27

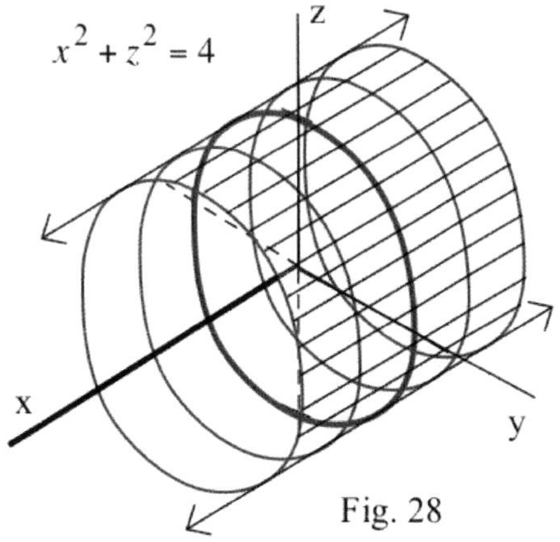

Fig. 28

Practice 5: The graph, a cylinder with radius 2 around the x–axis, is shown in Fig. 28. The dark circle is the graph of points satisfying $y^2 + z^2 = 4$ **and** $x = 0$.

Practice 6:

$\text{Dist}(A,B) = \sqrt{6^2 + 6^2 + 3^2} = 9$,

$\text{Dist}(A,C) = \sqrt{6^2 + 6^2 + 7^2} = 11$, and

$\text{Dist}(B,C) = \sqrt{0^2 + 0^2 + 4^2} = 4$. B and C are closest. A and C are farthest apart.

$4^2 + 9^2 \neq 11^2$ so the points do not form a right triangle.

Practice 7: (A) $(x-1)^2 + (y+5)^2 + (z-3)^2 = 10^2$. (B) $(x-7)^2 + (y+7)^2 + (z-0)^2 = 2^2$.

The distance between the centers is

$$\sqrt{(7-1)^2 + (-7+5)^2 + (0-3)^2} = \sqrt{49} = 7.$$

The radius of sphere A is larger than the distance between the centers plus the radius of sphere B so sphere B is inside sphere A (Fig. 29). The minimum distance between a point on A and a point on B is $10 - (5 + 2 + 2) = 1$. The maximum distance is $10 + (5 + 2 + 2) = 19$.

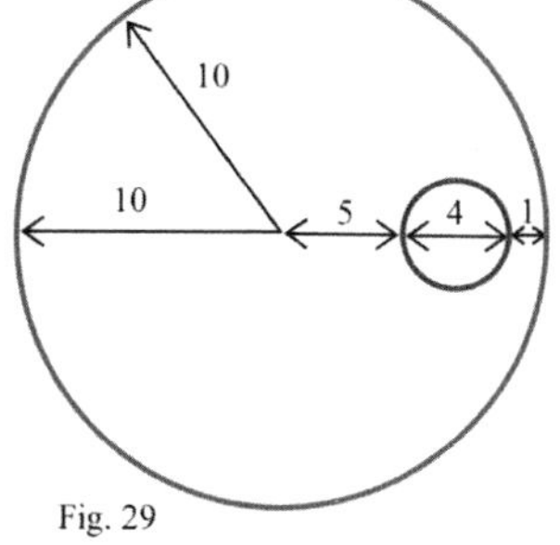

Fig. 29

Practice 8: A point $P = (v,w,x,y,z)$ is on the sphere if and only if the distance from P to the center $(3, 5, 0, -2, 4)$ is 8. Using the distance formula, and squaring each side, we have

$$(v-3)^2 + (w-5)^2 + (x-0)^2 + (y+2)^2 + (z-4)^2 = 8^2 .$$

11.3 VECTORS IN THREE DIMENSIONS

Once you understand the 3–dimensional coordinate system, 3–dimensional vectors are a straightforward extension of vectors in two dimensions. Vectors in three dimensions are more difficult to visualize and sketch, but all of the 2–dimensional algebraic techniques extend very naturally, with just one more component.

A vector in any setting is a **quantity that has both a direction and a magnitude**, and in three dimensions vectors can be represented geometrically as directed line segments, (arrows). The vector **V** given by the line segment from the starting point (tail) P = (1, 2, 3) to the ending point (head) Q = (4, 1, 4) is shown in Fig. 1. The vector **V** from P to Q is represented algebraically by the ordered triple enclosed in "bent" brackets: $\mathbf{V} = \langle 3, -1, 1 \rangle$ with each component representing the displacement from P to Q. (We continue to reserve the "round brackets" ( ) to represent points.) Fig. 1 also shows several other geometric representations of vector **V**.

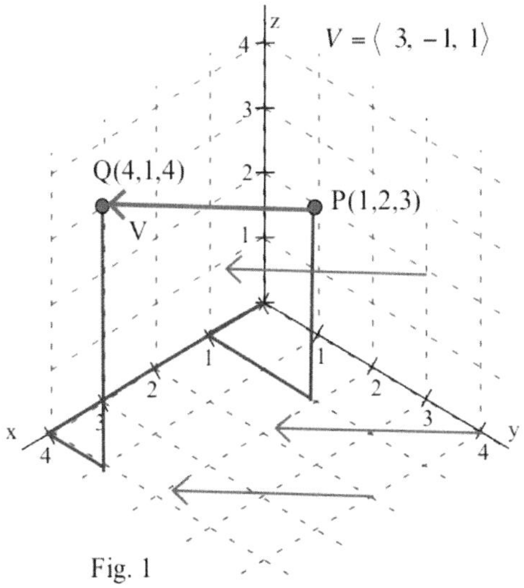

Fig. 1

Definitions and Properties

The following definitions, arithmetic operations, and properties of vectors in 3–dimensional space are straightforward generalizations from two dimensions.

> **Definition: Equality of Vectors**
>
> Geometrically, two vectors are equal if their lengths are equal and their directions are the same.
>
> Algebraically, two vectors are equal if their respective components are equal:
>
> if $\mathbf{U} = \langle a, b, c \rangle$ and $\mathbf{V} = \langle x, y, z \rangle$,
>
> then $\mathbf{U} = \mathbf{V}$ if and only if $a = x$, $b = y$, and $c = z$.

The definitions of scalar multiplication, vector addition and vector subtraction in three dimensions are similar to the definitions in two dimensions, but each vector has one more component.

> **Definitions**: Vector Arithmetic
>
> If $\mathbf{A} = \langle a_1, a_2, a_3 \rangle$ and $\mathbf{B} = \langle b_1, b_2, b_3 \rangle$ are vectors and k is a scalar, then
>
> then $k\mathbf{A} = \langle ka_1, ka_2, ka_3 \rangle$
>
> $\mathbf{A} + \mathbf{B} = \langle a_1 + b_1, a_2 + b_2, a_3 + b_3 \rangle$
>
> $\mathbf{A} - \mathbf{B} = \mathbf{A} + (-1)\mathbf{B} = \langle a_1 - b_1, a_2 - b_2, a_3 - b_3 \rangle$.

Example 1: For $\mathbf{A} = \langle 3, -4, 2 \rangle$ and $\mathbf{B} = \langle 5, 1, -3 \rangle$, calculate $\mathbf{C} = 3\mathbf{B}$, $\mathbf{D} = 2\mathbf{A} + 3\mathbf{B}$, and $\mathbf{E} = 5\mathbf{A} - 2\mathbf{B}$.

Solution: $\mathbf{C} = 3\mathbf{B} = 3\langle 5, 1, -3 \rangle = \langle 15, 3, -9 \rangle$.

$\mathbf{D} = 2\mathbf{A} + 3\mathbf{B} = 2\langle 3, -4, 2 \rangle + 3\langle 5, 1, -3 \rangle = \langle 6+15, -8+3, 4-9 \rangle = \langle 21, -5, -5 \rangle$.

$\mathbf{E} = 5\mathbf{A} - 2\mathbf{B} = 5\langle 3, -4, 2 \rangle - 2\langle 5, 1, -3 \rangle = \langle 5, -22, 16 \rangle$.

Practice 1: For $\mathbf{A} = \langle 5, -4, 1 \rangle$ and $\mathbf{B} = \langle 2, -3, 4 \rangle$, calculate $\mathbf{C} = 5\mathbf{A}$, $\mathbf{D} = 3\mathbf{A} + 4\mathbf{B}$, and $\mathbf{E} = 2\mathbf{B} - 3\mathbf{A}$.

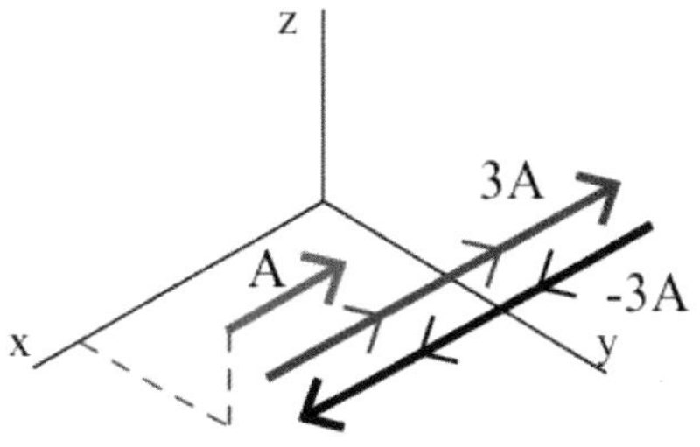

Fig. 2

Each of the given vector arithmetic operations also has a geometric interpretation:

Multiplying by a scalar k results in a vector that is $|k|$ times as long as the original vector, $|k\mathbf{A}| = |k|\,|\mathbf{A}|$. If k is positive, then $\mathbf{A}$ and $k\mathbf{A}$ have the same direction. If k is negative, then $\mathbf{A}$ and $k\mathbf{A}$ point in opposite directions (Fig. 2)

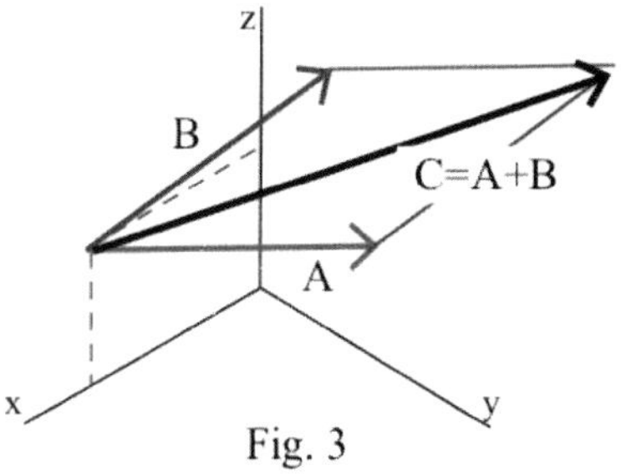

Fig. 3

The sum $\mathbf{C} = \mathbf{A} + \mathbf{B}$ vector can be found geometrically by using the parallelogram or head–to–tail methods described in Section 11.1. (Fig. 3)

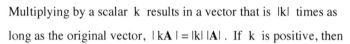

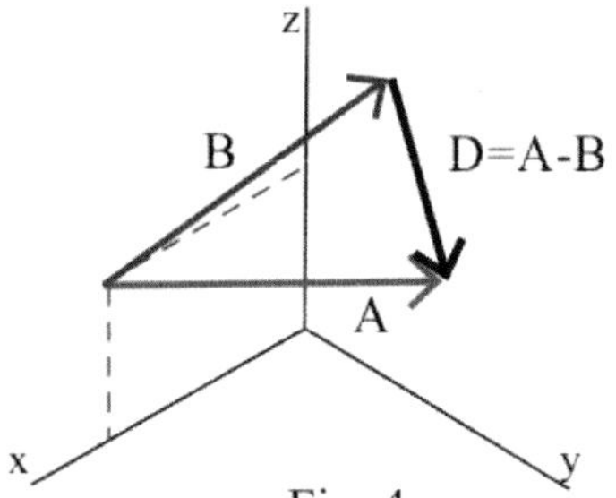

Fig. 4

The difference $\mathbf{D} = \mathbf{A} - \mathbf{B}$ vector can be sketched by adding $\mathbf{A}$ and $-\mathbf{B}$ or by drawing $\mathbf{A}$ and $\mathbf{B}$ with a common starting point and then drawing the line segment from the head of $\mathbf{B}$ to the head of $\mathbf{A}$. (Fig. 4)

Because it is more difficult to make precise drawings and measurements in three dimensions, the geometric methods are seldom used to perform vector arithmetic in three dimensions. These geometric interpretations are still very powerful and are important for understanding the meaning of various arithmetic operations and for understanding how certain algorithms are developed.

Visualizing vector arithmetic in three dimensions: If you took the time in Section 11.2 to build a small model of a 3–dimensional coordinate system, you can use it now to see and handle some 3–dimensional vectors. Sharpened pencils or skewer sticks make good physical "vectors."

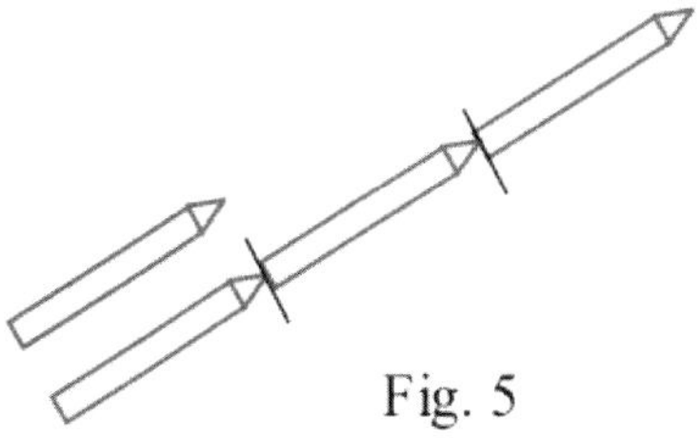

Fig. 5

Visualizing the vectors **A** and 3**A** is easy — just tape together a short pencil and one three times as long (Fig. 5). How could you modify this arrangement to illustrate **A** and –3**A**?

Addition and subtraction are more difficult, but the following vectors fit together nicely:
$\mathbf{C} = \mathbf{A} + \mathbf{B}$ with $\mathbf{A} = \langle 2, 3, 6 \rangle$ and $\mathbf{B} = \langle 4, 0, 0 \rangle$ (Fig. 6), and $\mathbf{D} = \mathbf{A} - \mathbf{B}$ with $\mathbf{A} = \langle 4, 7, 4 \rangle$ and $\mathbf{B} = \langle 4, 2, 4 \rangle$.

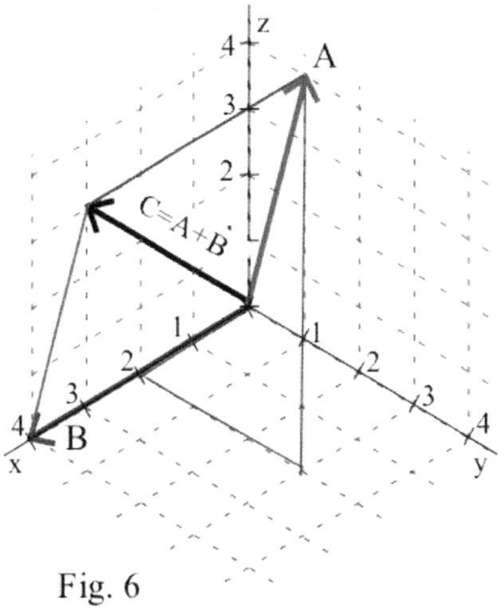

Fig. 6

The length of a vector in three dimensions follows directly from the formula for the distance between points in 3–dimensional space.

The **magnitude** or **length** of a vector $\mathbf{V} = \langle a, b, c \rangle$ is $|\mathbf{V}| = \sqrt{a^2 + b^2 + c^2}$.

Since the components a, b, and c of the vector $\mathbf{V} = \langle a, b, c \rangle$ represent the displacements of the ending point from the starting point in the x, y, and z directions, we can represent **V** as a line segment from the point (0, 0, 0) to the point (a, b, c). Then the length of **V** is the distance from the point (0, 0, 0) to the point (a, b, c): length = { distance from (0, 0, 0) to (a, b, c) } = $\sqrt{a^2 + b^2 + c^2}$.
The only vector in 3–dimensional space with magnitude 0 is the zero vector $\mathbf{0} = \langle 0, 0, 0 \rangle$. The zero vector has no specific direction.

Example 2: Determine the lengths of $\mathbf{A} = \langle 2, 8, 16 \rangle$, $\mathbf{B} = \langle -4, 8, 8 \rangle$, **C** = {vector represented by the line segment from (1,2,3) to (7,–1,9) } , $\mathbf{D} = \mathbf{A} - \mathbf{B}$, and $\mathbf{E} = \mathbf{B} + \mathbf{C}$.

Solution: $|\mathbf{A}| = \sqrt{2^2 + 8^2 + 16^2} = \sqrt{324} = 18$. $|\mathbf{B}| = \sqrt{(-4)^2 + 8^2 + 8^2} = \sqrt{144} = 12$.
$\mathbf{C} = \langle 7-1, -1-2, 9-3 \rangle = \langle 6, -3, 6 \rangle$ so $|\mathbf{C}| = \sqrt{81} = 9$.
$\mathbf{D} = \mathbf{A} - \mathbf{B} = \langle 2-(-4), 8-8, 16-8 \rangle = \langle 6, 0, 8 \rangle$ so $|\mathbf{D}| = \sqrt{100} = 10$.
$\mathbf{E} = \mathbf{B} + \mathbf{C} = \langle -4+6, 8+(-3), 8+6 \rangle = \langle 2, 5, 14 \rangle$ so $|\mathbf{E}| = \sqrt{225} = 15$.

Practice 2: Determine the lengths of $\mathbf{A} = \langle 2,3,6 \rangle$, $\mathbf{B} = \langle 2,1,2 \rangle$, and $\mathbf{C} = \mathbf{A} - 2\mathbf{B}$.

Definitions:

The **direction** of a nonzero vector **A** is the unit vector

$$\mathbf{U} = \frac{1}{|\mathbf{A}|}\mathbf{A} = \frac{\mathbf{A}}{|\mathbf{A}|}.$$

The **standard basis vectors** in the plane are

$$\mathbf{i} = \langle 1,0,0 \rangle, \mathbf{j} = \langle 0,1,0 \rangle, \text{ and } \mathbf{k} = \langle 0,0,1 \rangle. \text{ (Fig. 7)}$$

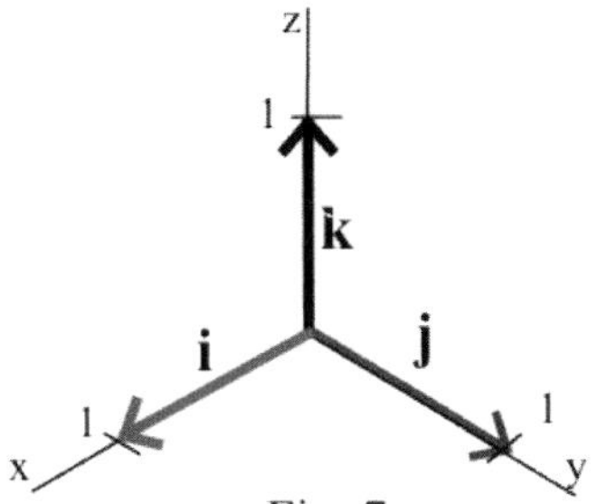

Fig. 7

Example 3: Determine the lengths and directions of $\mathbf{A} = 6\mathbf{i} + 2\mathbf{j} + 9\mathbf{k}$, $\mathbf{B} = 6\mathbf{i} + 3\mathbf{j} - 6\mathbf{k}$, $\mathbf{C} = 3\mathbf{A} + 2\mathbf{B}$, and $\mathbf{D} = \mathbf{B} - 2\mathbf{A}$.

Solution: $|\mathbf{A}| = \sqrt{6^2 + 2^2 + 9^2} = \sqrt{121} = 11$. Direction of **A** is $\frac{\mathbf{A}}{|\mathbf{A}|} = \frac{6}{11}\mathbf{i} + \frac{2}{11}\mathbf{j} + \frac{9}{11}\mathbf{k}$.

$|\mathbf{B}| = \sqrt{6^2 + 3^2 + (-6)^2} = \sqrt{81} = 9$. Direction of **B** is $\frac{\mathbf{B}}{|\mathbf{B}|} = \frac{2}{3}\mathbf{i} + \frac{1}{3}\mathbf{j} - \frac{2}{3}\mathbf{k}$.

$\mathbf{C} = 3\mathbf{A} + 2\mathbf{B} = 30\mathbf{i} + 12\mathbf{j} + 15\mathbf{k}$. $|\mathbf{C}| = \sqrt{1269} \approx 35.6$.

The direction of **C** is $\frac{\mathbf{C}}{|\mathbf{C}|} = \frac{30}{\sqrt{1269}}\mathbf{i} + \frac{12}{\sqrt{1269}}\mathbf{j} + \frac{15}{\sqrt{1269}}\mathbf{k} \approx 0.84\,\mathbf{i} + 0.34\,\mathbf{j} + 0.42\,\mathbf{k}$.

$\mathbf{D} = \mathbf{B} - 2\mathbf{A} = -6\mathbf{i} - 1\mathbf{j} - 24\mathbf{k}$. $|\mathbf{D}| = \sqrt{613} \approx 24.8$.

The direction of **D** is $\frac{\mathbf{D}}{|\mathbf{D}|} = \frac{-6}{\sqrt{613}}\mathbf{i} - \frac{1}{\sqrt{613}}\mathbf{j} - \frac{24}{\sqrt{613}}\mathbf{k} \approx 0.24\mathbf{i} - 0.04\mathbf{j} - 0.97\mathbf{k}$.

Practice 3: Determine the lengths and directions of $\mathbf{A} = 3\mathbf{i} + 2\mathbf{j} - 6\mathbf{k}$, $\mathbf{B} = 6\mathbf{j} - 8\mathbf{k}$, $\mathbf{C} = \mathbf{A} + 3\mathbf{B}$, and $\mathbf{D} = 2\mathbf{B} - 3\mathbf{A}$.

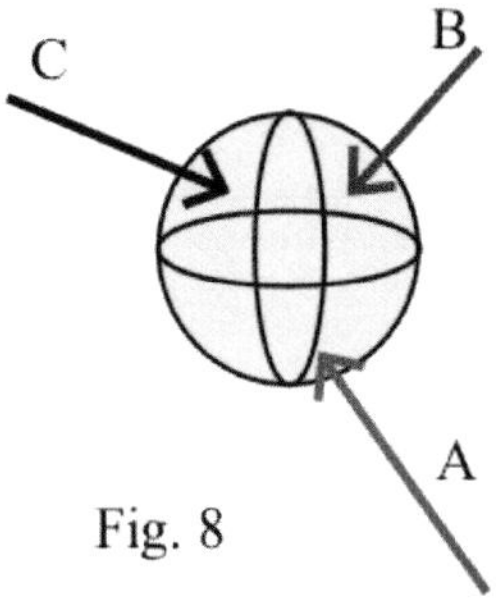

Fig. 8

Example 4: Three players are pushing on a ball, but the ball is not moving (Fig. 8). Player A is pushing with a force of 45 pounds in the direction $\frac{1}{9}\mathbf{i} - \frac{4}{9}\mathbf{j} + \frac{8}{9}\mathbf{k}$, and player B is pushing with a force of 60 pounds in the direction $\frac{4}{6}\mathbf{i} + \frac{4}{6}\mathbf{j} + \frac{2}{6}\mathbf{k}$. How hard and in what direction is player C pushing?

Solution: Since we know the magnitude and direction of the force vectors for players A and B, we can determine the force vector for each player:

$\mathbf{A} = \{\text{ magnitude }\}\{\text{ direction }\} = 45\{\frac{1}{9}\mathbf{i} - \frac{4}{9}\mathbf{j} + \frac{8}{9}\mathbf{k}\} = 5\mathbf{i} - 20\mathbf{j} + 40\mathbf{k}$, and

$\mathbf{B} = \{\text{ magnitude }\}\{\text{ direction }\} = 60\{\frac{4}{6}\mathbf{i} + \frac{4}{6}\mathbf{j} + \frac{2}{6}\mathbf{k}\} = 40\mathbf{i} + 40\mathbf{j} + 20\mathbf{k}$.

Pushing together, their force vector is $\mathbf{A} + \mathbf{B} = 45\mathbf{i} + 20\mathbf{j} + 60\mathbf{k}$. Since the ball is not moving, the force vectors of A, B, and C equal the zero vector, and we can solve for C's force vector:

$\mathbf{C} = -(\mathbf{A}+\mathbf{B}) = -45\mathbf{i} - 20\mathbf{j} - 60\mathbf{k}$.

C's force is $|\mathbf{C}| = \sqrt{(-45)^2 + (-20)^2 + (-60)^2} = \sqrt{6025} \approx 77.6$ pounds. C is pushing in

the direction $\frac{\mathbf{C}}{|\mathbf{C}|} = \frac{-45}{\sqrt{6025}}\mathbf{i} - \frac{20}{\sqrt{6025}}\mathbf{j} - \frac{60}{\sqrt{6025}}\mathbf{k} \approx -0.60\mathbf{i} - 0.26\mathbf{j} - 0.77\mathbf{k}$.

At this point, you should find the arithmetic of vectors in three dimensions is not much different or harder than in two dimensions. Angles, however, are another story, one we consider in Section 11.4. Fortunately, there is a straightforward process for determining the angle between two 3–dimensional vectors and it is useful in a variety of applications.

Beyond Three Dimensions

Just as we can represent points in 4, 5, or n–dimensional space, we can also work with n–dimensional vectors, $\langle a_1, a_2, a_3, \ldots a_n \rangle$. Even though it is no longer easy (or possible?) to work geometrically with these vectors, the arithmetic operations of scalar multiplication, vector addition, vector subtraction, and length are still defined component–by–component and are still easy.

Example 5: Write the vector **V** for the directed line segment from P = (1,2,–3,5,6) to Q = (5,–1,4,0,7), and find the length and direction of **V**.

Solution: $\mathbf{V} = \langle 5-1, -1-2, 4--3, 0-5, 7-6 \rangle = \langle 4, -3, 7, -5, 1 \rangle$.

$|\mathbf{V}| = \sqrt{4^2 + (-3)^2 + 7^2 + (-5)^2 + 1^2} = \sqrt{100} = 10$.

Direction of **V** is $\frac{\mathbf{V}}{|\mathbf{V}|} = \langle 0.4, -0.3, 0.7, -0.5, 0.1 \rangle$.

Practice 4: Calculate the lengths of $\mathbf{A} = \langle 4, 2, -5, 1, 0 \rangle$, $\mathbf{B} = \langle 6, 0, 2, -3, 6 \rangle$, and $\mathbf{C} = 2\mathbf{A} - 3\mathbf{B}$.

We can even determine formulas for some collections of points in higher dimensions.

Example 6: Find a formula for the set of points (w,x,y,z) in 4–dimensional space that are at a distance of 5 units from the point (5,3,–2,1). (This set is "4–sphere" with radius 5 and center (5,3,–2,1).)

Solution: We want the distance from (w,x,y,z) to (5,3,–2,1), $\sqrt{(w-5)^2 + (x-3)^2 + (y+2)^2 + (z-1)^2}$, to be 5 so

$$\sqrt{(w-5)^2 + (x-3)^2 + (y+2)^2 + (z-1)^2} = 5 \text{ and}$$

$$(w-5)^2 + (x-3)^2 + (y+2)^2 + (z-1)^2 = 25 .$$

PROBLEMS

In problems 1 – 4 the vectors **A** and **B** are shown. Sketch and label $\mathbf{C} = 2\mathbf{A}$, $\mathbf{D} = -\mathbf{B}$, and $\mathbf{E} = \mathbf{A} - \mathbf{B}$.

1. See Fig. 9.
2. See Fig. 10.
3. See Fig. 11.
4. See Fig. 12.

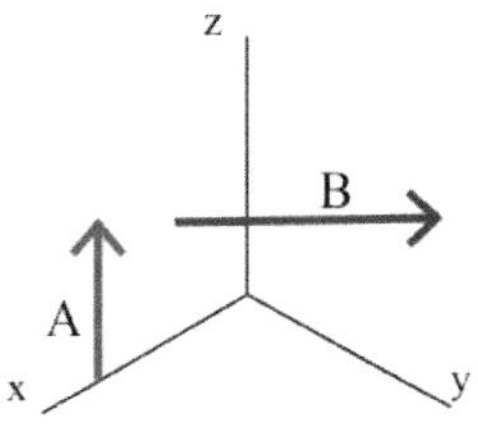

Fig. 9

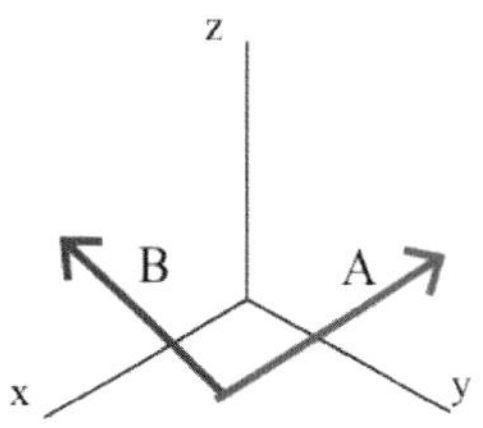

Fig. 10

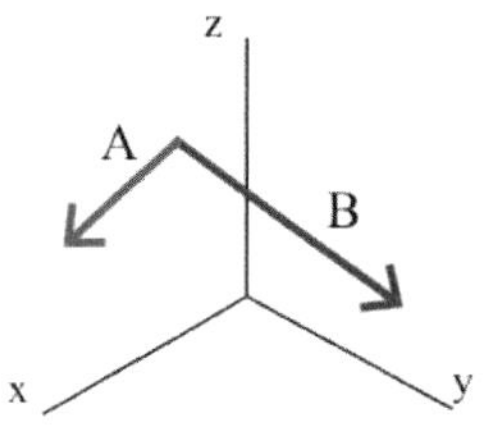

Fig. 11

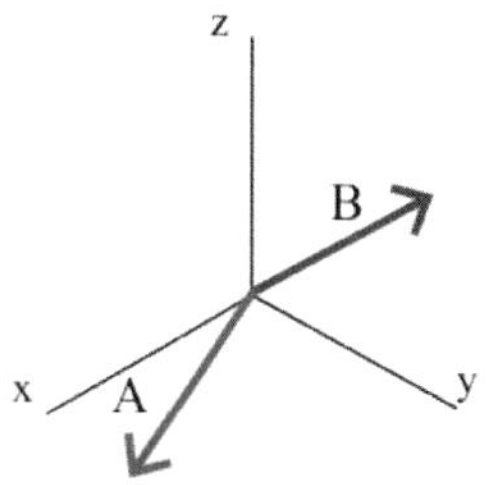

Fig. 12

In problems 5 – 12, vectors **U** and **V** are given. Calculate $\mathbf{W} = 2\mathbf{U} + \mathbf{V}$, $|\mathbf{U}|$, $|\mathbf{V}|$, and $|\mathbf{W}|$, and the directions of **U**, **V**, and **W**.

5. $\mathbf{U} = \langle 2, 3, 6 \rangle$, $\mathbf{V} = \langle 2, -9, 6 \rangle$
6. $\mathbf{U} = \langle 6, 3, 6 \rangle$, $\mathbf{V} = \langle 2, 4, 4 \rangle$
7. $\mathbf{U} = \langle 5, 2, 14 \rangle$, $\mathbf{V} = \langle 4, -7, 4 \rangle$
8. $\mathbf{U} = \langle 8, 4, 1 \rangle$, $\mathbf{V} = \langle 4, 4, -2 \rangle$
9. $\mathbf{U} = 9\mathbf{i} + 6\mathbf{j} + 2\mathbf{k}$, $\mathbf{V} = 3\mathbf{i} + 6\mathbf{j} - 6\mathbf{k}$
10. $\mathbf{U} = 24\mathbf{i} + 2\mathbf{j} + 24\mathbf{k}$, $\mathbf{V} = 10\mathbf{i} - 25\mathbf{j} + 2\mathbf{k}$
11. $\mathbf{U} = 10\mathbf{i} + 11\mathbf{j} + 2\mathbf{k}$, $\mathbf{V} = 6\mathbf{i} + 3\mathbf{j} - 6\mathbf{k}$
12. $\mathbf{U} = 8\mathbf{i} + 1\mathbf{j} + 4\mathbf{k}$, $\mathbf{V} = 3\mathbf{i} + 12\mathbf{j} + 4\mathbf{k}$
13. $\mathbf{A} = \langle 3, 6, -2 \rangle$, $\mathbf{B} = \langle 5, 0, -4 \rangle$. Find **C** so $\mathbf{A} + \mathbf{B} + \mathbf{C} = \mathbf{0}$.
14. $\mathbf{A} = \langle 1, -3, 7 \rangle$, $\mathbf{B} = \langle -5, 2, -3 \rangle$. Find **C** so $\mathbf{A} + \mathbf{B} + \mathbf{C} = \mathbf{0}$.
15. $\mathbf{A} = \langle 9, \pi, -3 \rangle$, $\mathbf{B} = \langle e, 7, 4 \rangle$, and $\mathbf{C} = \langle -6, 2, 0 \rangle$. Find **D** so $\mathbf{A} + \mathbf{B} + \mathbf{C} + \mathbf{D} = \mathbf{0}$.
16. $\mathbf{A} = \langle -4, 3, 1 \rangle$, $\mathbf{B} = \langle 5, 5, -\pi \rangle$, and $\mathbf{C} = \langle 0, -2, 1 \rangle$. Find **D** so $\mathbf{A} + \mathbf{B} + \mathbf{C} + \mathbf{D} = \mathbf{0}$.
17. Which of the following vectors has the smallest magnitude and which has the largest: $\mathbf{A} = \langle 6, 9, 2, 3 \rangle$, $\mathbf{B} = \langle -8, 0, 6, 0 \rangle$, $\mathbf{C} = \langle -6, 3, 3, 6 \rangle$, and $\mathbf{D} = \langle 9, 1, 8, -5 \rangle$?
18. Which of the following vectors has the smallest magnitude and which has the largest: $\mathbf{A} = \langle 11, 2, -10, 1 \rangle$, $\mathbf{B} = \langle -5, 14, 2, 6 \rangle$, $\mathbf{C} = \langle 2, 8, 16, -5 \rangle$, and $\mathbf{D} = \langle 12, 3, 7, 3 \rangle$?
19. $\mathbf{A} = \langle 2, 4, 3 \rangle$. Sketch **A** and find the "shadows" of **A** on the coordinate planes (e.g., on the xy, xz, and yz planes).

19. $\mathbf{A} = \langle 2, 4, 3 \rangle$. Sketch **A** and find the "shadows" of **A** on the coordinate planes (e.g., on the xy, xz, and yz planes).

20. $\mathbf{B} = \langle 4, 1, 2 \rangle$. Sketch **B** and find the "shadows" of **B** on the coordinate planes (e.g., on the xy, xz, and yz planes).

21. $\mathbf{C} = \langle 5, 2, 3 \rangle$. Sketch **C** and find the "shadows" of **C** on the coordinate planes.

22. $\mathbf{D} = \langle 3, 4, 0 \rangle$. Sketch **D** and find the "shadows" of **D** on the coordinate planes.

23. $\mathbf{A} = \langle 1, 0, 0 \rangle$. Sketch **A** and find three nonparallel vectors that are perpendicular to **A** . How many vectors are perpendicular to **A** ?

24. $\mathbf{B} = \langle 0, 0, 1 \rangle$. Sketch **B** and find three nonparallel vectors that are perpendicular to **B** . How many vectors are perpendicular to **B** ?

25. $\mathbf{C} = \langle 1, 2, 0 \rangle$. Sketch **C** and find two nonparallel vectors that are perpendicular to **C** .

26. $\mathbf{C} = \langle 0, 2, 3 \rangle$. Sketch **C** and find two nonparallel vectors that are perpendicular to **C** .

In problems 27 – 30, you are asked to sketch smooth curves that go through given points with given directions. Later, in Section 12.1, we will discuss how to find parametric equations for curves that satisfy conditions of this type.

27. Sketch a smooth curve that goes through the point (0,0,1) with direction $\langle 1, 0, 0 \rangle$ and then bends and goes through the point (0,1,0) with direction $\langle 0, 1, 0 \rangle$.

28. Sketch a smooth curve that goes through the point (0,0,1) with direction $\langle 0.8, 0, 0.6 \rangle$ and then bends and goes through the point (0,2,0) with direction $\langle 0, 0, 1 \rangle$.

29. Sketch a smooth curve that goes through the point (2,0,0) with direction $\langle 0, 0, 1 \rangle$ and then bends and goes through the point (0,1,2) with direction $\langle -1, 0, 0 \rangle$.

30. Sketch a smooth curve that goes through the point (4, 1, 2) with direction $\langle 0, 0, -1 \rangle$ and then bends and goes through the point (0,0,0) with direction $\langle 0, -1, 0 \rangle$.

31. Find a linear equation for x(t) so $x(0) = 3$ and $x(1) = 7$, for y(t) so $y(0) = 5$ and $y(1) = 4$, and for z(t) so $z(0) = 1$ and $z(1) = 1$.

32. Find a linear equation for x(t) so $x(0) = 1$ and $x(1) = 5$, for y(t) so $y(0) = 2$ and $y(1) = 0$, and for z(t) so $z(0) = 3$ and $z(1) = 5$.

33. Find a linear equation for x(t) so $x(0) = 2$ and $x(1) = 5$, for y(t) so $y(0) = 3$ and $y(1) = 3$, and for z(t) so $z(0) = 6$ and $z(1) = 1$.

34. Find a linear equation for x(t) so $x(0) = 0$ and $x(1) = -3$, for y(t) so $y(0) = -2$ and $y(1) = 1$, and for z(t) so $z(0) = 4$ and $z(1) = 1$.

Practice Answers

Practice 1: $\mathbf{C} = 5\mathbf{A} = 5\langle 5, -4, 1 \rangle = \langle 25, -20, 5 \rangle$.

$\mathbf{D} = 3\mathbf{A} + 4\mathbf{B} = 3\langle 5, -4, 1 \rangle + 4\langle 2, -3, 4 \rangle = \langle 15+8, -12-12, 3+16 \rangle = \langle 23, -24, 19 \rangle$.

$\mathbf{E} = 2\mathbf{B} - 3\mathbf{A} = 2\langle 2, -3, 4 \rangle - 3\langle 5, -4, 1 \rangle = \langle 4-15, -6+12, 8-3 \rangle = \langle -11, 6, 5 \rangle$.

Practice 2: $|\mathbf{A}| = \sqrt{2^2 + 3^2 + 6^2} = \sqrt{49} = 7$. $|\mathbf{B}| = \sqrt{2^2 + 1^2 + 2^2} = \sqrt{9} = 3$.

$\mathbf{C} = \mathbf{A} - 2\mathbf{B} = \langle 2, 3, 6 \rangle - 2\langle 2, 1, 2 \rangle = \langle -2, 1, 2 \rangle$ so $|\mathbf{C}| = \sqrt{9} = 3$.

Practice 3: $|\mathbf{A}| = \sqrt{3^2 + 2^2 + (-6)^2} = \sqrt{49} = 7$. Direction of $\mathbf{A}$ is $\frac{\mathbf{A}}{|\mathbf{A}|} = \frac{3}{7}\mathbf{i} + \frac{2}{7}\mathbf{j} - \frac{6}{7}\mathbf{k}$.

$|\mathbf{B}| = \sqrt{0^2 + 6^2 + (-8)^2} = \sqrt{100} = 10$. Direction of $\mathbf{B}$ is $\frac{\mathbf{B}}{|\mathbf{B}|} = \frac{6}{10}\mathbf{j} - \frac{8}{10}\mathbf{k}$.

$\mathbf{C} = \mathbf{A} + 3\mathbf{B} = 3\mathbf{i} + 20\mathbf{j} - 30\mathbf{k}$. $|\mathbf{C}| = \sqrt{1309} \approx 36.18$.

Direction of $\mathbf{C}$ is $\frac{\mathbf{C}}{|\mathbf{C}|} = \frac{3}{\sqrt{1309}}\mathbf{i} + \frac{20}{\sqrt{1309}}\mathbf{j} - \frac{30}{\sqrt{1309}}\mathbf{k} \approx 0.08\,\mathbf{i} + 0.55\,\mathbf{j} - 0.83\,\mathbf{k}$.

$\mathbf{D} = 2\mathbf{B} - 3\mathbf{A} = -9\mathbf{i} + 6\mathbf{j} + 2\mathbf{k}$. $|\mathbf{D}| = \sqrt{121} = 11$.

Direction of $\mathbf{D}$ is $\frac{\mathbf{D}}{|\mathbf{D}|} = \frac{-9}{11}\mathbf{i} + \frac{6}{11}\mathbf{j} + \frac{2}{11}\mathbf{k} \approx -0.82\mathbf{i} + 0.55\mathbf{j} + 0.18\mathbf{k}$.

Practice 4: $|\mathbf{A}| = \sqrt{4^2 + 2^2 + (-5)^2 + 1^2 + 0} = \sqrt{46} \approx 6.8$.

$|\mathbf{B}| = \sqrt{6^2 + 0^2 + 2^2 + (-3)^2 + 6^2} = \sqrt{85} \approx 9.2$.

$\mathbf{C} = 2\mathbf{A} - 3\mathbf{B} = 2\langle 4, 2, -5, 1, 0 \rangle - 3\langle 6, 0, 2, -3, 6 \rangle = \langle -10, 4, -16, 11, -18 \rangle$.

$|\mathbf{C}| = \sqrt{(-10)^2 + 4^2 + (-16)^2 + 11^2 + (-18)^2} = \sqrt{817} \approx 28.6$.

11.4 DOT PRODUCT

In the previous sections we looked at the meaning of vectors in two and three dimensions, but the only operations we used were addition and subtraction of vectors and multiplication by a scalar. Some of the applications of 2–dimensional vectors used the angles that the vectors made with the coordinate axes and with each other, but, so far, in three dimensions we have not used angles. This section addresses both of those situations. It introduces a way to multiply two vectors, in two and three dimensions, called the dot product, and this dot product provides us with a relatively easy way to determine angles between vectors. Section 11.5 introduces a different method of multiplying two vectors, the cross product, in three dimensions that has other useful applications.

Since we will soon have three different types of multiplications for a vector (scalar, dot, and cross), it is important that you distinguish among them and call each multiplication operation by its full name.

Definition: Dot Product

Two dimensions: The **dot product** of $\mathbf{A} = \langle a_1, a_2 \rangle$ and $\mathbf{B} = \langle b_1, b_2 \rangle$

is $\mathbf{A}\bullet\mathbf{B} = a_1b_1 + a_2b_2$.

Three dimensions: The **dot product** of $\mathbf{A} = \langle a_1, a_2, a_3 \rangle$ and $\mathbf{B} = \langle b_1, b_2, b_3 \rangle$

is $\mathbf{A}\bullet\mathbf{B} = a_1b_1 + a_2b_2 + a_3b_3$.

Both vectors in the dot product must have the same number of components, and the result of the dot product $\mathbf{U}\bullet\mathbf{V}$ is a scalar.

Example 1: For $\mathbf{A} = \langle 4, 1, 8 \rangle$ and $\mathbf{B} = \langle 2, -4, 4 \rangle$, calculate $\mathbf{A}\bullet\mathbf{B}$, $\mathbf{A}\bullet\mathbf{A}$, $\mathbf{B}\bullet\mathbf{B}$, and $(\mathbf{A}-\mathbf{B})\bullet(\mathbf{A}+2\mathbf{B})$.

Solution: $\mathbf{A}\bullet\mathbf{B} = \langle 4, 1, 8 \rangle\bullet\langle 2, -4, 4 \rangle = (4)(2) + (1)(-4) + (8)(4) = 8 - 4 + 32 = 36$.

$\mathbf{A}\bullet\mathbf{A} = \langle 4, 1, 8 \rangle\bullet\langle 4, 1, 8 \rangle = (4)(4) + (1)(1) + (8)(8) = 81$.

$\mathbf{B}\bullet\mathbf{B} = \langle 2, -4, 4 \rangle\bullet\langle 2, -4, 4 \rangle = (2)(2) + (-4)(-4) + (4)(4) = 36$.

You should notice that $\mathbf{A}\bullet\mathbf{A} = |\mathbf{A}|^2$ and $\mathbf{B}\bullet\mathbf{B} = |\mathbf{B}|^2$.

Finally, $\mathbf{A}-\mathbf{B} = \langle 2, 5, 4 \rangle$ and $\mathbf{A} + 2\mathbf{B} = \langle 8, -7, 16 \rangle$

so $(\mathbf{A}-\mathbf{B})\bullet(\mathbf{A}+2\mathbf{B}) = \langle 2, 5, 4 \rangle\bullet\langle 8, -7, 16 \rangle = (2)(8) + (5)(-7) + (4)(16) = 45$.

Practice 1: For $\mathbf{U} = \langle 2, 6, -3 \rangle$ and $\mathbf{V} = \langle -1, 2, 2 \rangle$, calculate $\mathbf{U}\bullet\mathbf{V}$, $\mathbf{U}\bullet\mathbf{U}$, $\mathbf{V}\bullet\mathbf{V}$, $\mathbf{U}\bullet(\mathbf{U}+\mathbf{V})$, and $\mathbf{U}\bullet\mathbf{U} + \mathbf{U}\bullet\mathbf{V}$. Does $\mathbf{U}\bullet\mathbf{U} = |\mathbf{U}|^2$? Does $\mathbf{U}\bullet\mathbf{V} = \mathbf{V}\bullet\mathbf{U}$?

As you might have noticed in Example 1 and Practice 1, the dot product seems to have some of the properties of ordinary multiplication of numbers.

Properties of the Dot Product:

(1) $\mathbf{A}\bullet\mathbf{A} = |\mathbf{A}|^2$

(2) $\mathbf{A}\bullet\mathbf{B} = \mathbf{B}\bullet\mathbf{A}$

(3) $k(\mathbf{A}\bullet\mathbf{B}) = (k\mathbf{A})\bullet\mathbf{B} = \mathbf{A}\bullet(k\mathbf{B})$

(4) $\mathbf{A}\bullet(\mathbf{B}+\mathbf{C}) = \mathbf{A}\bullet\mathbf{B} + \mathbf{A}\bullet\mathbf{C}$

All of these properties can be proved using the definition of the dot product.

Proof of (1): If $\mathbf{A} = \langle a_1, a_2, a_3 \rangle$ then $\mathbf{A}\bullet\mathbf{A} = (a_1)^2 + (a_2)^2 + (a_3)^2$ and

$$|\mathbf{A}|^2 = \left(\sqrt{(a_1)^2 + (a_2)^2 + (a_3)^2}\right)^2 = (a_1)^2 + (a_2)^2 + (a_3)^2 \text{ so } \mathbf{A}\bullet\mathbf{A} = |\mathbf{A}|^2 .$$

Proof of (3): $k(\mathbf{A}\bullet\mathbf{B}) = k(a_1b_1 + a_2b_2 + a_3b_3) = ka_1b_1 + ka_2b_2 + ka_3b_3$.

$(k\mathbf{A})\bullet\mathbf{B} = \langle ka_1, ka_2, ka_3 \rangle\bullet\langle b_1, b_2, b_3 \rangle = ka_1b_1 + ka_2b_2 + ka_3b_3$. And

$\mathbf{A}\bullet(k\mathbf{B}) = \langle a_1, a_2, a_3 \rangle\bullet\langle kb_1, kb_2, kb_3 \rangle = a_1(kb_1) + a_2(kb_2) + a_3(kb_3)$

$= ka_1b_1 + ka_2b_2 + ka_3b_3$ so $k(\mathbf{A}\bullet\mathbf{B}) = (k\mathbf{A})\bullet\mathbf{B} = \mathbf{A}\bullet(k\mathbf{B})$

Practice 2: Prove Property (2) for 3–dimensional vectors.

The next result about dot products is very important, and much of the usefulness of dot products follows from it. It enables us to easily determine the angle between two vectors in two or three (or more) dimensions.

Angle Property of Dot Products

$\mathbf{A}\bullet\mathbf{B} = |\mathbf{A}|\,|\mathbf{B}|\cos(\theta)$ where θ is the angle between **A** and **B**. Equivalently,

if **A** and **B** are nonzero vectors, then the angle θ between **A** and **B** satisfies $\cos(\theta) = \dfrac{\mathbf{A}\bullet\mathbf{B}}{|\mathbf{A}|\,|\mathbf{B}|}$.

Proof of the Angle Property: The proof uses the Law of Cosines and several of the properties of the dot product.

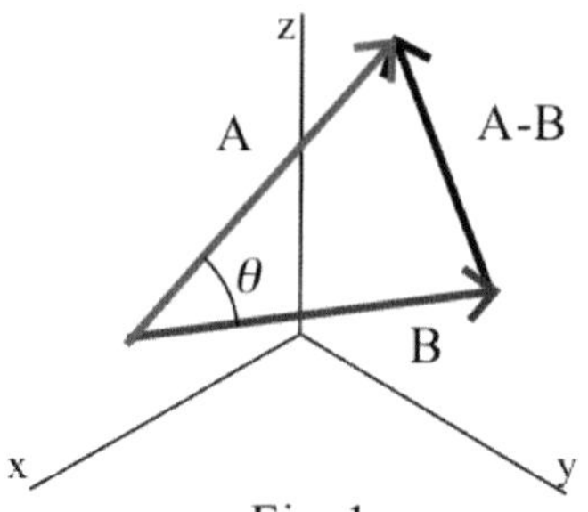

Fig. 1

The vectors **A**, **B** and **A–B** can be arranged to form a triangle (Fig. 1) with the angle θ between **A** and **B**. Applying the Law of Cosines to this triangle, we have

$$|\mathbf{A}-\mathbf{B}|^2 = |\mathbf{A}|^2 + |\mathbf{B}|^2 - 2|\mathbf{A}||\mathbf{B}|\cos(\theta) .$$

We can also use the properties of the dot product to expand $|\mathbf{A}-\mathbf{B}|^2$ in a different way: $|\mathbf{A}-\mathbf{B}|^2 = (\mathbf{A}-\mathbf{B})\bullet(\mathbf{A}-\mathbf{B})$

$$= \mathbf{A}\bullet\mathbf{A} - \mathbf{A}\bullet\mathbf{B} - \mathbf{B}\bullet\mathbf{A} + \mathbf{B}\bullet\mathbf{B} = |\mathbf{A}|^2 - 2\mathbf{A}\bullet\mathbf{B} + |\mathbf{B}|^2.$$

From these two representations for $|\mathbf{A}-\mathbf{B}|^2$ we have that

$$|\mathbf{A}|^2 - 2\mathbf{A}\bullet\mathbf{B} + |\mathbf{B}|^2 = |\mathbf{A}|^2 + |\mathbf{B}|^2 - 2|\mathbf{A}||\mathbf{B}|\cos(\theta) \text{ so}$$

$$-2\mathbf{A}\bullet\mathbf{B} = -2|\mathbf{A}||\mathbf{B}|\cos(\theta) \text{ and } \mathbf{A}\bullet\mathbf{B} = |\mathbf{A}|\,|\mathbf{B}|\cos(\theta).$$

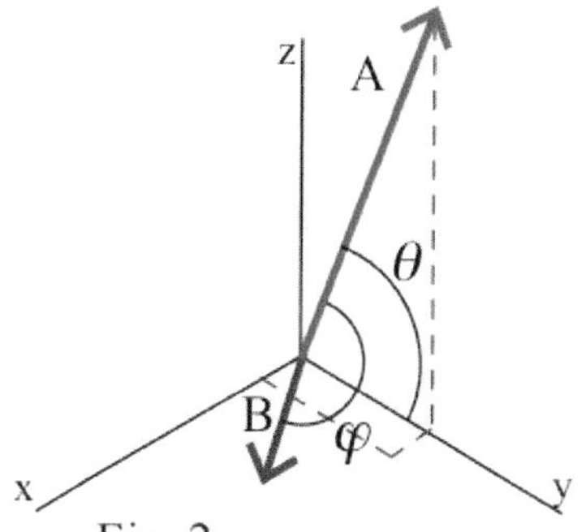

Fig. 2

Example 2: Let $\mathbf{A} = \langle 2, 5, 14 \rangle$ and $\mathbf{B} = \langle 2, 1, -2 \rangle$. Find the angles (a) between **A** and **B** and (b) between **A** and the positive y–axis (Fig. 2).

Solution: (a) $|\mathbf{A}| = \sqrt{4+25+196} = 15$, $|\mathbf{B}| = \sqrt{4+1+4} = 3$, $\mathbf{A}\bullet\mathbf{B} = 4+5-28 = -19$, and

$$\cos(\varphi) = \frac{\mathbf{A}\bullet\mathbf{B}}{|\mathbf{A}|\,|\mathbf{B}|} = \frac{-19}{(15)(3)} \approx -0.4222 \text{ so } \varphi \approx 2.01 \text{ or about } 115.0°.$$

(b) The basis vector $\mathbf{j} = \langle 0, 1, 0 \rangle$ points along the positive y–axis so the angle between **A** and the positive y–axis is the same as the angle between A and $\mathbf{j}$.

$|\mathbf{A}| = 15$, $|\mathbf{j}| = 1$, and $\mathbf{A}\bullet\mathbf{j} = (2)(0) + (5)(1) + (14)(0) = 5$ so

$$\cos(\theta) = \frac{\mathbf{A}\bullet\mathbf{j}}{|\mathbf{A}|\,|\mathbf{j}|} = \frac{5}{(15)(1)} \approx 0.333 \text{ so } \theta \approx 1.23 \text{ or about } 70.5°.$$

Practice 3: Let $\mathbf{A} = \langle 2, -6, 3 \rangle$, $\mathbf{B} = \langle 4, 8, -1 \rangle$ and $\mathbf{C} = \langle 3, 0, -4 \rangle$ and determine the angles between the vectors (a) **A** and **B**, (b) **A** and C, (c) **B** and the negative x–axis, and (d) **C** and the positive y–axis.

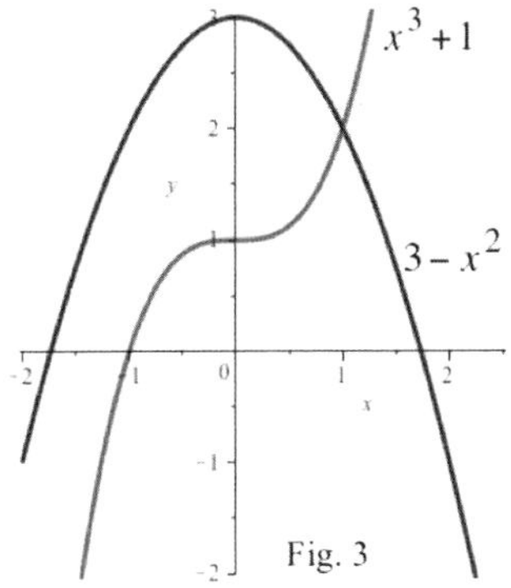

Fig. 3

Example 3: Find the angle of intersection of the graphs of $f(x) = x^3 + 1$ and $g(x) = 3 - x^2$ at the point (1, 2). (The angle of intersection is the angle between tangent vectors to the graphs at the point.) Fig. 3.

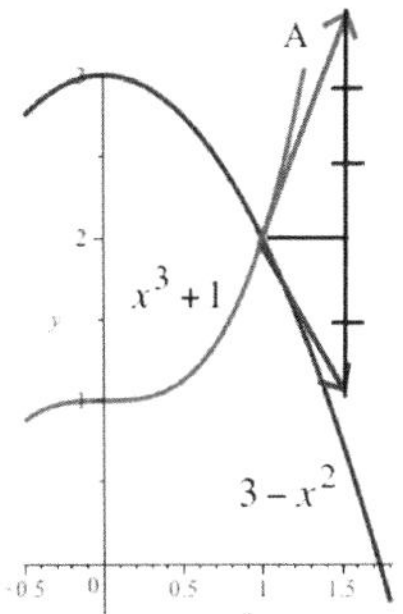

Fig. 4

Solution: $f'(x) = 3x^2$, so the slope of f at (1,2) is $f'(1) = 3(1)^2 = 3$: $\frac{\text{rise}}{\text{run}} = \frac{3}{1}$ and a vector, $\mathbf{A} = 1\mathbf{i} + 3\mathbf{j}$, with this slope is shown in Fig. 4. Similarly, $g'(x) = -2x$, so the slope of g at (1,2) is $g'(1) = -2(1) = -2$: $\frac{\text{rise}}{\text{run}} = \frac{-2}{1}$ and vector $\mathbf{B} = 1\mathbf{i} - 2\mathbf{j}$ has the same slope. Then $|\mathbf{A}| = \sqrt{10}$, $|\mathbf{B}| = \sqrt{5}$, and $\mathbf{A}\bullet\mathbf{B} = -5$, so

$$\cos(\theta) = \frac{-5}{\sqrt{10}\sqrt{5}} \approx -0.707. \text{ Then } \theta = \arccos(-0.707) \approx 2.356 \text{ (or } 135^{\circ}).$$

Note: If $y = f(x)$, then a tangent vector to the graph of f at the point $(x_0, f(x_0))$ is $\mathbf{T} = \langle 1, f'(x_0) \rangle$.
If a curve is given parametrically by $x(t)$ and $y(t)$,
then a tangent vector to the curve when $t = t_o$ is $\mathbf{T} = x'(t_o)\mathbf{i} + y'(t_o)\mathbf{j} = \langle x'(t_o), y'(t_o) \rangle$.

Some books define the dot product **A•B** as **A•B** = |**A**| |**B**| cos(θ) and then derive the definition we gave, **A•B** = $a_1b_1 + a_2b_2 + a_3b_3$, as a property. Either way, the pattern **A•B** = $a_1b_1 + a_2b_2 + a_3b_3$ is typically used to compute the dot product, and the angle property **A•B** = |**A**| |**B**| cos(θ) is typically used to help us see what the dot product measures and to help us derive and simplify some vector algorithms.

Criteria for A and B to be Perpendicular

Let **A** and **B** be nonzero vectors. **A** and **B** are perpendicular if and only if **A•B** = 0.

Proof: If **A** and **B** are perpendicular, then $\theta = \pm \pi/2$ so **A•B** = |**A**| |**B**| cos(θ) = |**A**| |**B**| (0) = 0.
If 0 = **A•B** = |**A**| |**B**| cos(θ) , then cos(θ) = 0 so $\theta = \pm \pi/2$ ($\pm 2\pi n$) and **A** and **B** are perpendicular.

Example 4: (a) Find a vector perpendicular to **V** = $\langle 1, 2, 3 \rangle$.
(b) Find a vector perpendicular to the line $4x + 3y = 24$.

Solution: (a) We need a nonzero vector **N** so that **N•V** = 0 . If **N** = $\langle a, b, c \rangle$, then we want to find values for a, b, and c so that $a + 2b + 3c = 0$, and there are lots of values for a, b, and c that work: **N** = $\langle 0, -3, 2 \rangle$, $\langle 3, 0, -1 \rangle$, $\langle 2, -1, 0 \rangle$, $\langle 1, 1, -1 \rangle$, $\langle 1, -2, 1 \rangle$ and lots of others all give **N•V** = 0 . Fig. 5 illustrates why a single vector in three dimensions can have perpendicular vectors that point in an infinite number of different directions.

(b) The points P = (6, 0) and Q = (0, 8) both lie on the line, so the vector **V** from P to Q is parallel to the line, and **V** = $\langle 0-6, 8-0 \rangle = \langle -6, 8 \rangle$. If **N** is perpendicular to **V** , then **N** is perpendicular to the line, and **N** = $\langle 4, 3 \rangle$ is perpendicular to **V** since **N•V** = 0 . The vector **N** = $\langle 4, 3 \rangle$ is perpendicular to the line $4x + 3y = 24$. Every scalar multiple k**N** with $k \neq 0$ is also perpendicular to the line.

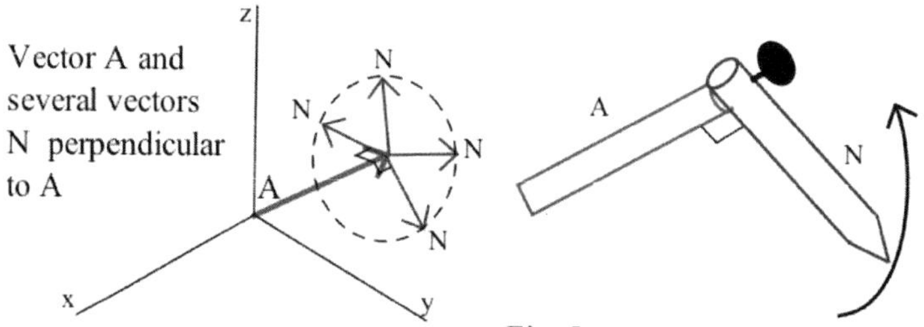

Fig. 5

Practice 4: Find a vector **N** perpendicular to the line $-5x + 2y = 30$. Is **N** = $\langle -5, 2 \rangle$ perpendicular to $-5x + 2y = 30$?

You might have noticed a pattern in the vectors perpendicular to the lines in Example 4 and Practice 4. The vector $\langle 4, 3 \rangle$ is perpendicular to the line $4x + 3y = 24$, and the vector $\langle -5, 2 \rangle$ is perpendicular to the line $-5x + 2y = 30$. The next result says this pattern is not an accident.

> **Finding a Vector Perpendicular to a Line**
>
> The vector $\mathbf{N} = a\mathbf{i} + b\mathbf{j}$ is perpendicular to the line $ax + by = c$.

Problem 66 asks you to prove this result.

Projection of a Vector onto a Vector

The length of a force vector tells us the amount of force in the direction of the vector, but sometimes we want to know the size of the force in another direction (Fig. 6). One of the examples in two dimensions (Fig. 7) involved finding the

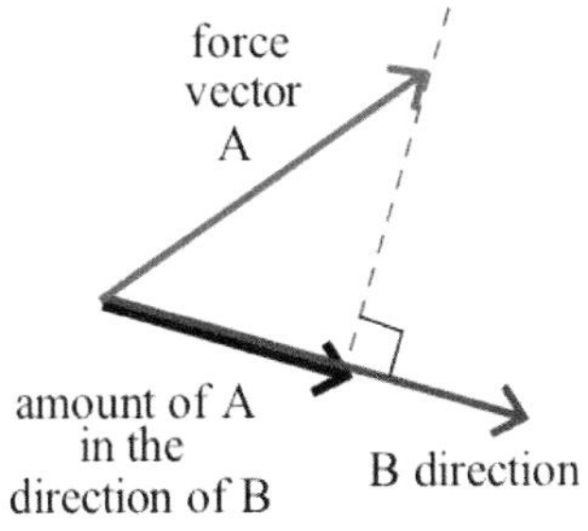

Fig. 6

amount of "horizontal force" obtained when we pulled on a box at an angle to the horizontal. Similar questions can also be asked if one of the directions is not horizontal (Fig. 8) and in three or more dimensions (Fig. 9). Since we now have a method for determining the angle between two vectors in three dimensions, the solutions are relatively straightforward. The vector representing the amount of a vector **A** in the direction of a vector **B** is called the "projection of **A** onto **B**" and is denoted as $\mathbf{Proj_B\,A}$.

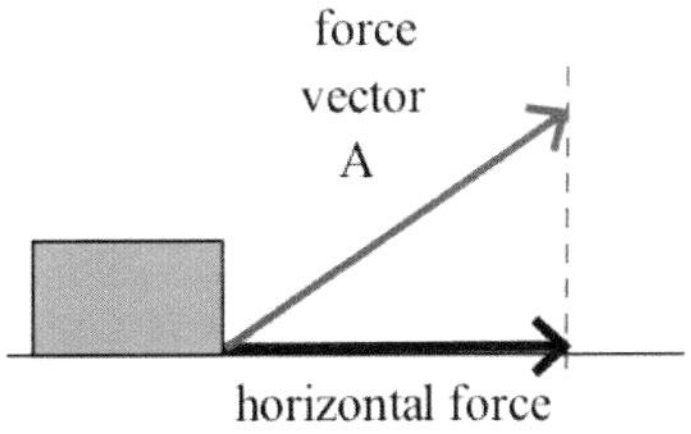

Fig. 7

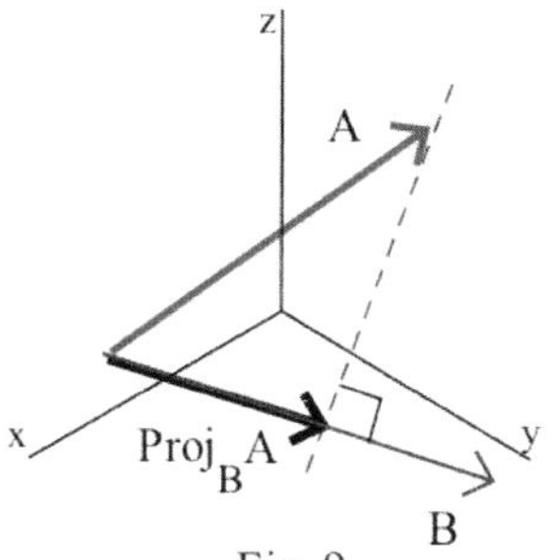

Fig. 9

Visualizing the projection of A onto B: Fig. 10 shows several geometric examples of the projection of a vector **A** onto a vector **B**. We arrange **A** and **B** to have the same starting point, draw a (dotted) line through the head of **A** and perpendicular to **B**, and form the projection of **A** onto **B** as the vector from the starting point of **B** to the point where the dotted line intersects **B** (or an extension of **B**). The projection of **A** onto **B** is a vector along the line of **B** — the direction of the projection of **A** onto **B** is either the same direction as **B**, **B**/|**B**|, or opposite the direction of **B**, –**B**/|**B**| .

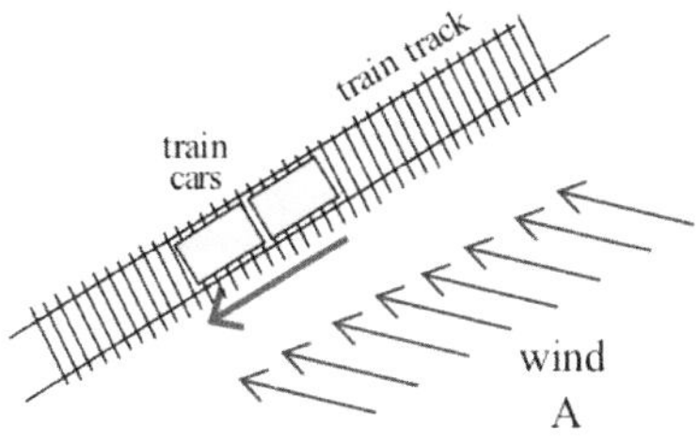

Fig. 8

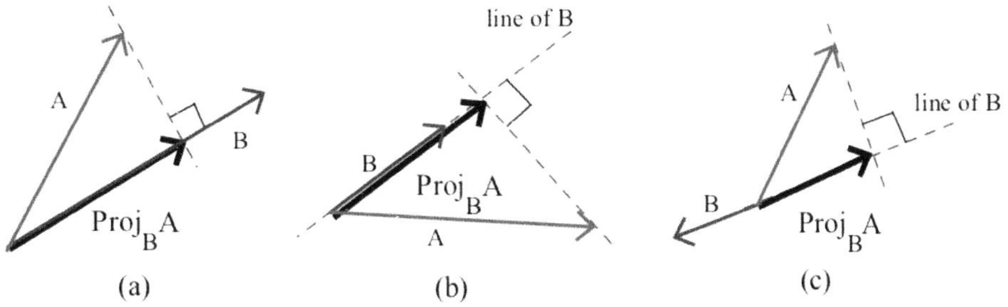

Fig. 10

Fig. 11 shows some examples of the projection of **B** onto **A**, $\mathbf{Proj_A\ B}$.

The resulting projection vector is always along the line of **A**.

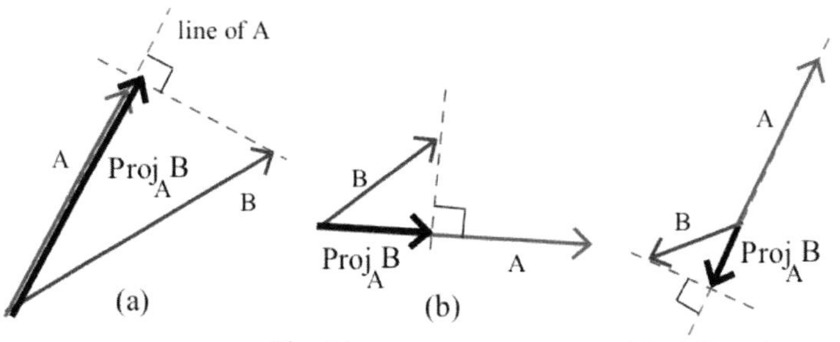

Fig. 11

Once we understand the geometric meaning of "the projection of **A** onto **B**," trigonometry enables us to determine the vector projection of **A** onto **B**: $\mathbf{Proj_B\ A}$ (Fig. 12), and its magnitude, $|\mathbf{Proj_B\ A}|$, called the scalar projection.

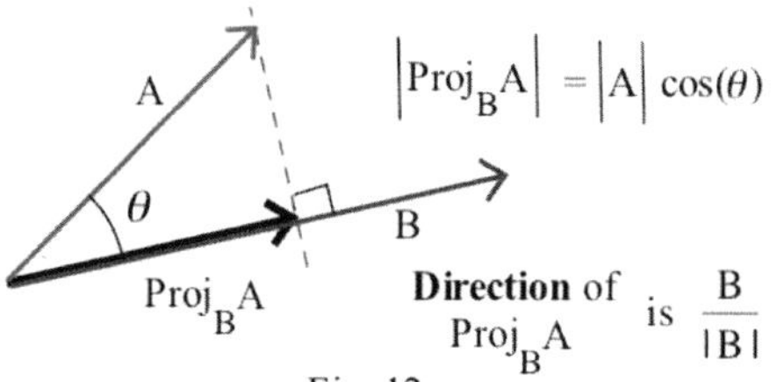

Fig. 12

Definitions: Vector Projection, Scalar Projection

The **vector projection of A onto B** is the

(magnitude of the projection) times (the direction of **B**): $\mathbf{Proj_B\ A} = (\ |\mathbf{A}|\cos(\theta)\ )\left(\frac{\mathbf{B}}{|\mathbf{B}|}\right)$

The **scalar projection of A onto B** is $|\mathbf{A}|\cos(\theta)$ where θ is the angle between **A** and **B**.

Note: "Projection of **A** onto **B**" usually means Vector Projection.

We can use properties of the dot product to simplify the calculation of projections.

Since θ is the angle between **A** and B, then $\cos(\theta) = \frac{\mathbf{A\bullet B}}{|\mathbf{A}||\mathbf{B}|}$ so the scalar projection of **A** onto **B** is

$|\mathbf{A}|\cos(\theta) = |\mathbf{A}|\frac{\mathbf{A\bullet B}}{|\mathbf{A}||\mathbf{B}|} = \frac{\mathbf{A\bullet B}}{|\mathbf{B}|}$. Putting this result into the definition of the vector projection of **A** onto **B**, we get

$$(\ |\mathbf{A}|\cos(\theta)\ )\left(\frac{\mathbf{B}}{|\mathbf{B}|}\right) = \frac{\mathbf{A\bullet B}}{|\mathbf{B}|}\left(\frac{\mathbf{B}}{|\mathbf{B}|}\right) = \frac{\mathbf{A\bullet B}}{|\mathbf{B}|^2}\mathbf{B}.$$

Calculating Scalar and Vector Projections

Vector projection of **A** onto **B** is $\mathbf{Proj_B\ A} = \left(\frac{\mathbf{A\bullet B}}{|\mathbf{B}|^2}\right)\mathbf{B}$ (a vector).

Scalar projection of **A** onto **B** = $\frac{\mathbf{A\bullet B}}{|\mathbf{B}|}$ (a scalar) = magnitude of $\mathbf{Proj_B\ A}$.

Example 5: For $\mathbf{A} = \langle 6, -2, 3 \rangle$ and $\mathbf{B} = \langle 4, 8, -1 \rangle$, calculate scalar and vector projections of
(a) **A** onto **B**, (b) **B** onto **A**, (c) **A** onto the positive x–axis,
(d) **A** onto the positive y–axis, and (e) **A** onto the positive z–axis.

Solution: $|\mathbf{A}| = 7$, $|\mathbf{B}| = 9$, and $\mathbf{A}\bullet\mathbf{B} = 24 - 16 - 3 = 5$.

(a) The scalar projection of **A** onto **B** is $\dfrac{\mathbf{A}\bullet\mathbf{B}}{|\mathbf{B}|} = \dfrac{5}{9}$.

The vector projection of **A** onto **B** is $\mathbf{Proj_B\, A} = \left(\dfrac{\mathbf{A}\bullet\mathbf{B}}{|\mathbf{B}|^2}\right)\mathbf{B} = \left(\dfrac{5}{81}\right)\langle 4, 8, -1 \rangle = \left\langle \dfrac{20}{81}, \dfrac{40}{81}, \dfrac{-5}{81} \right\rangle$.

(b) Scalar projection of **B** onto **A** is $\dfrac{\mathbf{A}\bullet\mathbf{B}}{|\mathbf{A}|} = \dfrac{5}{7}$. Vector projection is $\left(\dfrac{\mathbf{A}\bullet\mathbf{B}}{|\mathbf{A}|^2}\right)\mathbf{A} = \left\langle \dfrac{30}{49}, \dfrac{-10}{49}, \dfrac{15}{49} \right\rangle$.

(c) **i** has the same direction as the positive x–axis. Scalar projection of **A** onto **i** is $\dfrac{\mathbf{A}\bullet\mathbf{i}}{|\mathbf{i}|} = 6$.

Vector projection of **A** onto **i** is $\mathbf{Proj_i\, A} = \left(\dfrac{\mathbf{A}\bullet\mathbf{i}}{|\mathbf{i}|^2}\right)\mathbf{i} = \langle 6, 0, 0 \rangle$.

(d) and (e) The scalar projections of **A** onto **j** and **k** are –2 and 3, respectively.
The vector projections of **A** onto **j** and **k** are $\langle 0, -2, 0 \rangle$ and $\langle 0, 0, 3 \rangle$, respectively.

Practice 5: For $\mathbf{U} = \langle 9, -2, 6 \rangle$ and $\mathbf{V} = \langle 1, 2, -2 \rangle$, calculate the scalar and vector projections of
(a) **U** onto **V**, (b) **V** onto **U + V** , and (c) **V** onto the positive y–axis.

Applications

Projections are useful in a number of situations. The two examples given here illustrate only two of a variety of those uses. In the first example below, we use the geometric meaning of projection to derive a formula for the distance from a point to a line. In the second example, we illustrate how projections can be used to calculate work.

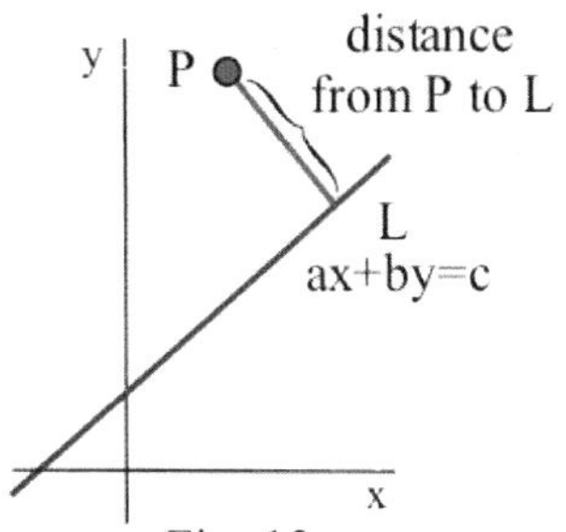

Fig. 13

Example 6: Find the distance from the point $P = (p, q)$ to a line $ax + by = c$. (Fig. 13)

Solution: Using vectors and projections, finding this distance can be broken into several simple steps (Fig. 14). The algorithm:

(1) Find a vector **N** perpendicular to the line: $\mathbf{N} = \langle a, b \rangle$.(Fig. 14a)

(2) Find a point Q on the line and call it (x_0, y_0) . (Fig. 14b)

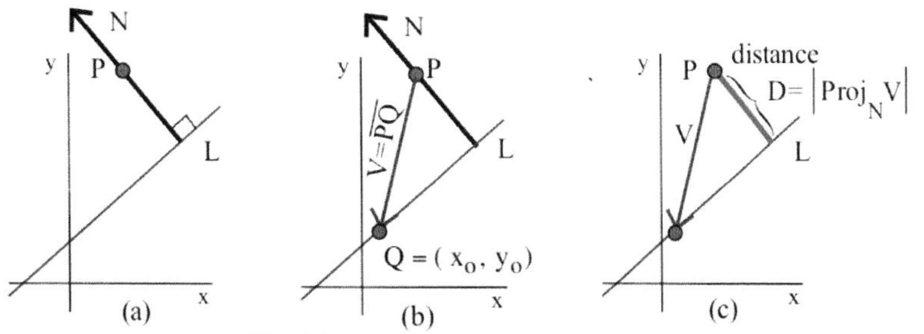

Fig. 14

(3) Form the vector **V** from P to Q: $\mathbf{V} = \langle p - x_0, q - y_0 \rangle$.

(4) Find the absolute value of the scalar projection of **V** onto **N**: $\left| \frac{\mathbf{V} \bullet \mathbf{N}}{|\mathbf{N}|} \right|$ (Fig. 14c)

The formula: The distance from P to the line is this absolute value of the scalar projection:

$$\text{distance} = \left| \frac{\mathbf{V} \bullet \mathbf{N}}{|\mathbf{N}|} \right| = \left| \frac{a(p - x_0) + b(q - y_0)}{\sqrt{a^2 + b^2}} \right|$$

$$= \frac{| ap + bq - (ax_0 + by_0) |}{\sqrt{a^2 + b^2}} = \frac{| ap + bq - c |}{\sqrt{a^2 + b^2}} \quad \text{since } ax_0 + by_0 = c.$$

Practice 6: Step through the algorithm in Example 6 to find the distance from the point $P = (2, 7)$ to the line $3x - 4y = 20$.

The projection of a force vector onto a vector with a different direction tells us the amount of the force in that other direction, a useful result to know for solving work problems.

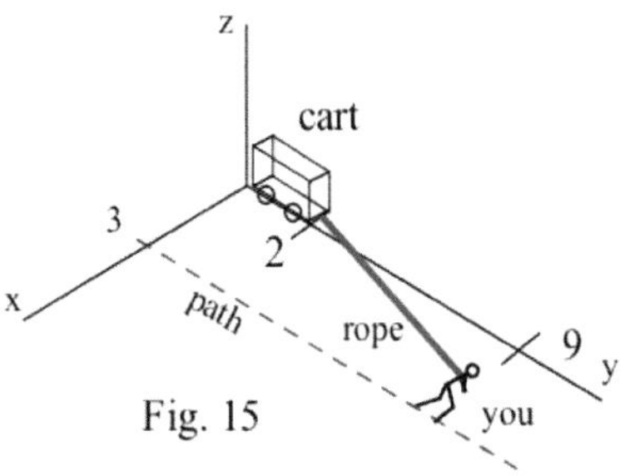

Fig. 15

Example 7: A cart moves on a track located on the y–axis, and you are pulling on a rope with a force of 70 pounds. Find the amount of work that you do in moving the cart from the point $P = (0, 2, 0)$ to the point $Q = (0, 9, 0)$ if you have the rope over your shoulder 4 feet above the ground and walk 12 feet in front of the cart along a path 3 feet to the side of the y–axis (Fig. 15).

Solution: Work = (distance moved)(force in the direction of the movement)

First, we can convert the given information into a vector form: the force vector has a magnitude of 70 pounds in the direction $\langle \frac{3}{13}, \frac{12}{13}, \frac{4}{13} \rangle$ so the force vector is $\mathbf{F} = \langle \frac{210}{13}, \frac{840}{13}, \frac{280}{13} \rangle$. The movement of the cart is in the direction of the vector from P to Q, $\mathbf{D} = Q - P = \langle 0, 7, 0 \rangle$, so the amount of work is

work = (distance moved)(magnitude of the force in the direction of the movement)

= (length of PQ)(scalar projection of F onto direction of PQ)

$$= (|\mathbf{D}|)\left(\frac{\mathbf{F} \bullet \mathbf{V}}{|\mathbf{V}|} \right) = \mathbf{F} \bullet \mathbf{V} = \langle \frac{210}{13}, \frac{840}{13}, \frac{280}{13} \rangle \bullet \langle 0, 7, 0 \rangle = \frac{5880}{13} \approx 452.3 \text{ foot–pounds.}$$

Work

If a constant force vector **F** moves an object from a point P to a point Q,

then the amount of work done is Work = **F**•**D** where **D** is the displacement vector from P to Q.

Practice 7: In Example 7, determine the amount of work done if you walk 6 feet to the side of the y–axis. (All of the other distances are the same as in Example 7.)

In Chapter 13, we discuss how to determine the amount of work done if the force is variable or if the object moves along a curved path.

Beyond Three Dimensions

If **A** and **B** have the same number of components, then we can define their dot product, the angle between them, and the projection of one onto the other with the same patterns as for two and three dimensions.

Definitions:

If $\mathbf{A} = \langle a_1, a_2, \ldots, a_n \rangle$ and $\mathbf{B} = \langle b_1, b_2, \ldots, b_n \rangle$ are nonzero vectors in n–dimensional space,

then $\mathbf{A}\bullet\mathbf{B} = a_1b_1 + a_2b_2 + \ldots + a_nb_n$,

the angle θ between **A** and **B** satisfies $\cos(\theta) = \dfrac{\mathbf{A}\bullet\mathbf{B}}{|\mathbf{A}||\mathbf{B}|}$, and

the vector projection of A onto B is $\mathbf{Proj_B\ A} = \left(\dfrac{\mathbf{A}\bullet\mathbf{B}}{|\mathbf{B}|^2}\right)\mathbf{B}$.

Now, even though we may not visualize 4 or 5–dimensional vectors, we can calculate the dot product and the angle between two vectors.

Practice 8: The psychological profiles for you and a friend were $\mathbf{Y} = \langle 5, 1, -7, 5 \rangle$ and $\mathbf{F} = \langle 6, 10, 8, 5 \rangle$ for the four personality categories measured by the profile. Should you say you and your friend are "very alike" ($\theta < 30°$), "somewhat alike" ($30° \le \theta < 60°$), "different" ($60° \le \theta \le 120°$), "somewhat opposite" ($120° < \theta \le 150°$). or "very opposite" ($150° < \theta \le 180°$)? What about you and and another friend with the profile $\mathbf{A} = \langle 10, -4, -10, 3 \rangle$?

This section has involved very little "calculus," but the ideas of dot products and projections of vectors are very powerful and useful, and they will be used often as we develop "vector calculus" in later chapters.

PROBLEMS

1. $\mathbf{A} = \langle 1, 2, 3 \rangle$, $\mathbf{B} = \langle -2, 4, -1 \rangle$. Calculate $\mathbf{A}\bullet\mathbf{B}$, $\mathbf{B}\bullet\mathbf{A}$, $\mathbf{A}\bullet\mathbf{A}$, $\mathbf{A}\bullet(\mathbf{B}+\mathbf{A})$, and $(2\mathbf{A}+3\mathbf{B})\bullet(\mathbf{A}-2\mathbf{B})$.

2. $\mathbf{A} = \langle 6, -1, 2 \rangle$, $\mathbf{B} = \langle 2, 4, -3 \rangle$. Calculate $\mathbf{A}\bullet\mathbf{B}$, $\mathbf{B}\bullet\mathbf{A}$, $\mathbf{A}\bullet\mathbf{A}$, $\mathbf{A}\bullet(\mathbf{B}+\mathbf{A})$, and $(2\mathbf{A}+3\mathbf{B})\bullet(\mathbf{A}-2\mathbf{B})$.

3. $\mathbf{U} = \langle 6, -1, 2 \rangle$, $\mathbf{V} = \langle 2, 4, -3 \rangle$. Calculate $\mathbf{U}\bullet\mathbf{V}$, $\mathbf{U}\bullet\mathbf{U}$, $\mathbf{U}\bullet\mathbf{i}$, $\mathbf{U}\bullet\mathbf{j}$, $\mathbf{U}\bullet\mathbf{k}$, and $(\mathbf{V} + \mathbf{i})\bullet\mathbf{U}$.

4. $\mathbf{U} = \langle -3, 3, 2 \rangle$, $\mathbf{V} = \langle 2, 4, -3 \rangle$. Calculate $\mathbf{U}\bullet\mathbf{V}$, $\mathbf{U}\bullet\mathbf{U}$, $\mathbf{V}\bullet\mathbf{i}$, $\mathbf{V}\bullet\mathbf{j}$, $\mathbf{V}\bullet\mathbf{k}$, $(\mathbf{V} + \mathbf{k})\bullet\mathbf{U}$.

5. $\mathbf{S} = 2\mathbf{i} - 4\mathbf{j} + \mathbf{k}$, $\mathbf{T} = 3\mathbf{i} + \mathbf{j} - 5\mathbf{k}$, $\mathbf{U} = \mathbf{i} + 3\mathbf{j} + 2\mathbf{k}$. Calculate $\mathbf{S}\bullet\mathbf{T}$, $\mathbf{T}\bullet\mathbf{U}$, $\mathbf{T}\bullet\mathbf{T}$, $(\mathbf{S} + \mathbf{T})\bullet(\mathbf{S} - \mathbf{T})$, and $(\mathbf{S}\bullet\mathbf{T})\,\mathbf{U}$.

6. $\mathbf{S} = \mathbf{i} - 3\mathbf{j} + 2\mathbf{k}$, $\mathbf{T} = 5\mathbf{i} + 3\mathbf{j} - 2\mathbf{k}$, $\mathbf{U} = 2\mathbf{i} + 4\mathbf{j} + 2\mathbf{k}$. Calculate $\mathbf{S}\bullet\mathbf{T}$, $\mathbf{S}\bullet\mathbf{U}$, $\mathbf{S}\bullet\mathbf{S}$, $(\mathbf{T} + \mathbf{U})\bullet(\mathbf{T} - \mathbf{U})$, and $\mathbf{S}(\mathbf{T}\bullet\mathbf{U})$.

In problems 7 – 18, calculate the angle between the given vectors. Also calculate the angles between the first vector and each of the coordinate axes.

7. $\mathbf{A} = \langle 1, 2, 3 \rangle$, $\mathbf{B} = \langle -2, 4, -1 \rangle$.

8. $\mathbf{A} = \langle 6, -1, 2 \rangle$, $\mathbf{B} = \langle 2, 4, -3 \rangle$.

9. $\mathbf{U} = \langle 6, -1, 2 \rangle$, $\mathbf{V} = \langle 2, 4, -3 \rangle$.

10. $\mathbf{U} = \langle -3, 3, 2 \rangle$, $\mathbf{V} = \langle 2, 4, -3 \rangle$.

11. $\mathbf{S} = 2\mathbf{i} - 4\mathbf{j} + \mathbf{k}$, $\mathbf{T} = 3\mathbf{i} + \mathbf{j} - 5\mathbf{k}$.

12. $\mathbf{S} = \mathbf{i} - 3\mathbf{j} + 2\mathbf{k}$, $\mathbf{T} = 5\mathbf{i} + 3\mathbf{j} - 2\mathbf{k}$.

13. $\mathbf{A} = 2\mathbf{i} - 3\mathbf{j} + 5\mathbf{k}$, $\mathbf{B} = -5\mathbf{i} + 0\mathbf{j} + 2\mathbf{k}$.

14. $\mathbf{A} = 4\mathbf{i} - 3\mathbf{j} + 2\mathbf{k}$, $\mathbf{B} = 3\mathbf{i} + 2\mathbf{j} - 3\mathbf{k}$.

15. $\mathbf{A} = \langle 5, -2, 0 \rangle$, $\mathbf{B} = \langle -3, 4, 0 \rangle$.

16. $\mathbf{A} = \langle 5, 0, 0 \rangle$, $\mathbf{B} = \langle 0, 4, -3 \rangle$.

17. $\mathbf{U} = \langle 1, 0, 3 \rangle$, $\mathbf{V} = \langle -2, 0, 1 \rangle$.

18. $\mathbf{U} = \langle 0, 1, 2 \rangle$, $\mathbf{V} = \langle 2, 4, 0 \rangle$.

In problems 19 – 28, determine the angle of intersection of the graphs of the given functions at the given point (i.e., determine the angle between the vectors tangent to the functions at the given point).

19. $f(x) = x^2 + 3x - 2$, $g(x) = 3x - 1$ at $(1, 2)$

20. $f(x) = x^2 + 3x - 2$, $g(x) = 3 - x^2$ at $(1, 2)$

21. $f(x) = e^x$, $g(x) = \cos(x)$ at $(0, 1)$

22. $f(x) = \sin(x)$, $g(x) = \cos(x)$ at $\left(\pi/4, \frac{\sqrt{2}}{2}\right)$

23. $f(x) = 1 + \arctan(3x)$, $g(x) = \ln(e + x)$ at $(0, 1)$

24. $f(x) = \sin(x^2)$, $g(x) = 1 - \cos(x^2)$ at $(0, 0)$

25. f: $x = 3t$, $y = t^2 + t - 1$; g: $x = 2t + 1$, $y = 2t - 1$ at the point when $t = 1$

26. f: $x = \sin(t)$, $y = \cos(t)$; g: $x = t^3$, $y = t^2 + 1$ at the point when $t = 0$

27. f: $x = e^t + 3$, $y = 2 + \cos(5t)$; g: $x = t^2 + 3t + 4$, $y = 3 + \ln(1 - t)$ at the point when $t = 0$

28. f: $x = 2 + \cos(5t)$, $y = e^t + 3$; g: $x = 3 + \ln(1 - t)$, $y = t^2 + 3t + 4$ at the point when $t = 0$

In problems 29 – 48, find a vector **N** that is perpendicular to the given vector or line. (Typically there are several correct answers.)

29. $\mathbf{A} = \langle 1, -2, 0 \rangle$

30. $\mathbf{B} = \langle -5, 0, 3 \rangle$

31. $\mathbf{C} = \langle 7, 3 \rangle$

32. $\mathbf{D} = \langle 7, -3 \rangle$

33. $\mathbf{E} = \langle 2, -1, 3 \rangle$

34. $\mathbf{S} = \langle 1, 2, 5 \rangle$

35. $\mathbf{T} = 3\mathbf{i} + \mathbf{j} - 5\mathbf{k}$

36. $\mathbf{U} = \mathbf{i} + 3\mathbf{j} + 2\mathbf{k}$

37. $\mathbf{V} = 2\mathbf{i} - 4\mathbf{j} + \mathbf{k}$

38. $\mathbf{W} = -6\mathbf{i} - 4\mathbf{j} + \mathbf{k}$

39. $\mathbf{A} = 2\mathbf{i} - 3\mathbf{j}$

40. $\mathbf{B} = 3\mathbf{i} + 2\mathbf{j}$

41. $\mathbf{C} = 3\mathbf{i} - 2\mathbf{j}$

42. $x + y = 6$

43. $3x + 2y = 6$

44. $5x - 3y = 30$

45. $x - 4y = 8$

46. $5x + y = 10$

47. $y = 3$

48. $x = 2$

In problems 49 – 52 sketch $\mathbf{Proj_B\ A}$.

49. **A** and **B** in Fig. 16. 50. **A** and **B** in Fig. 17. 51. **A** and **B** in Fig. 18. 52. **A** and **B** in Fig. 19.

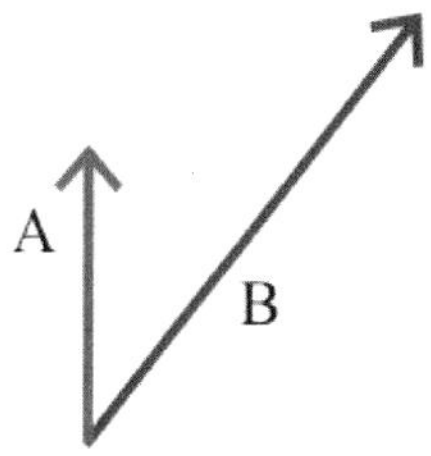

Fig. 16

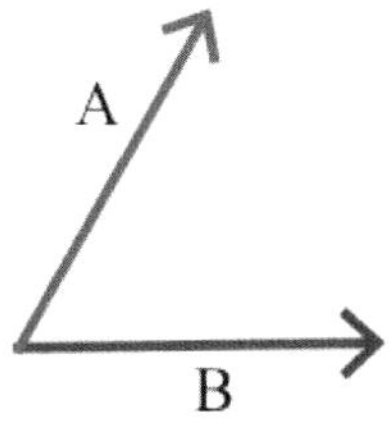

Fig. 17

Fig. 18

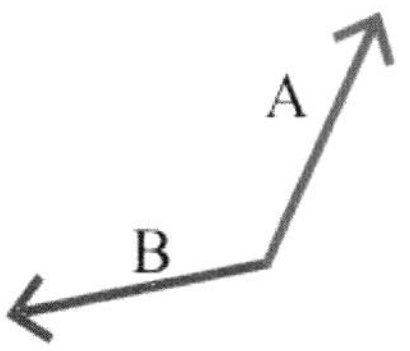

Fig. 19

In problems 53 – 56 sketch $\mathbf{Proj_A\ B}$.

53. **A** and **B** in Fig. 16. 54. **A** and **B** in Fig. 17. 55. **A** and **B** in Fig. 18. 56. **A** and **B** in Fig. 19.

In problems 57 – 63, calculate $\mathbf{Proj_B\ A}$ and $\mathbf{Proj_A\ B}$.

57. $\mathbf{A} = \langle 1, -2, 0 \rangle$, $\mathbf{B} = \langle -5, 0, 3 \rangle$

58. $\mathbf{A} = \langle 1, 2, 3 \rangle$, $\mathbf{B} = \langle -2, 4, -1 \rangle$

59. $\mathbf{A} = 2\mathbf{i} - 3\mathbf{j} + 5\mathbf{k}$, $\mathbf{B} = -5\mathbf{i} + 0\mathbf{j} + 2\mathbf{k}$

60. $\mathbf{A} = 4\mathbf{i} - 3\mathbf{j} + 2\mathbf{k}$, $\mathbf{B} = 3\mathbf{i} + 2\mathbf{j} - 3\mathbf{k}$

61. $\mathbf{A} = \langle 5, 0, 0 \rangle$, $\mathbf{B} = \langle 0, 4, -3 \rangle$

62. $\mathbf{A} = \langle 2, -1, 3 \rangle$, $\mathbf{B} = \langle 1, 2, 5 \rangle$

63. $\mathbf{A} = \langle 1, -2, 3 \rangle$, $\mathbf{B} = \mathbf{j}$

64. Suppose **A** and **B** have the same length. Which has the larger magnitude: $\mathbf{Proj_B\ A}$ or $\mathbf{Proj_A\ B}$? Justify your answer.

65. Suppose $|\mathbf{A}| = 3\,|\mathbf{B}|$. Which has the larger magnitude: $\mathbf{Proj_B\ A}$ or $\mathbf{Proj_A\ B}$? Justify your answer.

66. Prove that the vector $\mathbf{N} = a\mathbf{i} + b\mathbf{j}$ is perpendicular to the line $ax + by = c$.
(Suggestion: Pick any two points $P = (x_0, y_0)$ and $Q = (x_1, y_1)$ on the line. Then the vector **V** with starting point P and ending point Q has the same direction as the line, and $\mathbf{V} = \langle x_1 - x_0, y_1 - y0 \rangle$. Now show that **N** is perpendicular to **V** .)

In problems 67 – 72, calculate the distance from the given point to the given line using (a) the algorithm of Example 6, and (b) the formula found in Example 6.

67. $(1, 3), y = 7 - x$ 68. $(-2, 1), 3x - 2y = 6$ 69. $(5, 3), 3y = 3x + 7$

70. $(-3, -4), y = 3x + 2$ 71. $(0, 0), 4x + 3y = 7$ 72. $(0, 0), ax + by = c$

73. Fig. 20 shows the position of a road and a house. How close is the road to the house (minimum distance)?

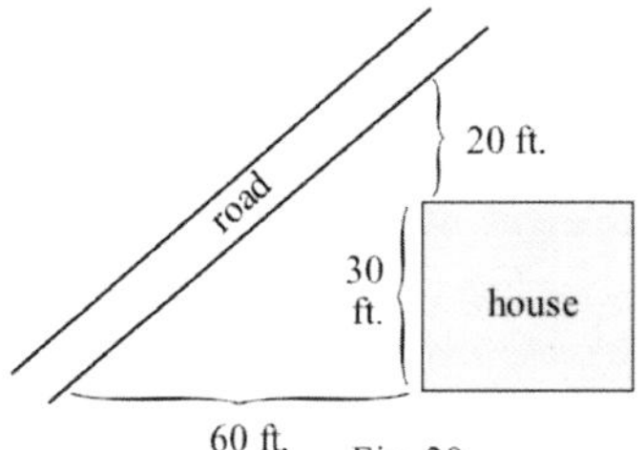

Fig. 20

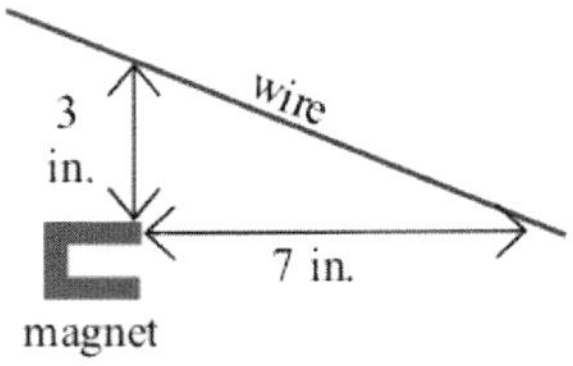

Fig. 21

74. In Fig. 21, how close is the wire to the magnet?

75. A person is standing directly below the electrical transmission wires in Fig. 22. Assuming the wires are so taut that they follow a straight line, how close do they come to the person's head?

76. The **direction cosines** of a vector $\mathbf{A} = a_1\mathbf{i} + a_2\mathbf{j} + a_3\mathbf{k}$ are the cosines of the angles **A** makes with each of the coordinate vectors $\mathbf{i}, \mathbf{j}$, and $\mathbf{k}$: if A makes angles θ_x, θ_y, and θ_z with the x, y, and z axes, respectively, then the direction cosines of **A** are $\cos(\theta_x)$, $\cos(\theta_y)$, and $\cos(\theta_z)$.

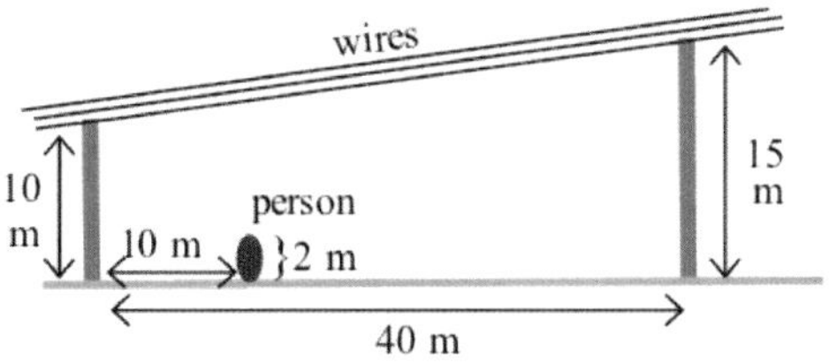

Fig. 22

Show that $\cos(\theta_x) = \frac{a_1}{|\mathbf{A}|}, \cos(\theta_y) = \frac{a_2}{|\mathbf{A}|}, \cos(\theta_z) = \frac{a_3}{|\mathbf{A}|}$,

and that $\cos^2(\theta_x) + \cos^2(\theta_y) + \cos^2(\theta_z) = 1$ for every nonzero vector **A**.

77. A car moves on a track located on the y–axis, and you are pulling on a rope with a force of 50 pounds. Find the amount of work that you do in moving the cart from the origin to the point $Q = (0, 10, 0)$ if you have the rope over your shoulder 4 feet above the ground and walk 10 feet in front of the cart along a path 5 feet to the side of the y–axis.

78. Redo Problem 77 assuming that you are now pulling on the rope with a force of 100 pounds.

79. A wind blowing parallel to the y–axis exerts a force of 8 pounds on a kite. How much work does the wind do in moving the kite in a straight line from the point $(20, 30, 40)$ to the point $(50, 90, 150)$.

80. Redo problem 79 assuming that the wind is blowing parallel to the x–axis.

81. How much work does the person in Fig. 23 do moving the box 20 feet along the ground?

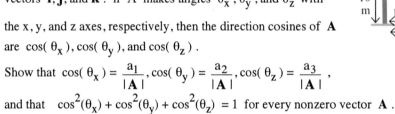

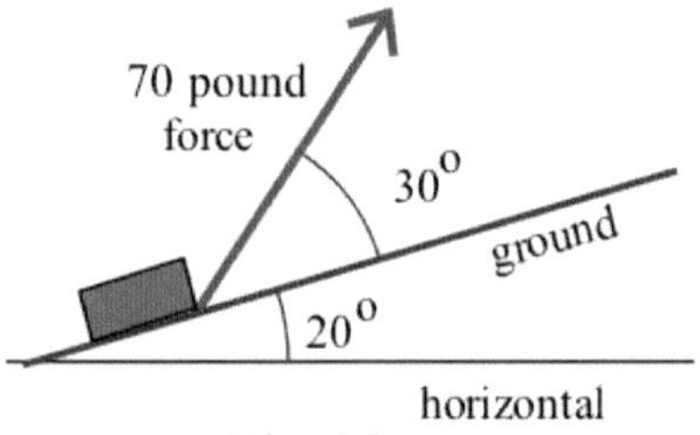

Fig. 23

Beyond Three Dimensions

83. $\mathbf{A} = \langle 1, 2, 3, 4 \rangle$, $\mathbf{B} = \langle -2, 4, -1, 3 \rangle$. Calculate $\mathbf{A} \bullet \mathbf{B}$, $\mathbf{B} \bullet \mathbf{A}$, and the angle between $\mathbf{A}$ and $\mathbf{B}$.

84. $\mathbf{A} = \langle -3, 4, 5, -1 \rangle$, $\mathbf{B} = \langle -2, 4, -2, -4 \rangle$. Calculate $\mathbf{A} \bullet \mathbf{B}$, $\mathbf{B} \bullet \mathbf{A}$, and the angle between $\mathbf{A}$ and $\mathbf{B}$.

In problems 85 – 88, a particular personality profile assigns a number between –1 and + 1 to each person on each of five personality characteristics. Using the categories ("very alike," "somewhat alike," , etc.) of Practice 8, determine the correct category for the pair of personality profiles given in each problem.

85. $\mathbf{A} = \langle 1, -0.2, 0.3, -0.6, 0.4 \rangle$, $\mathbf{B} = \langle -0.5, 0.3, 0.3, -0.1, 0.2 \rangle$

86. $\mathbf{A} = \langle 0.2, -0.3, -0.3, -0.7, 0.6 \rangle$, $\mathbf{B} = \langle 0.4, -0.2, -0.4, -0.5, 0.8 \rangle$

87. $\mathbf{A} = \langle 0.1, 0.2, 0.3, 0.4, 0.5 \rangle$, $\mathbf{B} = \langle 0.3, 0.4, 0.5, 0.6, 0.7 \rangle$

88. $\mathbf{A} = \langle 0.8, 0.3, -0.5, 0.2, 0.6 \rangle$, $\mathbf{B} = \langle -0.2, -0.4, -0.6, -0.4, 0.1 \rangle$

89. Prove the Parallelogram Law for vectors: $|\mathbf{A} + \mathbf{B}|^2 + |\mathbf{A} - \mathbf{B}|^2 \quad 2|\mathbf{A}|^2 + 2|\mathbf{B}|^2$

Practice Answers

Practice 1: $\mathbf{U} \bullet \mathbf{V} = (2)(-1) + (6)(2) + (-3)(2) = 4$. $\mathbf{U} \bullet \mathbf{U} = (2)(2) + (6)(6) + (-3)(-3) = 49$.

$\mathbf{V} \bullet \mathbf{V} = (-1)(-1) + (2)(2) + (2)(2) = 9$.

$\mathbf{U} \bullet (\mathbf{U}+\mathbf{V}) = \langle 2, 6, -3 \rangle \langle 1, 8, -1 \rangle = 53$. $\mathbf{U} \bullet \mathbf{U} + \mathbf{U} \bullet \mathbf{V} = 49 + 4 = 53$.

$|\mathbf{U}|^2 = \left(\sqrt{2^2 + 6^2 + (-3)^2} \right)2 = 2^2 + 6^2 + (-3)^2 = 49 = \mathbf{U} \bullet \mathbf{U}$.

$\mathbf{V} \bullet \mathbf{U} = (-1)(2) + (2)(6) + (2)(-3) = 4$ so $\mathbf{U} \bullet \mathbf{V} = \mathbf{V} \bullet \mathbf{U}$.

Practice 2: $\mathbf{A} \bullet \mathbf{B} = a_1 b_1 + a_2 b_2 + a_3 b_3 = b_1 a_1 + b_2 a_2 + b_3 a_3 = \mathbf{B} \bullet \mathbf{A}$.

Practice 3: $|\mathbf{A}| = \sqrt{4+36+9} = 7$, $|\mathbf{B}| = \sqrt{16+64+1} = 9$, and $|\mathbf{C}| = \sqrt{9+0+16} = 5$.

(a) $\cos(\theta) = \dfrac{\mathbf{A} \bullet \mathbf{B}}{|\mathbf{A}|\,|\mathbf{B}|} = \dfrac{8-48-3}{(7)(9)} = \dfrac{-43}{63} \approx -0.683$ so $\theta \approx 2.32$ (radians) or $133.04°$.

(b) $\cos(\theta) = \dfrac{\mathbf{A} \bullet \mathbf{C}}{|\mathbf{A}|\,|\mathbf{C}|} = \dfrac{6+0-12}{(7)(5)} = \dfrac{-6}{35} \approx -0.171$ so $\theta \approx 1.74$ or $99.87°$.

(c) $\cos(\theta) = \dfrac{\mathbf{B} \bullet (-\mathbf{i})}{|\mathbf{B}|\,|-\mathbf{i}|} = \dfrac{-4+0+0}{(9)(1)} = \dfrac{-4}{9} \approx -0.444$ so $\theta \approx 2.03$ or $116.36°$.

(d) $\cos(\theta) = \dfrac{\mathbf{C} \bullet \mathbf{j}}{|\mathbf{C}|\,|\mathbf{j}|} = \dfrac{0+0+0}{(5)(1)} = 0$ so $\theta = \pi/2$ or $90°$. $\mathbf{C}$ and $\mathbf{j}$ are perpendicular.

Practice 4: First we need to find two points on the line and then use those two points to form a vector parallel to the line: $P = (-6, 0)$ and $Q = (0, 15)$ are on the line, and $\mathbf{V} = \langle 0-(-6), 15-0 \rangle = \langle 6, 15 \rangle$ is parallel to the line. $\mathbf{N} \bullet \mathbf{V} = \langle -5, 2 \rangle \bullet \langle 6, 15 \rangle = -30 + 30 = 0$ so $\mathbf{N} = \langle -5, 2 \rangle$ is perpendicular to $-5x + 2y = 30$. Every scalar multiple $k\mathbf{N}$ with $k \neq 0$ is also perpendicular to the line.

Practice 5: (a) $|\mathbf{U}| = 11, |\mathbf{V}| = 3, \mathbf{U} \bullet \mathbf{V} = -7$. scalar projection of **U** onto **V** is $\dfrac{\mathbf{U} \bullet \mathbf{V}}{|\mathbf{V}|} = \dfrac{-7}{3}$.

Vector projection of **U** onto **V** is $\left(\dfrac{\mathbf{U} \bullet \mathbf{V}}{|\mathbf{V}|^2} \right) \mathbf{V} = \left\langle \dfrac{-7}{9}, \dfrac{-14}{9}, \dfrac{14}{9} \right\rangle$.

(b) $\mathbf{U} + \mathbf{V} = \langle 10, 0, 4 \rangle$, $|\mathbf{U} + \mathbf{V}| = \sqrt{116}$, $\mathbf{V} \bullet (\mathbf{U} + \mathbf{V}) = 10+0-8 = 2$.

Scalar projection of **V** onto **U** + **V** is $\dfrac{\mathbf{V} \bullet (\mathbf{U} + \mathbf{V})}{|\mathbf{U} + \mathbf{V}|} = \dfrac{2}{\sqrt{116}} \approx 0.19$.

Vector projection of **V** onto **U** + **V** is $\left(\dfrac{\mathbf{V} \bullet (\mathbf{U} + \mathbf{V})}{|\mathbf{U} + \mathbf{V}|^2} \right) (\mathbf{U} + \mathbf{V}) = \left\langle \dfrac{5}{29}, 0, \dfrac{2}{29} \right\rangle$.

(c) Scalar projection of **V** onto the positive y–axis is $\dfrac{\mathbf{V} \bullet \mathbf{j}}{|\mathbf{j}|} = 2$.

Vector projection of **V** onto **j** is $\left(\dfrac{\mathbf{V} \bullet \mathbf{j}}{|\mathbf{j}|^2} \right) \mathbf{j} = \langle 0, 2, 0 \rangle$.

Practice 6: (1) $\mathbf{N} = \langle 3, -4 \rangle$. Take $Q = (0, -5)$ (any other point on the line also works -- try one).

(3) **V** is the vector from P to Q: $\mathbf{V} = \langle -2, -12 \rangle$.

(4) scalar projection of **V** onto **N** is $\dfrac{\mathbf{V} \bullet \mathbf{N}}{|\mathbf{N}|} = \dfrac{-6 + 48}{\sqrt{9 + 16}} = \dfrac{42}{5} = 8.4$, the distance from the point to the line.

We get the same answer, but perhaps less understanding, using the formula:

$$\frac{|ap + bq - c|}{\sqrt{a^2 + b^2}} = \frac{|(3)(2) + (-4)(7) - 20|}{\sqrt{9 + 16}} = \frac{|6 - 28 - 20|}{5} = \frac{42}{5}.$$

Practice 7: The direction of the force vector is $\langle 6, 12, 4 \rangle / |\langle 6, 12, 4 \rangle| = \left\langle \dfrac{6}{14}, \dfrac{12}{14}, \dfrac{4}{14} \right\rangle$ so $\mathbf{F} = 70 \left\langle \dfrac{6}{14}, \dfrac{12}{14}, \dfrac{4}{14} \right\rangle = \langle 30, 60, 20 \rangle$. Then the work done is

Work = $\mathbf{F} \bullet \mathbf{D} = \langle 30, 60, 20 \rangle \bullet \langle 0, 7, 0 \rangle = 0 + (60 \text{ pounds})(7 \text{ feet}) + 0 = 420$ foot–pounds.

Practice 8: $|\mathbf{Y}| = \sqrt{5^2 + 1^2 + (-7)^2 + 5^2} = \sqrt{100} = 10$, $|\mathbf{F}| = 15$, and $|\mathbf{A}| = 15$.

For **Y** and **F**, $\cos(\theta) = \dfrac{\mathbf{Y} \bullet \mathbf{F}}{|\mathbf{Y}||\mathbf{F}|} = \dfrac{30+10-56+25}{(10)(15)} = \dfrac{9}{150}$ so $\theta \approx 86.6°$: "different."

For **Y** and **A**, $\cos(\theta) = \dfrac{\mathbf{Y} \bullet \mathbf{A}}{|\mathbf{Y}||\mathbf{A}|} = \dfrac{50-4+70+15}{(10)(15)} = \dfrac{131}{150}$ so $\theta \approx 29.2°$: "very alike."

11.5 CROSS PRODUCT

This section is the final one about the arithmetic of vectors, and it introduces a second type of vector–vector multiplication called the cross product. The material in this section and the previous sections is the foundation for the next several chapters on the calculus of vector–valued functions and functions of several variables, and all of the vector arithmetic is used extensively.

The dot product of two vectors results in a scalar, a number related to the magnitudes of the two original vectors and to the angle between them, and the dot product is defined for two vectors in 2–dimensional, 3–dimension, and higher dimensional spaces. The cross product of a vector and a vector differs from the dot product in several significant ways: the cross product is only defined for two vectors in 3–dimensional space, and the cross product of two vectors is a vector. At first, the definition of the cross product given below may seem strange, but the resulting vector has some very useful properties as well as some unusual ones. The torque wrench described next illustrates some of the properties we get with the cross product.

Torque wrench: As you pull down on the torque wrench in Fig. 1, a force is applied to the bolt that twists it into the wall. This "twisting" force is the result of two vectors, the length and a direction of the wrench **A** and the magnitude and direction of the pulling force **B**. To model this "twisting into the wall" result of **A** and **B**, we want a vector **C** that points into the wall and depends on the magnitudes of **A** and **B** as well as on the angle between the wrench and the direction of the pull.

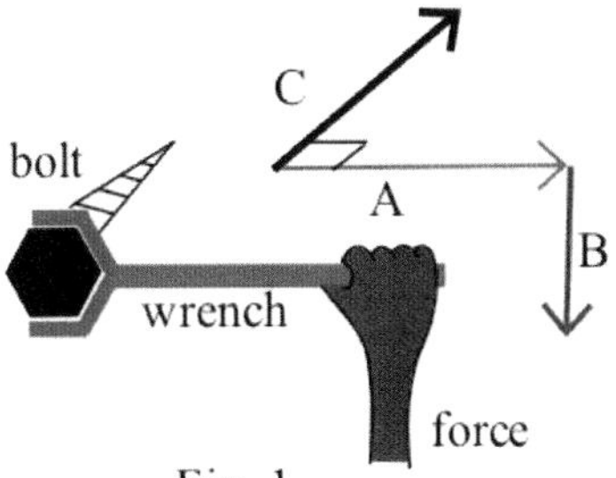

Fig. 1

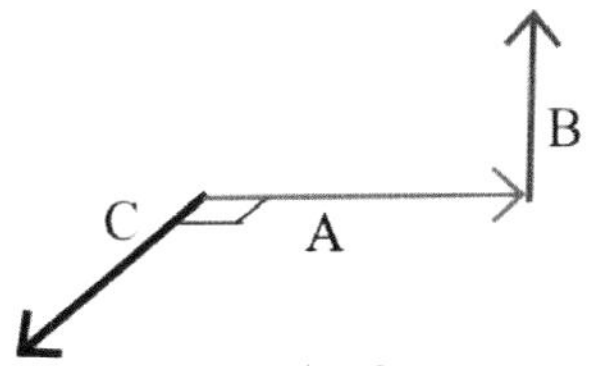

Fig. 2

If the pull is downward (Fig. 1), we want **C** to point into the page.

If the pull is upward (Fig. 2), we want **C** to point out of the page.

If the angle between **A** and **B** is close to $\pm 90^{\circ}$ (Fig. 3), we want the magnitude of **C** to be large.

If the angle between **A** and **B** is small (Fig. 4), we want the magnitude of **C** to be small.

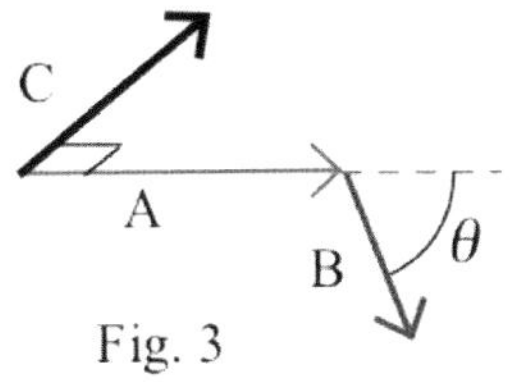

Fig. 3

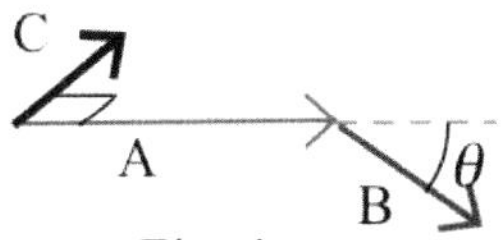

Fig. 4

There is a way to combine the vectors **A** and **B** to produce a vector **C** with the properties that model the torque wrench. This vector **C**, called the cross product of **A** and **B**, also turns out to be very useful when we discuss planes through given points and tangent planes to surfaces. It is not at all obvious from the definition given below for the cross product that the cross product has a relationship to torque wrenches, planes, or anything else of interest or use to us, but it does and we will investigate those applied and geometric properties. Try not to be repelled by the unusual definition — it leads to some lovely results.

Definition of the **Cross Product (Vector Product):**

For $\mathbf{A} = \langle a_1, a_2, a_3 \rangle$ and $\mathbf{B} = \langle b_1, b_2, b_3 \rangle$, the **cross product** of **A** and **B** is

$$\mathbf{A} \times \mathbf{B} = (a_2b_3 - a_3b_2)\mathbf{i} + (a_3b_1 - a_1b_3)\mathbf{j} + (a_1b_2 - a_2b_1)\mathbf{k}$$, a vector.

The symbols "**A** x **B**" are read "**A** cross **B**."

Many people find it difficult to remember a complicated formula like the definition of the cross product. Fortunately, there is an easy way to do so, but it requires a digression into the calculation of determinants.

Determinants

Determinants appear in a number of areas of mathematics, and you may have already seen them as part of Cramer's Rule for solving systems of equations. Here we only need them as a device for making it easier to remember and calculate and use the cross product.

Definition of the 2x2 ("two by two") Determinant: The determinant $\begin{vmatrix} a & b \\ c & d \end{vmatrix} = ad - bc$.

Some people prefer to remember the visual pattern in Fig. 5.

Example 1: Evaluate the determinants: $\begin{vmatrix} 1 & 4 \\ 3 & 5 \end{vmatrix}$, $\begin{vmatrix} x & y \\ -2 & 3 \end{vmatrix}$, and $\begin{vmatrix} \mathbf{i} & \mathbf{j} \\ 0 & 4 \end{vmatrix}$.

Solution: $\begin{vmatrix} 1 & 4 \\ 3 & 5 \end{vmatrix} = (1)(5) - (4)(3) = 5 - 12 = -7$.

$\begin{vmatrix} x & y \\ -2 & 3 \end{vmatrix} = (x)(3) - (y)(-2) = 3x + 2y$.

$\begin{vmatrix} \mathbf{i} & \mathbf{j} \\ 0 & 4 \end{vmatrix} = (\mathbf{i})(4) - (\mathbf{j})(0) = 4\mathbf{i}$.

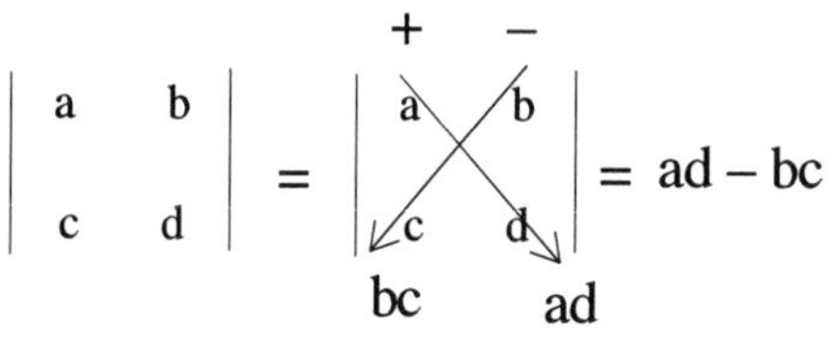

Fig. 5: Pattern for a 2x2 determinant

Practice 1: Evaluate the determinants: $\begin{vmatrix} -3 & 4 \\ 5 & 6 \end{vmatrix}$, $\begin{vmatrix} x & y \\ 0 & -3 \end{vmatrix}$, and $\begin{vmatrix} \mathbf{i} & \mathbf{j} \\ -4 & 3 \end{vmatrix}$.

A 3x3 determinant can be defined in terms of several 2x2 determinants.

Definition of the 3x3 Determinant: $\begin{vmatrix} a_1 & a_2 & a_3 \\ b_1 & b_2 & b_3 \\ c_1 & c_2 & c_3 \end{vmatrix} = a_1 \begin{vmatrix} b_2 & b_3 \\ c_2 & c_3 \end{vmatrix} - a_2 \begin{vmatrix} b_1 & b_3 \\ c_1 & c_3 \end{vmatrix} + a_3 \begin{vmatrix} b_1 & b_2 \\ c_1 & c_2 \end{vmatrix}$.

Many people prefer to remember the visual pattern in Fig. 6.

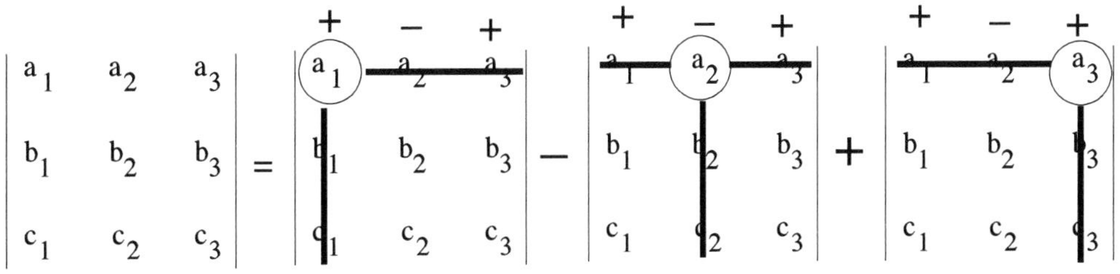

$$= +a_1 \begin{vmatrix} b_2 & b_3 \\ c_2 & c_3 \end{vmatrix} - a_2 \begin{vmatrix} b_1 & b_3 \\ c_1 & c_3 \end{vmatrix} + a_3 \begin{vmatrix} b_1 & b_2 \\ c_1 & c_2 \end{vmatrix}$$

Fig. 6: Pattern for a 3x3 determinant

In the 3x3 definition, the first 2x2 determinant, $\begin{vmatrix} b_2 & b_3 \\ c_2 & c_3 \end{vmatrix}$, is the part of the original 3x3 table after the first row and the first column have been removed, the row and column containing a_1. The second 2x2 determinant $\begin{vmatrix} b_1 & b_3 \\ c_1 & c_3 \end{vmatrix}$ is what remains of the original table after the first row and the second column have been removed, the row and column containing a_2. The third 2x2 determinant $\begin{vmatrix} b_1 & b_2 \\ c_1 & c_2 \end{vmatrix}$ is what remains of the original table after the first row and the third column have been removed, the row and column containing a_3. (Note: The leading signs attached to the three terms alternate: **+ – +**.)

Example 2: Evaluate the determinants $\begin{vmatrix} 2 & 3 & 5 \\ 0 & -4 & 1 \\ -3 & 4 & 0 \end{vmatrix}$ and $\begin{vmatrix} \mathbf{i} & \mathbf{j} & \mathbf{k} \\ 3 & 4 & 1 \\ 0 & -2 & 5 \end{vmatrix}$.

Solution: $\begin{vmatrix} 2 & 3 & 5 \\ 0 & -4 & 1 \\ -3 & 4 & 0 \end{vmatrix} = (2)\begin{vmatrix} -4 & 1 \\ 4 & 0 \end{vmatrix} - (3)\begin{vmatrix} 0 & 1 \\ -3 & 0 \end{vmatrix} + (5)\begin{vmatrix} 0 & -4 \\ -3 & 4 \end{vmatrix} = (2)(-4) - (3)(3) + (5)(-12) = -77.$

$$\begin{vmatrix} \mathbf{i} & \mathbf{j} & \mathbf{k} \\ 3 & 4 & 1 \\ 0 & -2 & 5 \end{vmatrix} = \mathbf{i}\begin{vmatrix} 4 & 1 \\ -2 & 5 \end{vmatrix} - \mathbf{j}\begin{vmatrix} 3 & 1 \\ 0 & 5 \end{vmatrix} + \mathbf{k}\begin{vmatrix} 3 & 4 \\ 0 & -2 \end{vmatrix} = +22\mathbf{i} - 15\mathbf{j} + (-6)\mathbf{k}.$$

Practice 2: Evaluate the determinants $\begin{vmatrix} 3 & 5 & 0 \\ 1 & 4 & -1 \\ -2 & 0 & 6 \end{vmatrix}$ and $\begin{vmatrix} \mathbf{i} & \mathbf{j} & \mathbf{k} \\ 2 & -1 & 3 \\ 4 & 0 & 5 \end{vmatrix}$.

The original definition of the cross product can now be rewritten using the determinant notation.

Determinant Form of the Cross Product Definition:

For $\mathbf{A} = \langle a_1, a_2, a_3 \rangle$ and $\mathbf{B} = \langle b_1, b_2, b_3 \rangle$, $\mathbf{A} \times \mathbf{B} = \begin{vmatrix} \mathbf{i} & \mathbf{j} & \mathbf{k} \\ a_1 & a_2 & a_3 \\ b_1 & b_2 & b_3 \end{vmatrix}$.

The second determinant in Example 2 represents the cross product $\mathbf{A} \times \mathbf{B}$ for $\mathbf{A} = \langle 3, 4, 1 \rangle$ and $\mathbf{B} = \langle 0, -2, 5 \rangle$, and the second determinant in Practice 2 is the cross product $\mathbf{A} \times \mathbf{B}$ for $\mathbf{A} = \langle 2, -1, 3 \rangle$ and $\mathbf{B} = \langle 4, 0, 5 \rangle$.

The cross products of various pairs of the basis vectors $\mathbf{i}, \mathbf{j}$, and $\mathbf{k}$ are relatively easy to evaluate, and they begin to illustrate some of the properties of cross products.

Example 3: Use the determinant form of the definition of the cross product to evaluate (a) $\mathbf{i} \times \mathbf{j}$, (b) $\mathbf{j} \times \mathbf{i}$, and (c) $\mathbf{i} \times \mathbf{i}$.

Solution:

$$\mathbf{i} \times \mathbf{j} = \begin{vmatrix} \mathbf{i} & \mathbf{j} & \mathbf{k} \\ 1 & 0 & 0 \\ 0 & 1 & 0 \end{vmatrix} = \mathbf{i} \begin{vmatrix} 0 & 0 \\ 1 & 0 \end{vmatrix} - \mathbf{j} \begin{vmatrix} 1 & 0 \\ 0 & 0 \end{vmatrix} + \mathbf{k} \begin{vmatrix} 1 & 0 \\ 0 & 1 \end{vmatrix} = 0\mathbf{i} + 0\mathbf{j} + 1\mathbf{k} = \mathbf{k}.$$

$$\mathbf{j} \times \mathbf{i} = \begin{vmatrix} \mathbf{i} & \mathbf{j} & \mathbf{k} \\ 0 & 1 & 0 \\ 1 & 0 & 0 \end{vmatrix} = \mathbf{i} \begin{vmatrix} 1 & 0 \\ 0 & 0 \end{vmatrix} - \mathbf{j} \begin{vmatrix} 0 & 0 \\ 1 & 0 \end{vmatrix} + \mathbf{k} \begin{vmatrix} 0 & 1 \\ 1 & 0 \end{vmatrix} = 0\mathbf{i} + 0\mathbf{j} + (-1)\mathbf{k} = -\mathbf{k}.$$

$$\mathbf{i} \times \mathbf{i} = \begin{vmatrix} \mathbf{i} & \mathbf{j} & \mathbf{k} \\ 1 & 0 & 0 \\ 1 & 0 & 0 \end{vmatrix} = \mathbf{i} \begin{vmatrix} 0 & 0 \\ 0 & 0 \end{vmatrix} - \mathbf{j} \begin{vmatrix} 1 & 0 \\ 1 & 0 \end{vmatrix} + \mathbf{k} \begin{vmatrix} 1 & 0 \\ 1 & 0 \end{vmatrix} = 0\mathbf{i} + 0\mathbf{j} + 0\mathbf{k} = \mathbf{0}.$$

You should note that $\mathbf{i} \times \mathbf{j} = \mathbf{k}$ is perpendicular to the vectors $\mathbf{i}$ and $\mathbf{j}$ and the xy–plane. Similarly, $\mathbf{j} \times \mathbf{i} = -\mathbf{k}$ is also perpendicular to $\mathbf{i}$ and $\mathbf{j}$ and the xy–plane.

Practice 3: Use the determinant form of the definition of the cross product to evaluate (a) $\mathbf{j} \times \mathbf{k}$, (b) $\mathbf{k} \times \mathbf{j}$, and (c) $\mathbf{j} \times \mathbf{j}$.

The cross products of pairs of basis vectors follow a simple pattern given below and in Fig. 7.

$\mathbf{i} \times \mathbf{j} = \mathbf{k}$	$\mathbf{j} \times \mathbf{i} = -\mathbf{k}$	$\mathbf{i} \times \mathbf{i} = \mathbf{0}$
$\mathbf{j} \times \mathbf{k} = \mathbf{i}$	$\mathbf{k} \times \mathbf{j} = -\mathbf{i}$	$\mathbf{j} \times \mathbf{j} = \mathbf{0}$
$\mathbf{k} \times \mathbf{i} = \mathbf{j}$	$\mathbf{i} \times \mathbf{k} = -\mathbf{j}$	$\mathbf{k} \times \mathbf{k} = \mathbf{0}$

Pattern for the Cross Product of the basis vectors

clockwise is +

counterclockwise is –

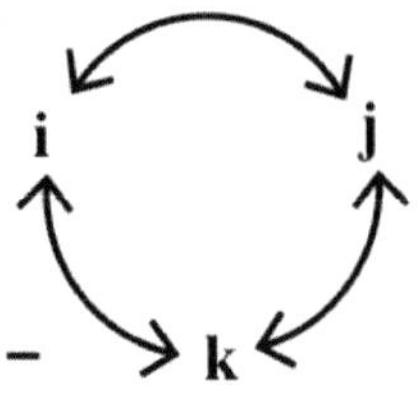

Fig. 7

Example 4: Evaluate **A x B** for $A = \langle 2, -3, 0 \rangle$ and $B = \langle 3, 1, -4 \rangle$.

Solution: $$\mathbf{A} \times \mathbf{B} = \begin{vmatrix} \mathbf{i} & \mathbf{j} & \mathbf{k} \\ 2 & -3 & 0 \\ 3 & 1 & -4 \end{vmatrix} = \mathbf{i}\begin{vmatrix} -3 & 0 \\ 1 & -4 \end{vmatrix} - \mathbf{j}\begin{vmatrix} 2 & 0 \\ 3 & -4 \end{vmatrix} + \mathbf{k}\begin{vmatrix} 2 & -3 \\ 3 & 1 \end{vmatrix} = 12\mathbf{i} + 8\mathbf{j} + 11\mathbf{k}.$$

Practice 4: Evaluate **A x B** for $A = \langle 3, 0, -5 \rangle$ and $B = \langle -2, 4, 1 \rangle$.

Properties of the Cross Product:

(a) $\mathbf{0} \times \mathbf{A} = \mathbf{A} \times \mathbf{0} = \mathbf{0}$

(b) $\mathbf{A} \times \mathbf{A} = \mathbf{0}$

(c) $\mathbf{A} \times \mathbf{B} = -\mathbf{B} \times \mathbf{A}$

(d) $k(\mathbf{A} \times \mathbf{B}) = k\mathbf{A} \times \mathbf{B} = \mathbf{A} \times k\mathbf{B}$

(e) $\mathbf{A} \times (\mathbf{B} + \mathbf{C}) = (\mathbf{A} \times \mathbf{B}) + (\mathbf{A} \times \mathbf{C})$

The proofs of all of these properties are straightforward applications of the definition of the cross product as is illustrated below for part (c). Proofs of (a) and (b) are given in the Appendix after the Practice Answers, and the proofs of (d) and (e) are left as exercises.

Proof of (c): If $\mathbf{A} = \langle a_1, a_2, a_3 \rangle$ and $\mathbf{B} = \langle b_1, b_2, b_3 \rangle$, then

$$\mathbf{B} \times \mathbf{A} = \begin{vmatrix} \mathbf{i} & \mathbf{j} & \mathbf{k} \\ b_1 & b_2 & b_3 \\ a_1 & a_2 & a_3 \end{vmatrix} = \mathbf{i}\begin{vmatrix} b_2 & b_3 \\ a_2 & a_3 \end{vmatrix} - \mathbf{j}\begin{vmatrix} b_1 & b_3 \\ a_1 & a_3 \end{vmatrix} + \mathbf{k}\begin{vmatrix} b_1 & b_2 \\ a_1 & a_2 \end{vmatrix}$$

$$= \mathbf{i}\,(a_3b_2 - a_2b_3) - \mathbf{j}\,(a_3b_1 - a_1b_3) + \mathbf{k}\,(a_2b_1 - a_1b_2).$$

$$\mathbf{A} \times \mathbf{B} = \begin{vmatrix} \mathbf{i} & \mathbf{j} & \mathbf{k} \\ a_1 & a_2 & a_3 \\ b_1 & b_2 & b_3 \end{vmatrix} = \mathbf{i}\begin{vmatrix} a_2 & a_3 \\ b_2 & b_3 \end{vmatrix} - \mathbf{j}\begin{vmatrix} a_1 & a_3 \\ b_1 & b_3 \end{vmatrix} + \mathbf{k}\begin{vmatrix} a_1 & a_2 \\ b_1 & b_2 \end{vmatrix}$$

$$= \mathbf{i}\,(a_2b_3 - a_3b_2) - \mathbf{j}\,(a_1b_3 - a_3b_1) + \mathbf{k}\,(a_1b_2 - a_2b_1) = -\mathbf{B} \times \mathbf{A}.$$

A vector has both a direction and a magnitude, and the direction and magnitude of the vector **A x B** each give us useful information: the direction of **A x B** is perpendicular to **A** and **B**, and the magnitude of **A x B** is the area of the parallelogram with sides **A** and **B**. These are the two properties of the cross product that get used most often in the later chapters, and they enable us to visualize **A x B**.

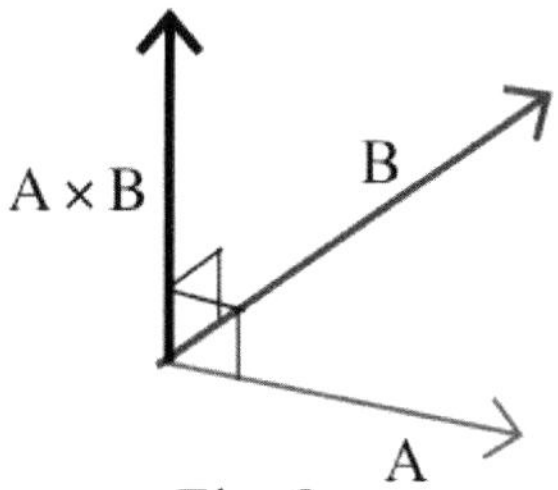

Fig. 8

The **direction** property of **A x B:**

A•(A x B) = 0 and **B•(A x B)** = 0

so **A x B** is perpendicular to **A** and to **B**. (Fig. 8)

Proof: $\mathbf{A}\bullet(\mathbf{A}\times\mathbf{B}) = \langle a_1, a_2, a_3 \rangle \bullet \langle a_2b_3 - a_3b_2, a_3b_1 - a_1b_3, a_1b_2 - a_2b_1 \rangle$

$= a_1a_2b_3 - a_1a_3b_2 + a_2a_3b_1 - a_2a_1b_3 + a_3a_1b_2 - a_3a_2b_1 = 0$.

The proof that **B•(A x B)** = 0 is similar .

The **magnitude** property of **A x B:**

If **A** and **B** are nonzero vectors with angle θ between them ($0 \le \theta \le \pi$),

then $|\mathbf{A}\times\mathbf{B}| = |\mathbf{A}||\mathbf{B}|\,|\sin(\theta)|$

= **area of the parallelogram** formed by the vectors **A** and **B**. (Fig. 9)

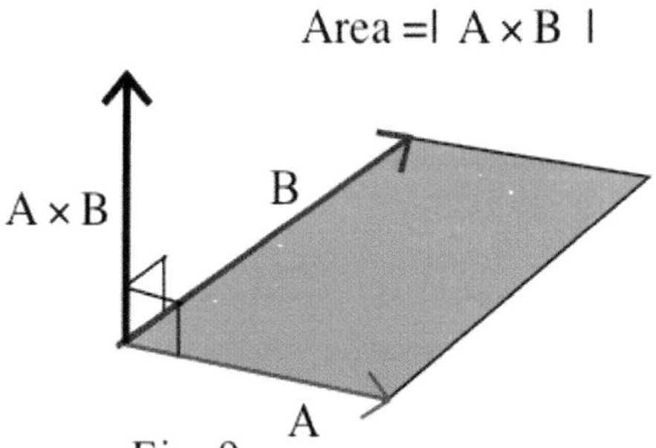

Fig. 9

The proof that $|\mathbf{A}\times\mathbf{B}| = |\mathbf{A}||\mathbf{B}|\,|\sin(\theta)|$ is algebraically complicated and is given in the Appendix after the Practice Answers. These properties of the direction and magnitude are sometimes used to define the cross product, and then the algebraic definition is derived from them.

Corollary to the magnitude property of **A x B:**

The **area of the triangle** formed by the vectors **A** and **B** is $\frac{1}{2}\,|\mathbf{A}\times\mathbf{B}|$.

Visualizing the cross product: The cross product **A x B** is perpendicular to both **A** and **B** and satisfies a "right hand rule" so we can visualize the direction of **A x B** using, of course, your right hand. If **A** and **B** are the indicated fingers (Fig. 10 and Fig. 11)) then your extended right thumb points in the direction of **A x B**.

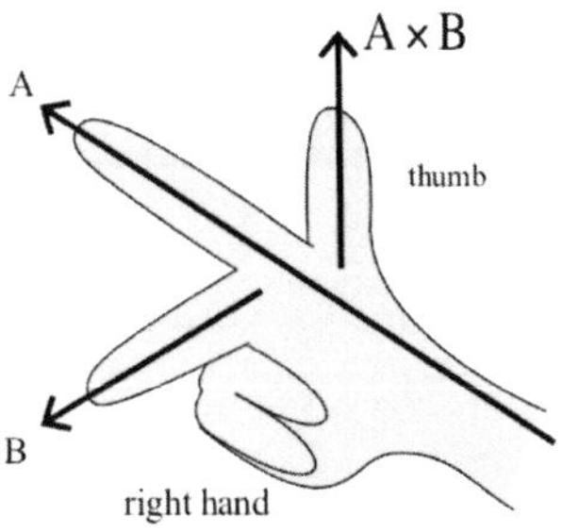

Fig. 10

The relative length of **A x B** can be estimated from the magnitude property of **A x B** : $|\mathbf{A}\times\mathbf{B}|$ = parallelogram area.

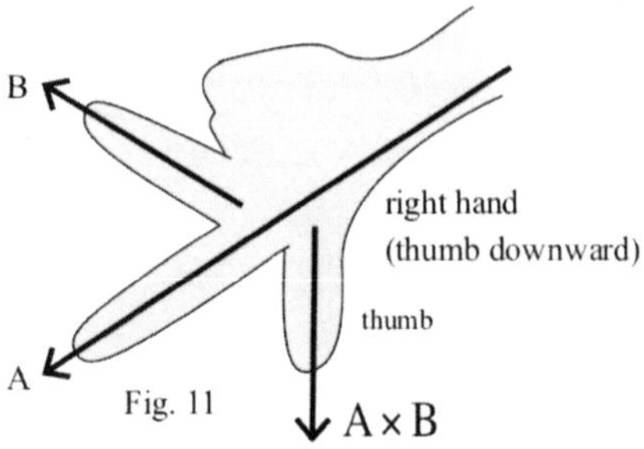

Fig. 11

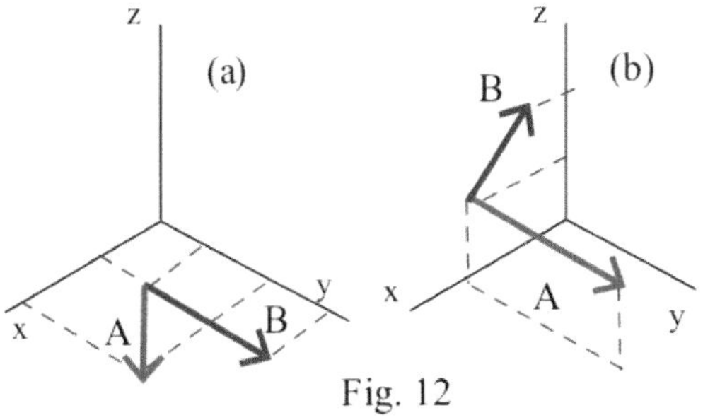

Fig. 12

Example 5: Fig. 12 shows two pairs of vectors **A** and **B**. For each pair sketch the direction of **A** x **B** . For which pair is | **A** x **B** | larger?

Solution: See Fig. 13. | **A** x **B** | is larger for (b) since the area of their parallelogram is larger in (b).

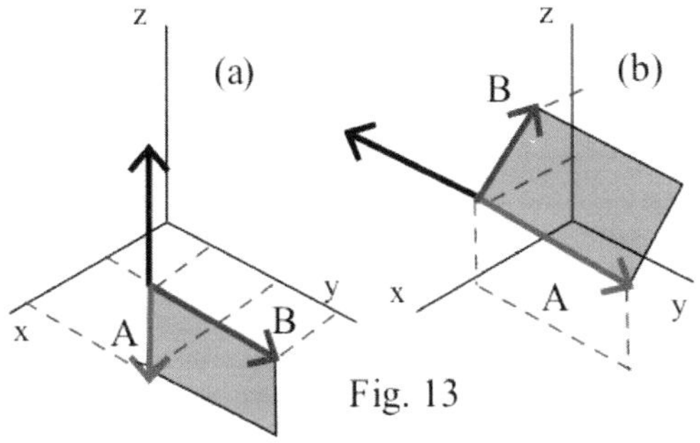

Fig. 13

Practice 5: Fig. 14 shows two pairs of vectors **A** and **B**. For each pair sketch the direction of **A** x **B** . For which pair is | **A** x **B** | larger?

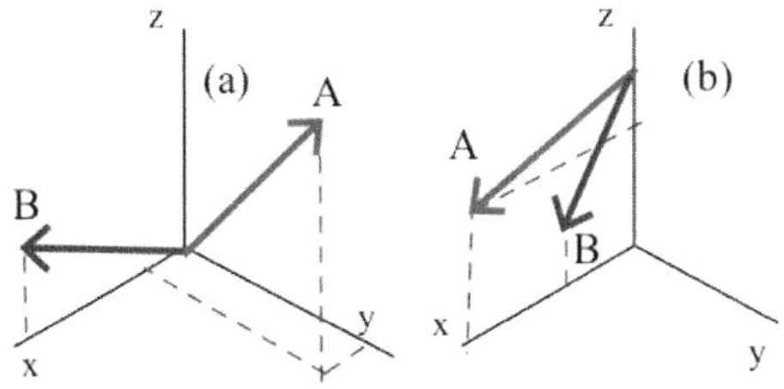

Fig. 14

Torque Wrench Revisited

Now we can use the ideas and properties of the cross product to analyze the original torque wrench problem.

Definition: The **torque vector** produced by a lever arm vector **A** and a force vector **B** is **A** x **B** .

The direction of the torque vector tells us whether the wrench is driving the bolt into the wall or pulling it out of the wall. The magnitude of the torque vector describes the strength of the tendency of the wrench to drive the bolt in or pull it out.

Example 6: Fig. 15 shows two force vectors acting at the end of a 10 inch wrench. Which vector produces the larger torque?

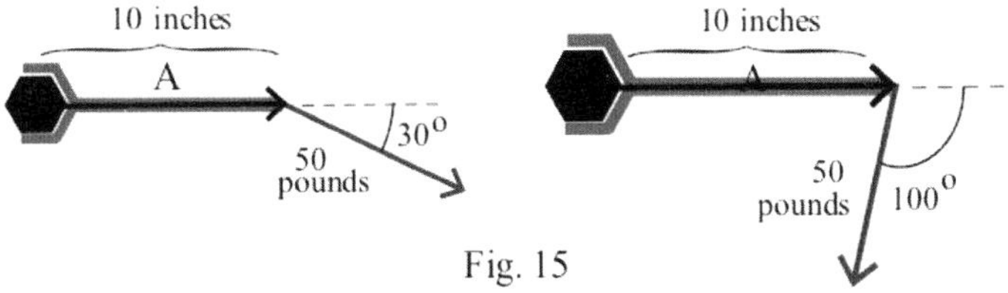

Fig. 15

Solution: $|\mathbf{B}| = 50$ pounds with $\theta = 30°$ so

$| \mathbf{A} \times \mathbf{B} | = |\mathbf{A}||\mathbf{B}| \, |\sin(\theta)|$

$= (10 \text{ inches})(50 \text{ pounds}) \, |\sin(30°)| \approx 250$ pound–inches of force.

$|\mathbf{C}| = 30$ pounds with $\theta = 100°$ so

$| \mathbf{A} \times \mathbf{C} | = |\mathbf{A}||\mathbf{C}| \, |\sin(\theta)| = (10 \text{ inches})(30 \text{ pounds}) | \sin(100°) | \approx 295.4$ pound–inches of force.

Vector **C** produces the larger torque: the smaller force, used intelligently, produced the larger result.

Practice 6: In Fig. 16 which force vector produces the larger torque?

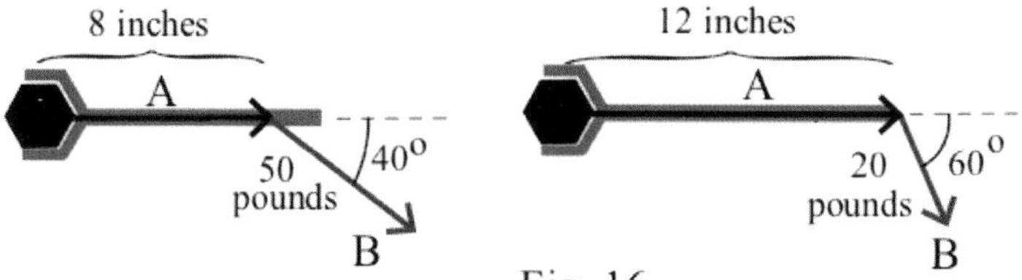

Fig. 16

The Triple Scalar Product A•(B x C)

The combination of the dot and cross products, **A•(B x C)** , for the three vectors **A**, **B**, and **C** in 3–dimensional space is called the triple scalar product because the result of these operations is a scalar: **A•(B x C)** = **A•**(vector) = scalar. And the magnitude of this scalar has a nice geometric meaning.

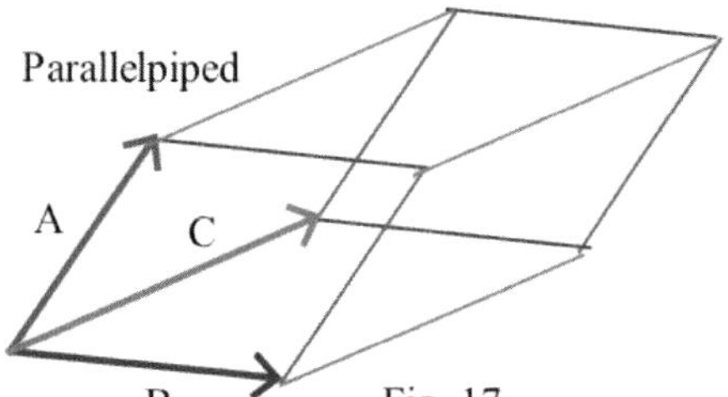

Fig. 17

> Geometric meaning of | **A•(B x C)** |
>
> For the vectors **A**, **B**, and **C** in 3–dimensional space (Fig.17)
>
> | **A•(B x C)** | = volume of the parallelpiped (box) with sides **A**, **B**, and **C** .

Proof: From the definition of the dot product,

A•(B x C) = | **B x C** | |**A**| cos(θ) so

| **A•(B x C)** | = | **B x C** | |**A**| |cos(θ)| .

Volume = (area of the base)(height),

and the area of the base of the box is | **B x C** | .

Since **B x C** is perpendicular to the base, the height h is the projection of **A** onto **B x C** (Fig. 18): h = |**A**| | cos(θ) |.

Then Volume = | **B x C**| |**A**| | cos(θ) | which we showed was equal to | **A•(B x C)** | .

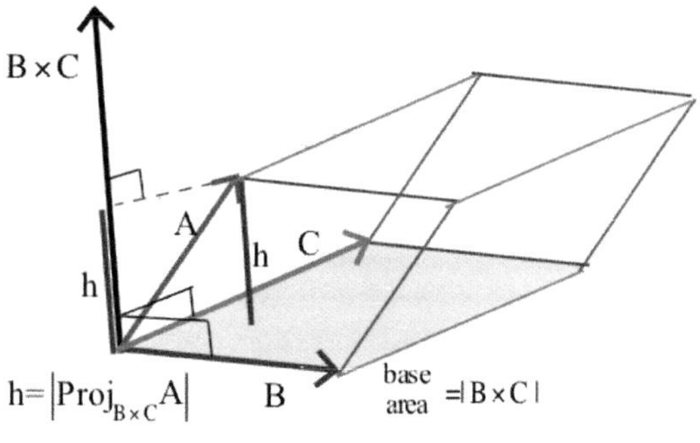

Fig. 18

Problem 46 asks you to show that the triple scalar product can be calculated as a 3x3 determinant.

Beyond Three Dimensions

The objects we examined in previous sections (points, distances, vectors, dot products, angles between vectors, projections) all had rather nice extensions to more than three dimensions. The cross product is different: the cross product **A x B** we have defined requires that **A** and **B** be 3–dimensional vectors, and there is no easy extension to vectors in more than three dimensions that preserves the properties of the cross product.

PROBLEMS

In problems 1 – 12, evaluate the determinants.

1. $\begin{vmatrix} 3 & 4 \\ 2 & 5 \end{vmatrix}$ 2. $\begin{vmatrix} 4 & -1 \\ 3 & 1 \end{vmatrix}$ 3. $\begin{vmatrix} x & 5 \\ y & 2 \end{vmatrix}$ 4. $\begin{vmatrix} 5 & a \\ b & 3 \end{vmatrix}$

5. $\begin{vmatrix} 1 & 0 \\ 0 & 1 \end{vmatrix}$ 6. $\begin{vmatrix} 0 & 1 \\ 1 & 0 \end{vmatrix}$ 7. $\begin{vmatrix} 1 & 3 & 2 \\ 0 & 5 & 2 \\ 1 & 1 & 0 \end{vmatrix}$ 8. $\begin{vmatrix} 2 & 3 & 0 \\ 1 & -3 & 2 \\ -1 & 0 & 4 \end{vmatrix}$

9. $\begin{vmatrix} x & y & z \\ 1 & 2 & 3 \\ 3 & 1 & 2 \end{vmatrix}$ 10. $\begin{vmatrix} a & b & c \\ 0 & 3 & 5 \\ 2 & 1 & 3 \end{vmatrix}$ 11. $\begin{vmatrix} 2 & 3 & 5 \\ 0 & -4 & 1 \\ -3 & 4 & 0 \end{vmatrix}$ 12. $\begin{vmatrix} x & 0 & 0 \\ 0 & 3 & 0 \\ 0 & 0 & 1-x \end{vmatrix}$

In problems 13 – 18, vectors **A** and **B** are given. Calculate (a) **A** x **B** , (b) (**A** x **B**)•**A**, (c) (**A** x **B**)•**B** , and (d) | **A** x **B** | .

13. $\mathbf{A} = \langle 3, 4, 5 \rangle$, $\mathbf{B} = \langle -1, 2, 0 \rangle$

14. $\mathbf{A} = \langle -2, 2, 2 \rangle$, $\mathbf{B} = \langle 3, 1, 2 \rangle$

15. $\mathbf{A} = \langle 1, -3, 2 \rangle$, $\mathbf{B} = \langle -2, 6, 4 \rangle$

16. $\mathbf{A} = \langle 6, 8, -2 \rangle$, $\mathbf{B} = \langle 3, 4, -1 \rangle$

17. $\mathbf{A} = 3\mathbf{i} - 1\mathbf{j} + 4\mathbf{k}$, $\mathbf{B} = 1\mathbf{i} - 2\mathbf{j} + 5\mathbf{k}$

18. $\mathbf{A} = 4\mathbf{i} - 1\mathbf{j} + 0\mathbf{k}$, $\mathbf{B} = 3\mathbf{i} - 2\mathbf{j} + 0\mathbf{k}$

In problems 19 – 22, state whether the result of the given calculation is a vector, a scalar, or is not defined.

19. **A**•(**B** x **C**) 20. **A** x (**B**•**C**) 21. (**A**•**B**) x (**A**•**C**) 22. **A**(**B** x **C**)

23. Prove property (d) of the Properties of the Cross Product: k(**A** x **B**) = k**A** x **B** = **A** x k**B** .

24. Prove property (e) of the Properties of the Cross Product: **A** x (**B** + **C**) = (**A** x **B**) + (**A** x **C**) .

25. Explain geometrically why **A** x **A** = **0** .

26. If |**A**| and |**B**| are fixed, what angle(s) between **A** and **B** maximizes | **A** x **B** | ? Why?

In problems 27 – 30, vectors **A** and **B** are given graphically. Sketch **A** x **B** .

27. See Fig. 19. 28. See Fig. 20. 29. See Fig. 21. 30. See Fig. 22.

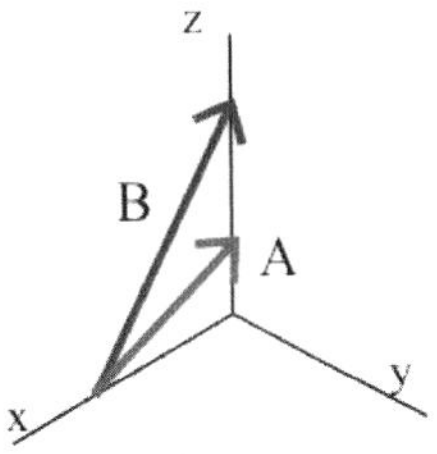

Fig. 19

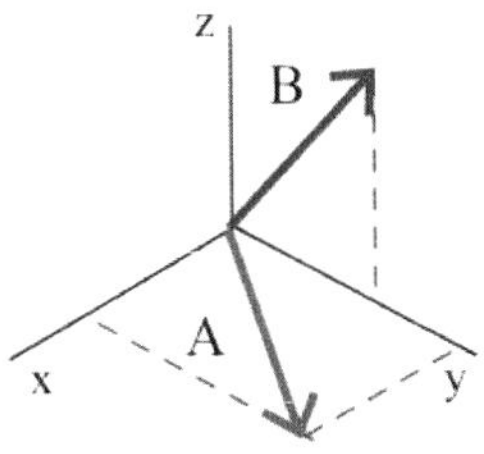

Fig. 20

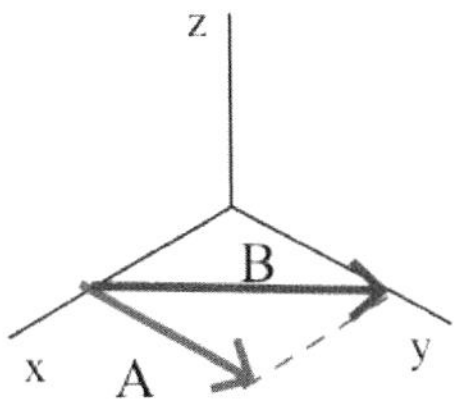

Fig. 21

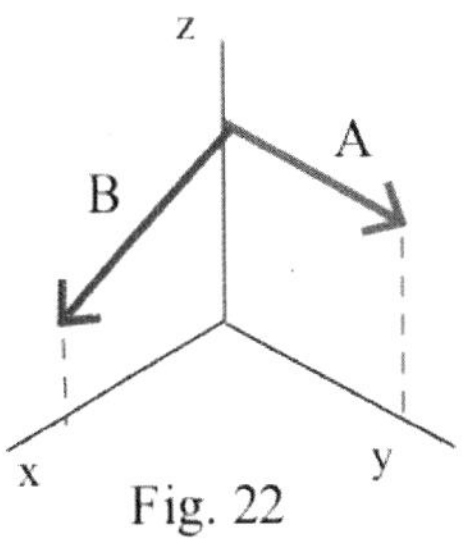

Fig. 22

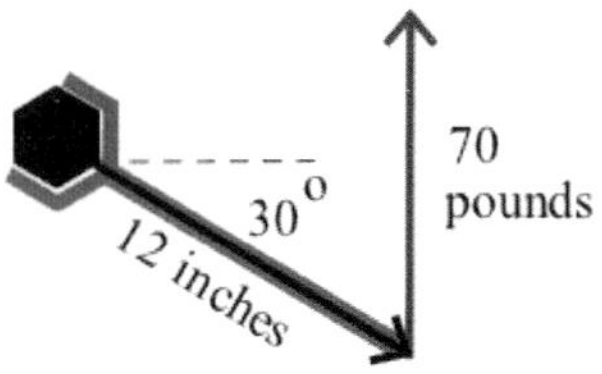

Fig. 23

31. (a) Calculate the torque produced by the wrench and force shown in Fig. 23.

(b) Calculate the torque produced by the wrench and force shown in Fig. 24.

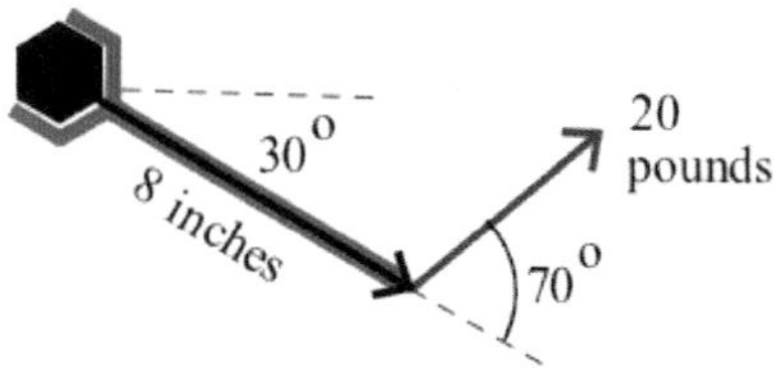

Fig. 24

32. When a tire nut is "frozen" (stuck), a pipe is sometimes put over the handle of the tire wrench (Fig. 25). Using the vocabulary and ideas of vectors, explain why this is effective.

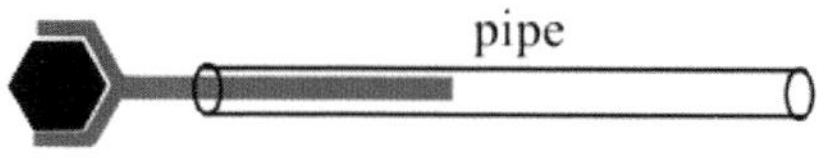

Fig. 25

33. Does the torque on wrench **A** produced by **B** plus the torque produced by **C** equal the torque produced by **B** + **C** ? Why or why not?

Areas

34. Sketch the parallelogram with sides $\mathbf{A} = \langle 5, 1, 0 \rangle$ and $\mathbf{B} = \langle 2, 4, 0 \rangle$ and find its area. Sketch and find the area of the triangle with sides **A** and **B** .

35. Sketch the parallelogram with sides $\mathbf{A} = \langle 1, 2, 0 \rangle$ and $\mathbf{B} = \langle 0, 4, 2 \rangle$ and find its area. Sketch and find the area of the triangle with sides **A** and **B** .

36. Sketch the triangle with vertices P = (4, 0, 1), Q =(1, 3, 1), and R = (2, 0, 5) and find its area.

37. Sketch the triangle with vertices P = (1, 4, –2), Q =(3, 5, 1), and R = (5, 2, 2) and find its area.

38. Sketch the triangle with vertices P = (a, 0, 0), Q =(0, b, 0), and R = (0, 0, c) and find its area.

Triple Scalar Products

39. Sketch a parallelpiped with edges $\mathbf{A} = \langle 2, 1, 0 \rangle$, $\mathbf{B} = \langle -1, 4, 1 \rangle$, $\mathbf{C} = \langle 1, 1, 2 \rangle$ and find its volume.

40. Sketch a parallelpiped with edges $\mathbf{A} = \langle 2, 0, 3 \rangle$, $\mathbf{B} = \langle 0, 4, 5 \rangle$, $\mathbf{C} = \langle 4, 3, 0 \rangle$ and find its volume.

41. Sketch a parallelpiped with edges $\mathbf{A} = \langle a, 0, 0 \rangle$, $\mathbf{B} = \langle 0, b, 0 \rangle$, $\mathbf{C} = \langle 0, 0, c \rangle$ and find its volume.

Use the result that "the volume of the tetrahedron with edges **A**, **B**, **C** (Fig. 26) is 1/6 the volume of the parallelpiped with the same edges" to find the areas of the tetrahedrons in problems 42 – 45.

42. Sketch the tetrahedron with vertices P = (0,0,0), Q =(3,1,0), R = (0,4,0), and S = (0,0,3) and find its volume.

43. Sketch the tetrahedron with vertices P = (1,0,2), Q =(3,1,2), R = (0,4,3), and S = (0,1,4) and find its volume.

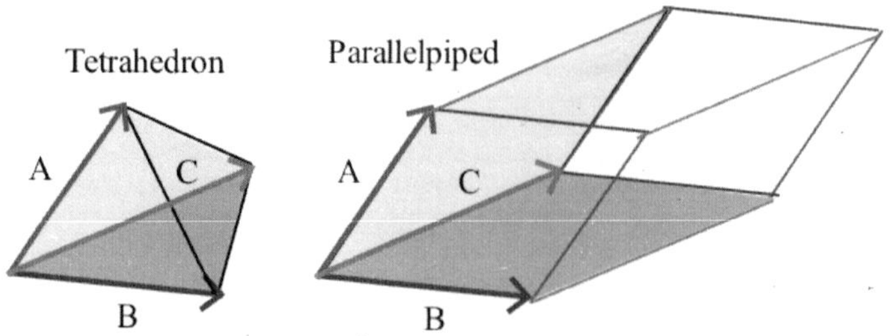

Tetrahedron volume = $\frac{1}{6}$ (Parallelpiped volume)

Fig. 26

44. Sketch the tetrahedron with vertices P = (0,0,0), Q =(2,0,0), R = (0,4,0), and S = (0,0,6) and find its volume.

45. Sketch the tetrahedron with vertices P = (0,0,0), Q =(a,0,0), R = (0,b,0), and S = (0,0,c) and find its volume.

46. Show that if $\mathbf{A} = \langle a_1, a_2, a_3 \rangle$, $\mathbf{B} = \langle b_1, b_2, b_3 \rangle$, and $\mathbf{C} = \langle c_1, c_2, c_3 \rangle$,

then $\mathbf{A} \bullet (\mathbf{B} \times \mathbf{C}) = \begin{vmatrix} a_1 & a_2 & a_3 \\ b_1 & b_2 & b_3 \\ c_1 & c_2 & c_3 \end{vmatrix}$.

Right Tetrahedrons

47. The four points (0,0,0), (2,0,0), (0,4,0), and (0,0,4) form a tetrahedron (Fig. 27) with four triangular faces. Find the areas A_{xy}, A_{xz}, A_{yz}, and A_{xyz} of the four triangular faces.

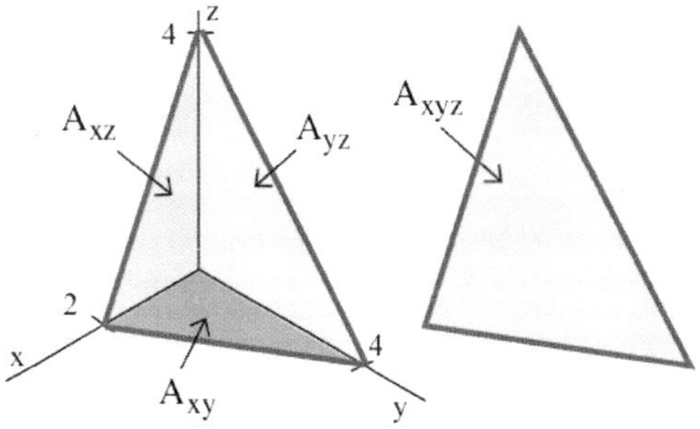

Fig. 27

48. The four points (0,0,0), (2,0,0), (0,4,0), and (0,0,6) form a tetrahedron (Fig. 28) with four triangular faces. Find the areas A_{xy}, A_{xz}, A_{yz}, and A_{xyz} of the four triangular faces.

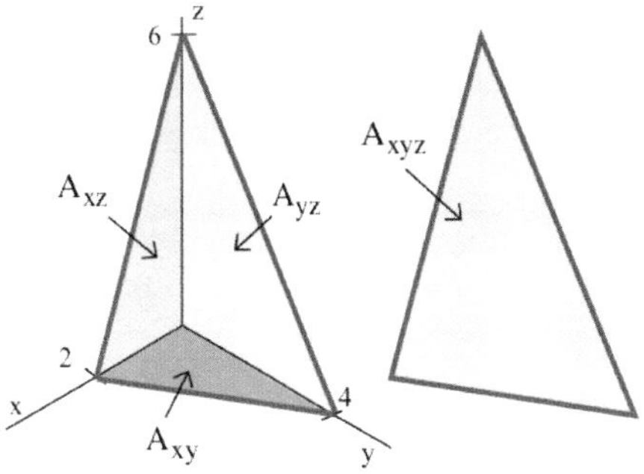

Fig. 28

49. Verify that the answers to problems 47 and 48 satisfy the relationship $(A_{xy})^2 + (A_{xz})^2 + (A_{yz})^2 = (A_{xyz})^2$.

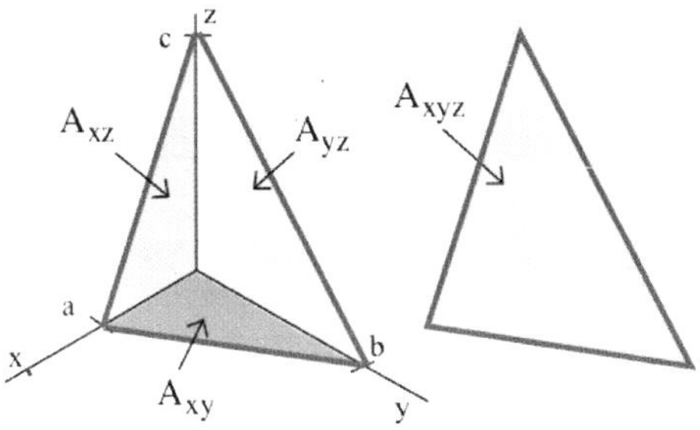

Fig. 29

50. For the right tetrahedron with vertices (0,0,0), (a,0,0), (0,b,0), and (0,0,c), determine the areas of the four triangular faces (Fig. 29) and prove the Pythagorean type result for areas of triangles in a right tetrahedron:

$$(A_{xy})^2 + (A_{xz})^2 + (A_{yz})^2 = (A_{xyz})^2 .$$

51. The Pythagorean pattern

$a^2 + b^2 = c^2$

can be thought of as relating a line segment C in two dimensions and its "shadows" a and b on the coordinate axes. Show that this "shadow" interpretation also holds for the area of a triangle in three dimensions and the areas of its "shadows" on the three coordinate planes (Fig. 30):

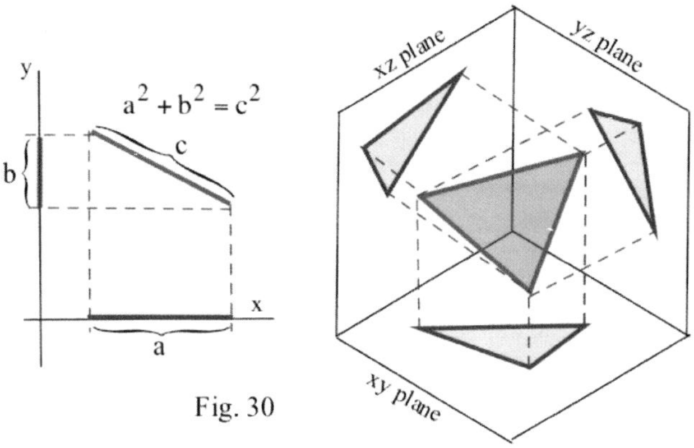

Fig. 30

$$(\text{area of dark triangle})^2 = (\text{xy shadow area})^2 + (\text{xz shadow area})^2 + (\text{yz shadow area})^2 .$$

Areas of Regions in the Plane

Among its several uses, the cross product also leads to a simple, easily programmed algorithm for finding the area of a "simple" (no edges cross) polygon in the plane, and this algorithm is used to approximate the areas of other regions as well.

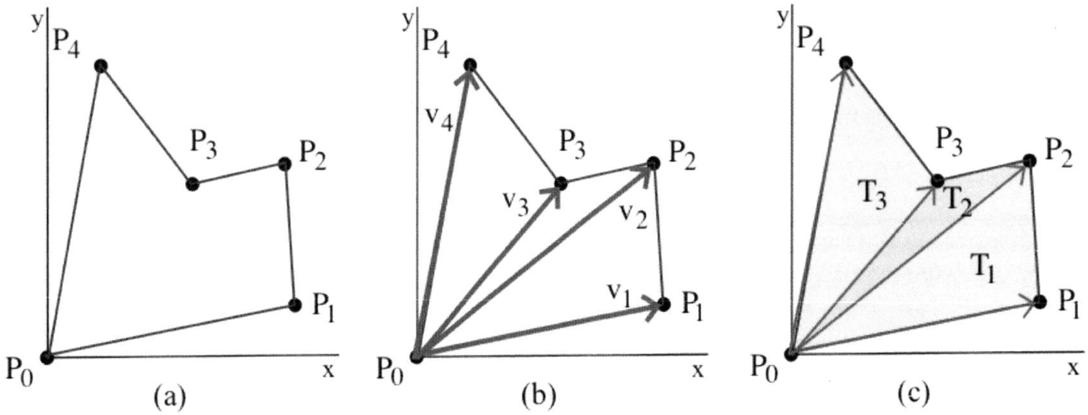

Fig. 31

Suppose $P_0 = (0, 0)$, $P_1 = (x_1, y_1)$, ... , $P_4 = (x_4, y_4)$ are 5 vertices of a simple polygon (Fig. 31a), with one vertex at the origin and then labeling the others as we travel counterclockwise around the polygon. Let $\mathbf{V}_1$ be the vector from P_0 to P_1, $\mathbf{V}_2$ from P_0 to P_2, ... (Fig. 31b). Then the area of the polygon in Fig. 31c is the sum of the 3 triangular areas T_1, T_2, and T_3 and each triangular area can be found using a cross product: $T_1 = \frac{1}{2} \,|\, \mathbf{V}_1 \times \mathbf{V}_2 \,|\, = \frac{1}{2}(x_1y_2 - x_2y_1)$,

$T_2 = \frac{1}{2} \,|\, \mathbf{V}_2 \times \mathbf{V}_3 \,|\, = \frac{1}{2}(x_2y_3 - x_3y_2)$, and $T_3 = \frac{1}{2} \,|\, \mathbf{V}_3 \times \mathbf{V}_4 \,|\, = \frac{1}{2}(x_3y_4 - x_4y_4)$.

Finally, the total area is the sum

$$\text{Area} = T_1 + T_2 + T_3$$

$$= \frac{1}{2}\{ (x_1y_2 - x_2y_1) + (x_2y_3 - x_3y_2) + (x_3y_4 - x_4y_3) \}$$

$$= \frac{1}{2}\sum_{k=1}^{n-1} (x_ky_{k+1} - x_{k+1}y_k) \quad \text{with } n = 4 .$$

The last summation formula works for polygons with at least three vertices. In fact, this algorithm is used by computers to report the area of a region traced by a cursor or stylus: the computer reads the (x,y) location of the cursor several times per second and uses the data and this algorithm to calculate the area of the region (as approximated by a many–sided polygon).

52. Use the given pattern to find the area of the rectangle with vertices (0,0), (2,0), (2,3), and (0,3). Does the pattern give the area of the rectangle?

53. Use the given pattern to find the area of the pentagon with vertices (0,0), (4,1), (5,3), (4,4), (2,4), and (1,3).

54. How can we modify the algorithm to handle the situation in which none of the vertices are at the origin? Show that your modification works for the rectangle with vertices (1,3), (3,3), (3,6), and (1,6).

Note: The cross product satisfies a "right hand rule" so if we go counterclockwise from **U** to **V** (Fig. 32a) then **U** x **V** is positive, and if we go clockwise from **U** to **V** (Fig. 32b) then **U** x **V** is negative.

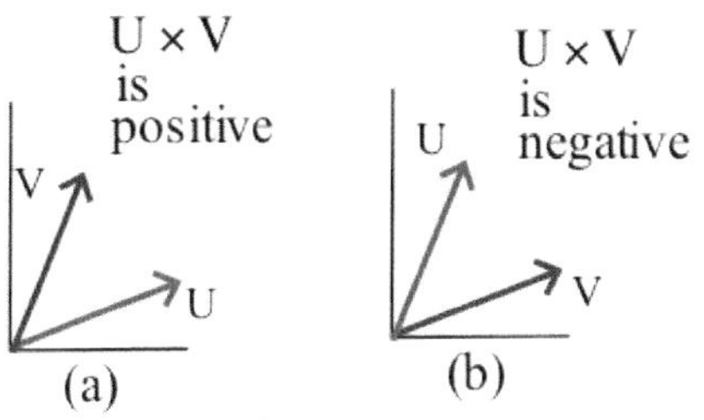

Fig. 32

55. In Fig. 33a, $\frac{1}{2}$ $\mathbf{V}_1$ x $\mathbf{V}_2$ gives the area of T_1 as a positive number; $\frac{1}{2}$ $\mathbf{V}_2$ x $\mathbf{V}_3$ gives the area of T_2 as a negative number; and $\frac{1}{2}$ $\mathbf{V}_3$ x $\mathbf{V}_4$ gives the area of T_3 as a positive number.

Explain geometrically how these positive and negative numbers "fit together" to give the correct area for the region in Fig. 33b.

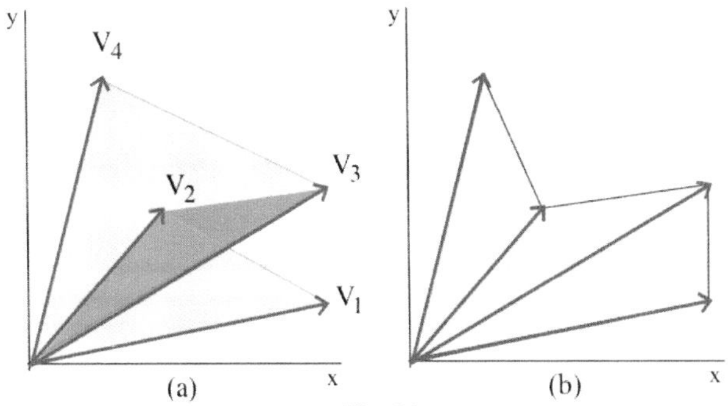

Fig. 33

Practice Answers

Practice 1: $\begin{vmatrix} -3 & 4 \\ 5 & 6 \end{vmatrix} = (-3)(6) - (4)(5) = -38.$ $\begin{vmatrix} x & y \\ 0 & -3 \end{vmatrix} = (x)(-3) - (y)(0) = -3x$.

$\begin{vmatrix} \mathbf{i} & \mathbf{j} \\ -4 & 3 \end{vmatrix} = (\mathbf{i})(3) - (\mathbf{j})(-4) = 3\mathbf{i} + 4\mathbf{j}$.

Practice 2: $\begin{vmatrix} 3 & 5 & 0 \\ 1 & 4 & -1 \\ -2 & 0 & 6 \end{vmatrix} = (3)\begin{vmatrix} 4 & -1 \\ 0 & 6 \end{vmatrix} - (5)\begin{vmatrix} 1 & -1 \\ -2 & 6 \end{vmatrix} + (0)\begin{vmatrix} 1 & 4 \\ -2 & 0 \end{vmatrix} = (3)(24) - (5)(4) + (0)(8) = 52.$

$\begin{vmatrix} \mathbf{i} & \mathbf{j} & \mathbf{k} \\ 2 & -1 & 3 \\ 4 & 0 & 5 \end{vmatrix} = \mathbf{i}\begin{vmatrix} -1 & 3 \\ 0 & 5 \end{vmatrix} - \mathbf{j}\begin{vmatrix} 2 & 3 \\ 4 & 5 \end{vmatrix} + \mathbf{k}\begin{vmatrix} 2 & -1 \\ 4 & 0 \end{vmatrix} = -5\mathbf{i} + 2\mathbf{j} + 4\mathbf{k}$.

Practice 3: $\mathbf{j} \times \mathbf{k} = \begin{vmatrix} \mathbf{i} & \mathbf{j} & \mathbf{k} \\ 0 & 1 & 0 \\ 0 & 0 & 1 \end{vmatrix} = \mathbf{i}\begin{vmatrix} 1 & 0 \\ 0 & 1 \end{vmatrix} - \mathbf{j}\begin{vmatrix} 0 & 0 \\ 0 & 1 \end{vmatrix} + \mathbf{k}\begin{vmatrix} 0 & 1 \\ 0 & 0 \end{vmatrix} = 1\mathbf{i} - 0\mathbf{j} + 0\mathbf{k} = \mathbf{i}$.

$\mathbf{k} \times \mathbf{j} = \begin{vmatrix} \mathbf{i} & \mathbf{j} & \mathbf{k} \\ 0 & 0 & 1 \\ 0 & 1 & 0 \end{vmatrix} = -\mathbf{i}$. $\mathbf{j} \times \mathbf{j} = \begin{vmatrix} \mathbf{i} & \mathbf{j} & \mathbf{k} \\ 0 & 1 & 0 \\ 0 & 1 & 0 \end{vmatrix} = \mathbf{0}$.

Practice 4: $\mathbf{A} \times \mathbf{B} = \begin{vmatrix} \mathbf{i} & \mathbf{j} & \mathbf{k} \\ 3 & 0 & -5 \\ -2 & 4 & 1 \end{vmatrix} = \mathbf{i}\begin{vmatrix} 0 & -5 \\ 4 & 1 \end{vmatrix} - \mathbf{j}\begin{vmatrix} 3 & -5 \\ -2 & 1 \end{vmatrix} + \mathbf{k}\begin{vmatrix} 3 & 0 \\ -2 & 4 \end{vmatrix} = 20\mathbf{i} + 7\mathbf{j} + 12\mathbf{k}$.

Practice 5: See Fig. 34. The pair in (a) produces the larger torque.

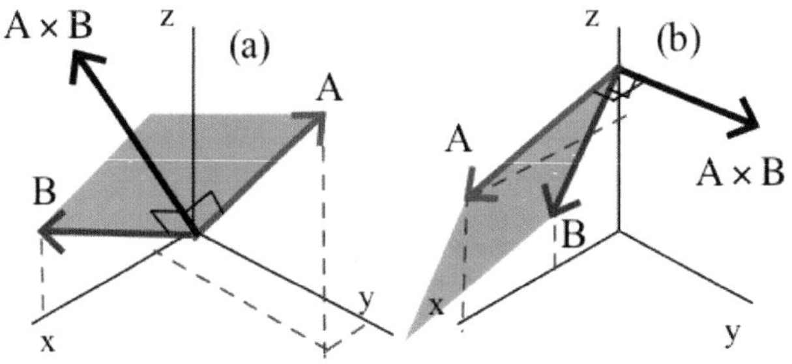

Fig. 34

Practice 6: For **B**,

| torque | = (8 inches)(50 pounds) sin(40°)

≈ 257.1 inch–pounds.

For **C**, | torque | = (12 inches)(20 pounds) sin(60°) ≈ 207.8 inch–pounds.

Force **B** produces the larger torque: sometimes strength is enough.

Appendix: Some Proofs

Proof of (a): $\mathbf{0} \times \mathbf{A} = \mathbf{A} \times \mathbf{0} = \mathbf{0}$. If $\mathbf{A} = \langle a_1, a_2, a_3 \rangle$, then

$$\mathbf{0} \times \mathbf{A} = \begin{vmatrix} \mathbf{i} & \mathbf{j} & \mathbf{k} \\ 0 & 0 & 0 \\ a_1 & a_2 & a_3 \end{vmatrix} = \mathbf{i} \begin{vmatrix} 0 & 0 \\ a_2 & a_3 \end{vmatrix} - \mathbf{j} \begin{vmatrix} 0 & 0 \\ a_1 & a_3 \end{vmatrix} + \mathbf{k} \begin{vmatrix} 0 & 0 \\ a_1 & a_2 \end{vmatrix} = 0\mathbf{i} - 0\mathbf{j} + 0\mathbf{k} .$$

The proof that $\mathbf{A} \times \mathbf{0} = \mathbf{0}$ is similar.

Proof of (b): $\mathbf{A} \times \mathbf{A} = \mathbf{0}$. If $\mathbf{A} = \langle a_1, a_2, a_3 \rangle$, then

$$\mathbf{A} \times \mathbf{A} = \begin{vmatrix} \mathbf{i} & \mathbf{j} & \mathbf{k} \\ a_1 & a_2 & a_3 \\ a_1 & a_2 & a_3 \end{vmatrix} = \mathbf{i} \begin{vmatrix} a_2 & a_3 \\ a_2 & a_3 \end{vmatrix} - \mathbf{j} \begin{vmatrix} a_1 & a_3 \\ a_1 & a_3 \end{vmatrix} + \mathbf{k} \begin{vmatrix} a_1 & a_2 \\ a_1 & a_2 \end{vmatrix} = 0\mathbf{i} - 0\mathbf{j} + 0\mathbf{k} = \mathbf{0} .$$

Proof that $| \mathbf{A} \times \mathbf{B} | = |\mathbf{A}||\mathbf{B}| |\sin(\theta)|$:

$$\mathbf{A} \times \mathbf{B} = (a_2b_3 - a_3b_2)\mathbf{i} + (a_3b_1 - a_1b_3)\mathbf{j} + (a_1b_2 - a_2b_1)\mathbf{k} \quad \text{so}$$

$$\begin{aligned} | \mathbf{A} \times \mathbf{B} |^2 &= (\mathbf{A} \times \mathbf{B})\bullet(\mathbf{A} \times \mathbf{B}) \\ &= (a_2b_3 - a_3b_2)^2 + (a_3b_1 - a_1b_3)^2 + (a_1b_2 - a_2b_1)^2 \\ &= a_2^2b_3^2 - 2a_2b_3a_3b_2 + a_3^2b_2^2 + a_3^2b_1^2 - 2a_3b_1a_1b_3 + a_1^2b_3^2 + a_1^2b_2^2 - 2a_1b_2a_2b_1 + a_2^2b_1^2 \\ &= (a_1^2 + a_2^2 + a_3^2)(b_1^2 + b_2^2 + b_3^2) - (a_1b_1 + a_2b_2 + a_3b_3)^2 \quad \text{(expand \& check)} \\ &= |\mathbf{A}|^2|\mathbf{B}|^2 - (\mathbf{A}\bullet\mathbf{B})^2 \\ &= |\mathbf{A}|^2|\mathbf{B}|^2 - (|\mathbf{A}||\mathbf{B}|\cos(\theta))^2 \quad \text{since } \mathbf{A}\bullet\mathbf{B} = |\mathbf{A}||\mathbf{B}|\cos(\theta) \\ &= |\mathbf{A}|^2|\mathbf{B}|^2 - |\mathbf{A}|^2|\mathbf{B}|^2\cos^2(\theta) \\ &= |\mathbf{A}|^2|\mathbf{B}|^2 \{ 1 - \cos^2(\theta) \} \\ &= |\mathbf{A}|^2|\mathbf{B}|^2 \sin^2(\theta) . \end{aligned}$$

Then, taking the square root of each side of $| \mathbf{A} \times \mathbf{B} |^2 = |\mathbf{A}|^2|\mathbf{B}|^2 \sin^2(\theta)$, we have

$$| \mathbf{A} \times \mathbf{B} | = |\mathbf{A}||\mathbf{B}| \, | \sin(\theta) | .$$

11.6 LINES AND PLANES IN THREE DIMENSIONS

In two dimensions the graph of a linear equation, either y=ax+b or parametric (x(t),y(t)), is always a straight line, and a point and slope or two points completely determine the line. In three dimensions the graph of a linear equation, either z=ax+by+c or parametric (x(t), y(t), z(t)), has more freedom, and it can be a line or a plane. In this section we examine the equations of lines and planes and their graphs in 3–dimensional space, discuss how to determine their equations from information known about them, and look at ways to determine intersections, distances, and angles in three dimensions. Lines and planes are the simplest graphs in three dimensions, and they are useful for a variety of geometric and algebraic applications. They are also the building blocks we need to find tangent lines to curves and tangent planes to surfaces (Fig. 1) in three dimensions.

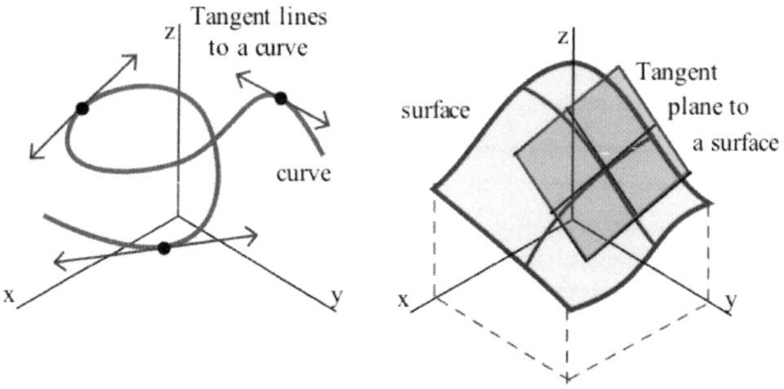

Fig. 1: Tangent lines and planes in three dimensions

Lines in Three Dimensions

Early in calculus we often used the point–slope formula to write the equation of the line through a given point $P = (x_0, y_0)$ with a given slope m: $y - y_0 = m(x - x_0)$. If the point $P = (x_0, y_0)$ is given and the direction of the line is parallel to a given vector $\mathbf{A} = \langle a, b \rangle$, then it is easier to use parametric equations to specify an equation for the line:

a point $Q = (x, y) \neq P$ is on the line if and only if $\dfrac{y - y_0}{x - x_0} = \dfrac{\text{rise}}{\text{run}} = \dfrac{b}{a} = \dfrac{b \cdot t}{a \cdot t}$ for some $t \neq 0$

so the equations $y - y_0 = b \cdot t$, and $x - x_0 = a \cdot t$ describe points on the line and

$x = x_0 + a \cdot t,\ y = y_0 + b \cdot t$.

> **Parametric Equation of a Line in Two Dimensions**
>
> Parametric equations for a line through the point $P = (x_0, y_0)$ and parallel to the vector $\mathbf{A} = \langle a, b \rangle$ are
>
> $x = x(t) = x_0 + a \cdot t$ and $y = y(t) = y_0 + b \cdot t$. (Fig. 2)

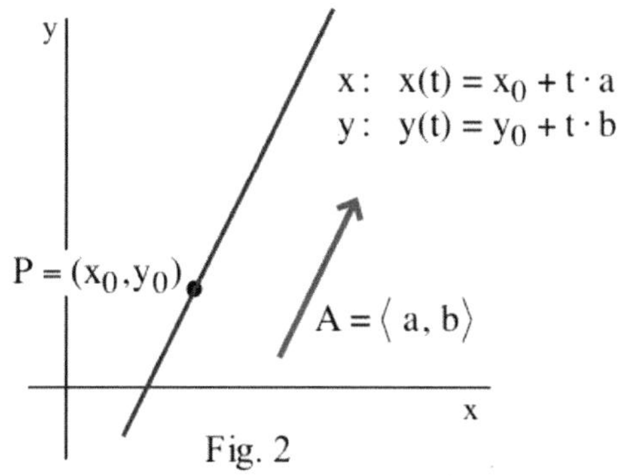

Fig. 2

Notice that the coordinates of the point (x_0, y_0) become the constant terms for x(t) and y(t), and the components of the vector $\langle a, b \rangle$ become the coefficients of the variable terms in the parametric equations. This same pattern is true for lines in three (and more) dimensions.

Example 1: Find parametric equations for the lines through the point $P = (1,2)$ that are (a) parallel to the vector $\mathbf{A} = \langle 3, 5 \rangle$, and (b) parallel to the vector $\mathbf{B} = \langle 6, 10 \rangle$. Then graph the two lines.

Solution: (a) $x(t) = 1 + 3t, y(t) = 2 + 5t$. (b) $x(t) = 1 + 6t, y(t) = 2 + 10t$.

The graphs of the lines are shown in Fig. 3. In this example, both sets of parametric equations have the same line when graphed. If we interpret t as time, and (x(t), y(t)) as the location of an object at time t, then the graph is a picture of all of the points on the path of the object (for all times). The objects in (a) and (b) have the same path, but they are at different points on the path at time t (except when $t = 0$).

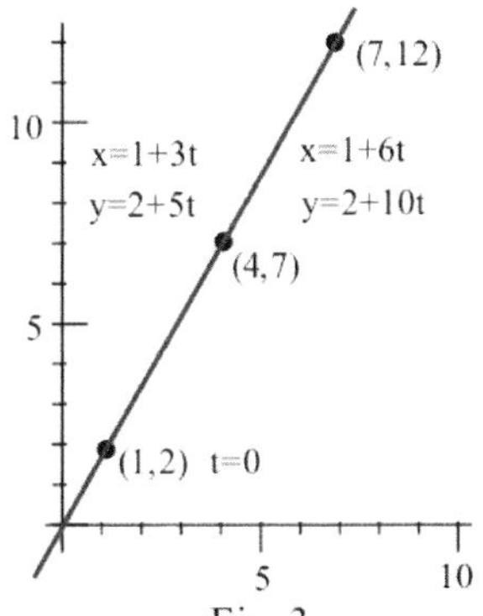

Fig. 3

Practice 1: Find parametric equations for the lines through the point $P = (3,-1)$ that are (a) parallel to the vector $\mathbf{A} = \langle 2, -4 \rangle$, and (b) parallel to the vector $\mathbf{B} = \langle 1, 5 \rangle$. Then graph the two lines.

The parametric pattern works for lines in three dimensions.

Parametric Equation of a Line in Three Dimensions

An equation of a line through the point $P = (x_0, y_0, z_0)$ and parallel to the vector $\mathbf{A} = \langle a, b, c \rangle$ is given by the parametric equations

$$x = x(t) = x_0 + at,$$
$$y = y(t) = y_0 + bt,$$
$$z = z(t) = z_0 + ct. \quad \text{(Fig. 4)}$$

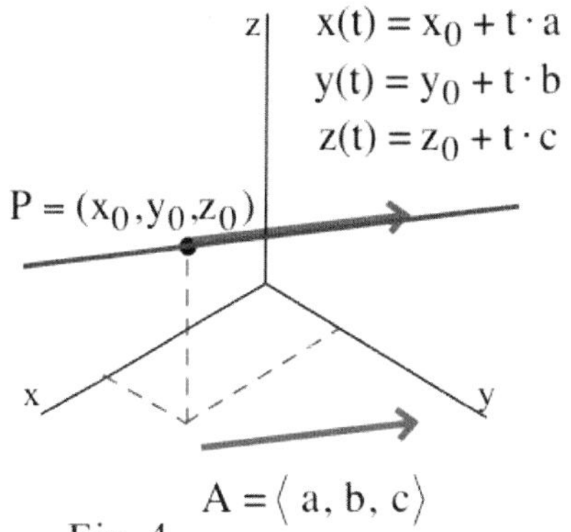

Fig. 4

Proof: To show that $P = (x_0, y_0, z_0)$ is on the line, put $t = 0$ and evaluate the three parametric equations: $x = x(0) = x_0 + 0, y = y(0) = y_0 + 0$, and $z = z(0) = z_0 + 0$.

To show that the line described by the parametric equations has the same direction as $\mathbf{A}$, we pick another point Q on the line and show that the vector from P to Q is parallel to $\mathbf{A}$.

Put $t = 1$, and let $Q = (x(1), y(1), z(1)) = (x_0 + a, y_0 + b, z_0 + c)$. Then the vector from P to Q is $\mathbf{V} = \langle (x_0 + a) - x_0, (y_0 + b) - y_0, (z_0 + c) - z_0 \rangle = \langle a, b, c \rangle = \mathbf{A}$ so the line described by the parametric equations is parallel to $\mathbf{A}$.

Example 2: Find parametric equations for the line (a) through the point $P = (3,0,2)$ and parallel to the vector $\mathbf{A} = \langle 1, 2, 0 \rangle$, and (b) through the two points $P = (3,0,2)$ and $Q = (5, -1, 1)$.

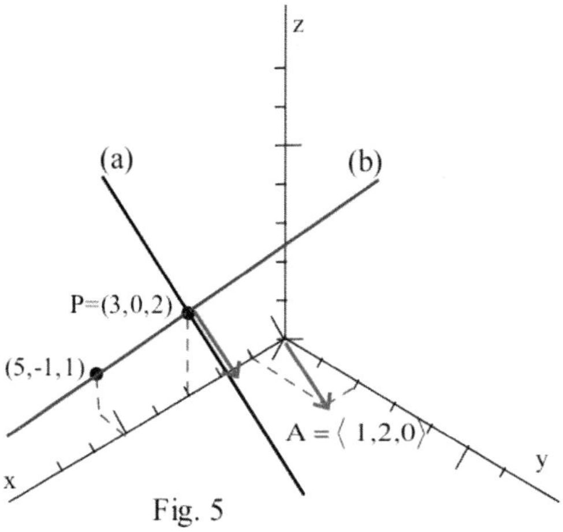

Fig. 5

Solution: (a) $x(t) = 3 + 1t, y(t) = 0 + 2t, z(t) = 2 + 0t$.

(b) We can use the two points to get a direction for the line. The direction of the line is parallel to the vector from P to Q, $\langle 2,-1,-1 \rangle$, so

$x(t) = 3 + 2t, y(t) = 0 - 1t , z(t) = 2 - 1t$.

The graphs of these two lines are shown in Fig. 5.

Practice 2: Find parametric equations for the line

(a) through the point $P = (2,-1,0)$ and parallel to the vector $\mathbf{A} = \langle 3,-4,1 \rangle$, and (b) through the two points $P = (3,0,2)$ and $Q = (2,5,4)$.

Given a point $P = (x_0, y_0, z_0)$ and a vector $\mathbf{A} = \langle a, b, c \rangle$, the point Q is on the line through P in the direction of **A** if and only if $Q = (x_0+at, y_0+bt, z_0+ct)$ for some value of t.

Once we are able to write the equations of lines in three dimensions, it is natural to ask about the points and angles of intersection of these lines.

Example 3: (a) Find the point of intersection of the lines K: $x_K = 2 + t, y_K = 3 + 2t, z_K = 0 + t$ and L: $x_L = 5 - t, y_L = 1 + 2t, z_L = -1 + t$.

(b) Find the angle of intersection of the lines K and L.

(c) Where does line L intersect the xy–plane?

Solution: (a) If we could find a value of t so that $x_K = x_L$, $y_K = y_L$, and $z_K = z_L$, that would say that the lines not only intersect at the point (x_K, y_K, z_K), but that the two objects were at that point at the same time. For these parametric equations, there is no value of t such that $x_K = x_L$, $y_K = y_L$, and $z_K = z_L$, but it is still possible for the lines to intersect, just not at the same "time."

To see if the lines go through a common point, but at different "times," we can change the parameter for the line L to "s" instead of "t," and represent L as $x_L = 5 - s, y_L = 1 + 2s, z_L = -1 + s$. Then the equations $x_K = x_L$, $y_K = y_L$, and $z_K = z_L$ become

$$x: 2 + t = 5 - s, \quad y: 3 + 2t = 1 + 2s, \text{ and } z: 0 + t = -1 + s .$$

Solving the first two equations for s and t, we get $s = 2$ and $t = 1$. These values also satisfy the equation for the z–coordinate, $0 + t = -1 + s$, so $x = 2 + t = 3, y = 3 + 2t = 5$, and $z = 0 + t = 1$. The point (3, 5, 1) lies on both lines. (If the values of s and t from the x–component and y–component equations do not satisfy the z–component equation, the lines do not intersect.)

(b) Since the lines intersect, we can find their angle of intersection. Line K is parallel to $\mathbf{A} = \langle 1, 2, 1 \rangle$, the coefficients of the t terms, and L is parallel to $\mathbf{B} = \langle -1, 2, 1 \rangle$, and the angle between the lines equals the angle between **A** and **B**:

$$\cos(\theta) = \frac{\mathbf{A} \bullet \mathbf{B}}{|\mathbf{A}||\mathbf{B}|} = \frac{4}{\sqrt{6}\sqrt{6}} = \frac{2}{3} \quad \text{so } \theta \approx 0.84 \text{ (about } 48.2°\text{)}.$$

(c) Every point on the xy–plane has z–coordinate equal to 0, so we can set $-1 + t = 0$ and solve for $t = 1$. When $t = 1$, then $x = 5 - (1) = 4$ and $y = 1 + 2(1) = 3$ so line L intersects the xy–plane at the point $(4, 3, 0)$.

Practice 3: If the pairs of lines in (a) or (b) intersect, find the point and angle of intersection.

(a) K: $x = 1 + t, y = 1 - 2t, z = -3 + 2t$ and L: $x = 8 + 4t, y = -4 + t, z = -5 - 8t$.

(b) K: $x = 1 + t, y = 1 - 2t, z = 2 + 2t$ and L: $x = 8 + 4t, y = -4 + t, z = 3 + t$.

(c) Where does L: $x = 8 + 4t, y = -4 + t, z = 3 + t$ intersect the yz–plane?

Practice 4: An arrow is shot from the point (1,2,3) and travels in a straight line in the direction $\langle 4, 5, 1 \rangle$. Will the arrow go over a 10 foot high wall built on the xy–plane along the line $y = 20$? (Fig. 6)

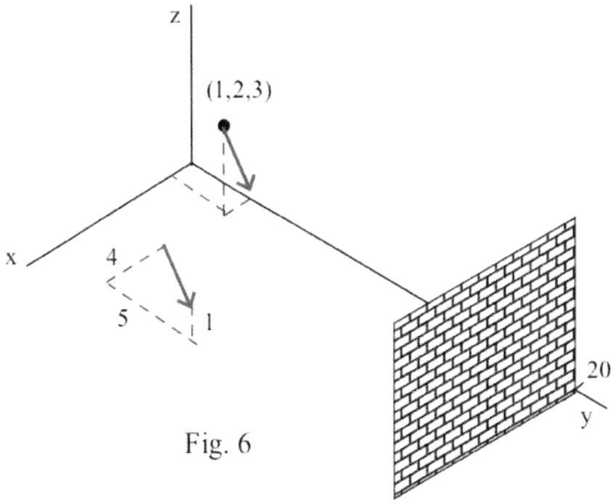

Fig. 6

Planes in Three Dimensions

The vectors in a plane point in infinitely many directions (Fig. 7) so, at first thought, you might think it would be more difficult to find an equation for a plane than for a line.

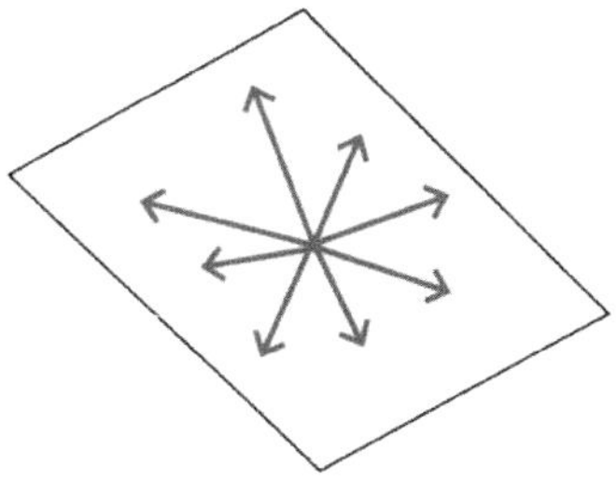

Fig. 7: Vectors in a plane

Fortunately, however, there is only one vector (and its scalar multiples) that is perpendicular to the plane (Fig. 8), and this "normal" vector makes the task easy.

Suppose $P = (x_0, y_0, z_0)$ is a point on the plane that has normal vector $\mathbf{N} = \langle a, b, c \rangle$. Let $Q = (x, y, z)$ be another point. Since **N** is perpendicular to every vector on the plane, the point Q is on the plane if and only if **N** is perpendicular to the vector from P to Q.

That is the idea that leads to an easy equation for the plane.

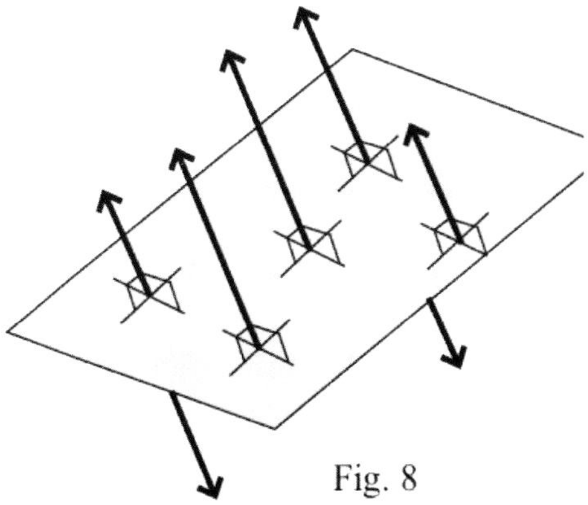

Fig. 8

Let $\mathbf{V}$ = vector from P to Q = $\langle x - x_0, y - y_0, z - z_0 \rangle$. Then Q is on the plane if and only if $\mathbf{V}$ is perpendicular to $\mathbf{N}$: $\mathbf{V} \bullet \mathbf{N} = 0$. But $\mathbf{V} \bullet \mathbf{N} = a(x - x_0) + b(y - y_0) + c(z - z_0)$, so the point Q is on the plane if and only if $a(x - x_0) + b(y - y_0) + c(z - z_0) = 0$.

Equation for a Plane in Three Dimensions:

(Point–Normal Form)

An equation for a plane through the point $P = (x_0, y_0, z_0)$ with normal vector $\mathbf{N} = \langle a, b, c \rangle$ is $a(x - x_0) + b(y - y_0) + c(z - z_0) = 0$. (Fig. 9)

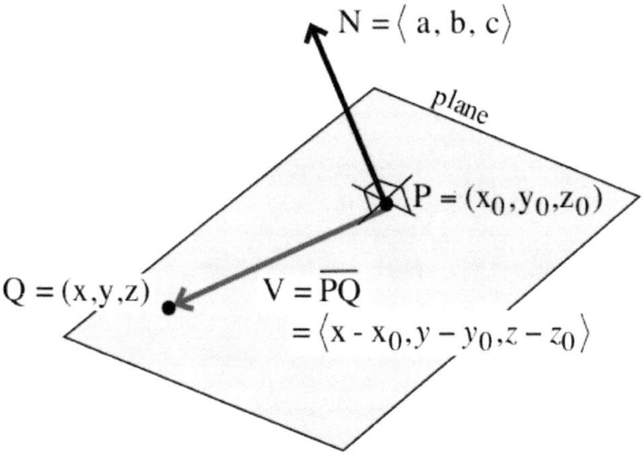

Equation of the plane
$a(x - x_0) + b(y - y_0) + c(z - z_0) = 0$
Fig. 9

The Point–Normal form is the fundamental pattern for the equation of a plane, and other information can usually be translated so the Point–Normal form can be used. If we have point P and two vectors $\mathbf{U}$ and $\mathbf{V}$ in the plane, then we can use the result that the cross product of two vectors is perpendicular to each of them to find a vector perpendicular to the plane: $\mathbf{N} = \mathbf{U} \times \mathbf{V}$. Once we have $\mathbf{N}$, we can use the Point–Normal form for the equation of the plane. If we have three points P, Q, and R, we can form the vectors $\mathbf{U}$ from P to Q and $\mathbf{V}$ from P to R, calculate the normal vector $\mathbf{N} = \mathbf{U} \times \mathbf{V}$ and then use the Point–Normal form for the equation for the plane.

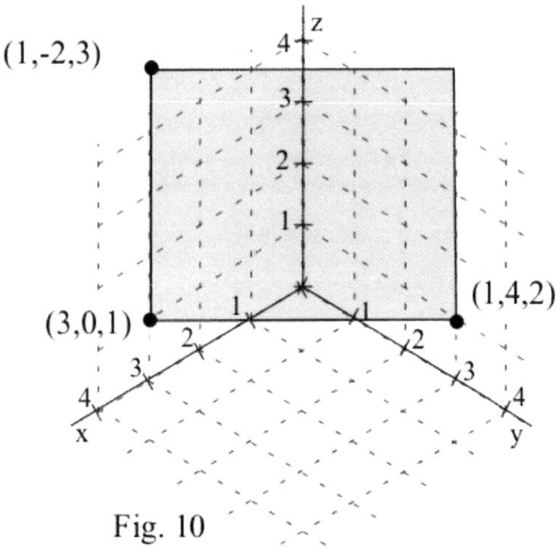

Fig. 10

Example 4: Find the equation of the plane:

(a) through the point $(1, -2, 3)$ and with normal vector $\mathbf{N} = \langle 5, 3, -4 \rangle$,

(b) through the points $P = (1, -2, 3)$, $Q = (3,0,1)$ and $R = (1,4,2)$. (Fig. 10)

Solution: (a) The point (x,y,z) is on the plane if and only if $5(x-1) + 3(y+2) - 4(z-3) = 0$, or, equivalently, $5x + 3y - 4z = -13$.

(b) Let $\mathbf{U}$ = vector from P to Q = $\langle 2, 2, -2 \rangle$ and $\mathbf{V}$ = vector from P to R = $\langle 0, 6, -1 \rangle$.

Then $\mathbf{N} = \mathbf{U} \times \mathbf{V} = \begin{vmatrix} \mathbf{i} & \mathbf{j} & \mathbf{k} \\ 2 & 2 & -2 \\ 0 & 6 & -1 \end{vmatrix} = \mathbf{i}\begin{vmatrix} 2 & -2 \\ 6 & -1 \end{vmatrix} - \mathbf{j}\begin{vmatrix} 2 & -2 \\ 0 & -1 \end{vmatrix} + \mathbf{k}\begin{vmatrix} 2 & 2 \\ 0 & 6 \end{vmatrix} = 10\mathbf{i} + 2\mathbf{j} + 12\mathbf{k}$.

The equation of the plane is $10(x-1) + 2(y+2) + 12(z-3) = 0$ or $10x + 2y + 12z = 42$ (or $5x+y+6z=21$).

Practice 5: Find the equation of the plane:

(a) through the point (4, 1, 0) and with normal vector $\mathbf{N} = \langle 3, -2, 6 \rangle$,

(b) determined by the lines K: x = 2 + t, y = 2 + 2t, z = 4 – t and L: x = 2 – 3t, y = 2 + t, z = 4 + 4t. (Fig. 11)
(The lines intersect at (2,2,4).)

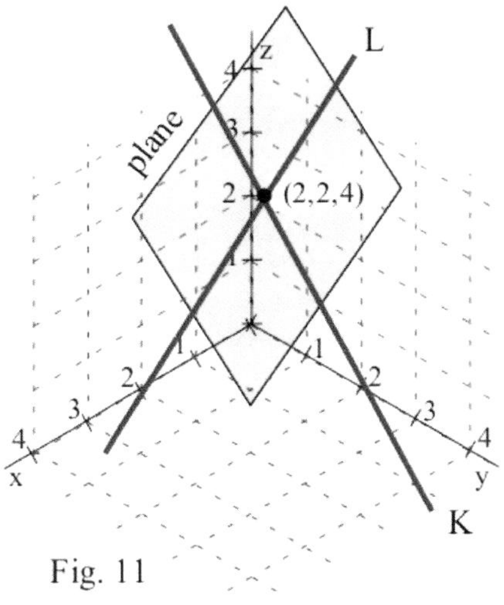

Fig. 11

Example 5: (a) Find a normal vector to the plane 3x – 2y + 4z = 12.

(b) Where does the plane 3x – 2y + 4z = 12 intersect each coordinate axis?

(c) Where does the line x = 3 + 5t , y = –4 + 2t , z = 1 – 2t intersect the plane 3x – 2y + 4z = 12?

Solution: (a) We can get one normal vector to the plane simply by using the coefficients of the variables of the equation of the plane: $\mathbf{N} = \langle 3, -2, 4 \rangle$. Any other nonzero vector is normal to the plane if and only if it is a nonzero scalar multiple of **N** .

(b) Every point on the x–axis has y=0 and z=0, so we can find where the plane intersects the x–axis by setting y and z equal to zero in the plane equation and solving for x: 3x – 2(0) + 4(0) = 12 so x = 4. The plane intersects the x–axis at (4,0,0). Similarly, the plane intersects the y–axis at (0,–6,0) and the z–axis at (0,0,3) .

(c) Substitute the parametric patterns for x, y, and z from the line into the equation for the plane and then solve for t: 3(3 + 5t) – 2(–4 + 2t) + 4(1 – 2t) = 12 so

9 + 15t + 8 – 4t + 4 – 8t = 12

21 + 3t = 12 and t = – 3.

Then x = 3 + 5(–3) = –12, y = –10, and z = 7. The point (–12, –10, 7) is on the line and on the plane.

Practice 6: (a) Find a normal vector to the plane 3x + 10y – 4z = 30.

(b) Where does the plane 3x + 10y – 4z = 30 intersect each coordinate axis?

(c) Where does the line x = 4 + t , y = –2 + 2t , z = 4 – t intersect the plane 3x + 10y + 4z = 30?

Normal vectors also provide us with a way to determine the angle between a line and a plane and the angle between two planes.

Angles of Intersection

The angle between a line and a plane (Fig. 12) is

$\pi/2$ – (angle between the line and the normal vector of the plane) .

The angle between two planes (Fig. 13) is the angle between the normal vectors of the planes.

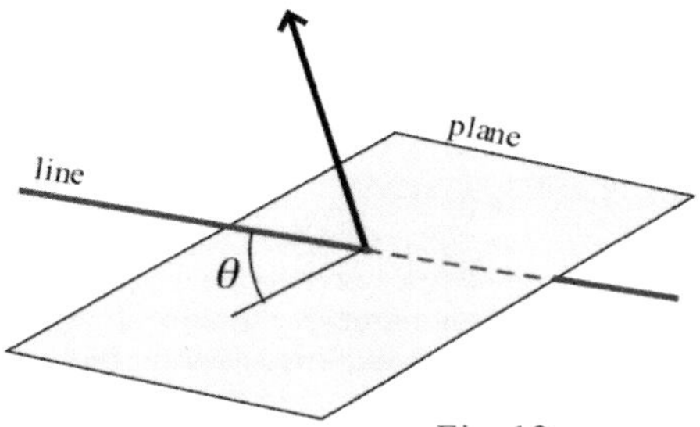

Fig. 12

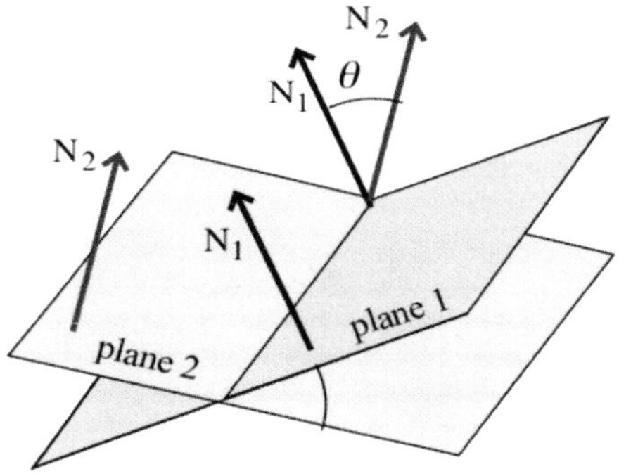

Fig. 13

If the normal vectors of two planes have different directions, then the planes intersect in a line (the set of points common to both planes is a line), and we can determine parametric equations for that line. One method is to eliminate the y variable from the equations of the planes and write x in terms of z: $x = x(z)$. Then we eliminate the x variable and write y in terms of z: $y = y(z)$. Finally, we treat the variable z as the parameter "t" in the parametric equations for the line, and we have $x = x(t), y = y(t), z = t$.

Example 6: Let P be the plane $12x + 7y - 3z = 43$ and Q be the plane $4x + 7y + 13z = 19$.

(a) Find a parametric equation representation of the line of intersection of the two planes.

(b) Find the angle between the planes.

Solution: (a) This is basically an algebra problem to use the equations of the two planes to solve for two of the variables in terms of the third variable. Then we can treat that third variable as the parameter t and write the equation of the line of intersection in parametric form. We can eliminate y and solve for x in terms of z: (equation for P) – (equation for Q) is $8x - 16z = 24$ so $x = 3 + 2z$. Then we can eliminate x and solve for y in terms of z: (equation for P) – 3(equation for Q) is $-14y - 42z = -14$ so $y = 1 - 3z$. Treating the variable z as our parameter "t" we have the line $\mathbf{x = 3 + 2t, y = 1 - 3t,}$ **and** $\mathbf{z = t}$.

As a check, when $t = 0$ then $x = 3, y = 1$, and $z = 0$, and the point $(3,1,0)$ lies on both planes. When $t = 1$, then $x = 5, y = -2$, and $z = 1$, and the point $(5,-2,1)$ also lies on both planes.

(b) The angle between the planes is the angle between the normal vectors of the planes.
$\mathbf{N}_P = \langle 12, 7, -3 \rangle, \mathbf{N}_Q = \langle 4, 7, 13 \rangle$ so

$$\cos(\theta) = \frac{\mathbf{N}_P \cdot \mathbf{N}_Q}{|\mathbf{N}_P||\mathbf{N}_Q|} = \frac{58}{\sqrt{202}\sqrt{234}} \approx 0.267 \text{ so } \theta \approx 1.30 \text{ (about } 74.5°) .$$

Practice 7: Let R be the plane $2x + 3y - z = 13$ and S be the plane $2x - y + 3z = 1$.

(a) Find the line of intersection of the two planes. (b) Find the angle between the planes.

Sometimes an alternate method is easier: find two points that lie on the intersection and then write the parametric equations for the line through those two points.

Example 7: Let P be the plane $7x - y - 11z = 10$ and Q be the plane $9x + y - 5z = 22$.

Find a parametric equation representation of the line of intersection of the two planes.

Solution: Set one variable equal to 0, say $z=0$, so we then have $7x - y = 10$ and $9x + y = 22$. Adding these two equations together gives $16x = 32$ so $x=2$ and then $y=4$ so one point on the intersection of the planes is $A = (2, 4, 0)$. Setting $x=0$, we have $-y - 11z = 10$ and $y - 5z = 22$. Adding these together gives $-16z = 32$ so $z=-2$ and then $y=12$ so another point on the intersection is $B = (0, 12, -2)$. A parametric equation representation of the line through A and B is $x(t) = 2 - 2t$, $y(t) = 4 + 8t$, $z(t) = -2t$. (We could have put $y=0$ and then found the point $C = (3, 0, 1)$. Check that point C satisfies the equations we just found.)

Distance Algorithms

Distances between objects are needed in a variety of applications in three dimensions:

- How close does the power line come to the school (Fig. 14)?
- How close can airplanes on two different linear flight paths come to each other (Fig. 15)?
- How far is the charged particle from the planar plate (Fig. 16)?

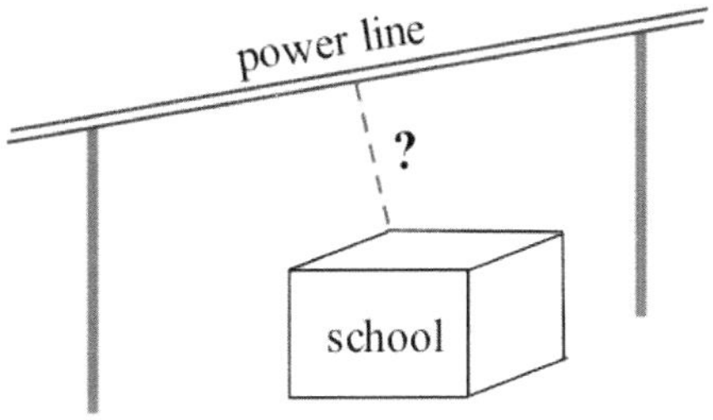

F

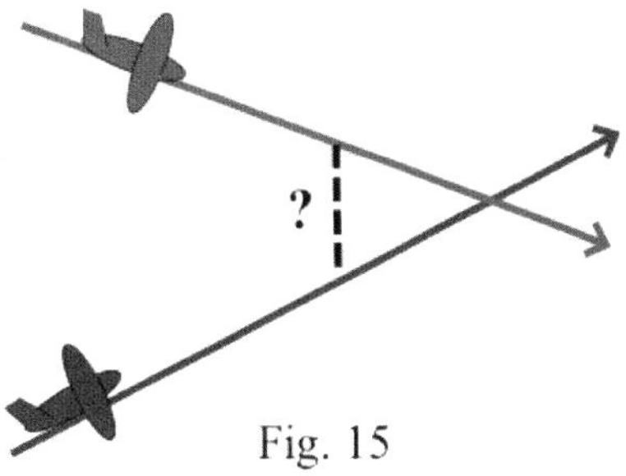

Fig. 15

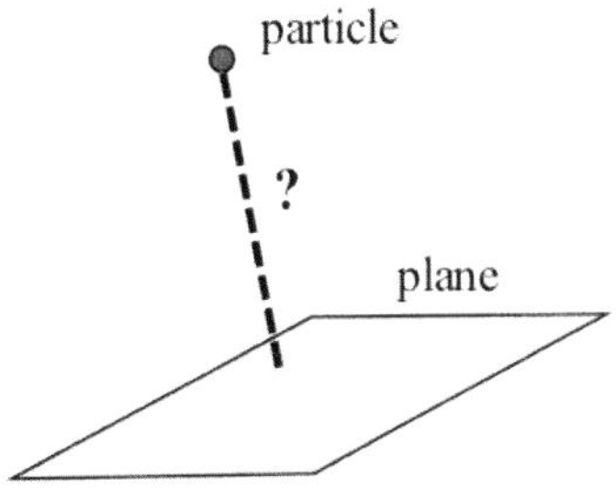

Fig. 16

Below we give a collection of distance algorithms, geometric constructions, and formulas for the distances between different types of objects. At first this may seem to be a large task, but most of the patterns use the idea of vector projection and follow from thinking about the geometry of the situations.

The distance from a point to a line in two dimensions was discussed in Section 11.3 and is presented again here because it uses the type of reasoning we need for three dimensions and because the resulting formula for two dimensions reappears for some distances in three dimensions.

Two dimensions: Distance from a **point** $P = (x_0, y_0)$ to a **line** $ax + by = c$ (Fig.17).

(1) Find **N** perpendicular to the line: $\mathbf{N} = \langle a, b \rangle$ works .

(2) Find a point Q on the line and form the vector **V** from P to Q .

(3) Distance = | Projection of **V** onto **N** |

$$= \frac{|\mathbf{V} \bullet \mathbf{N}|}{|\mathbf{N}|} = \frac{|ax_0 + by_0 - c|}{\sqrt{a^2 + b^2}} .$$

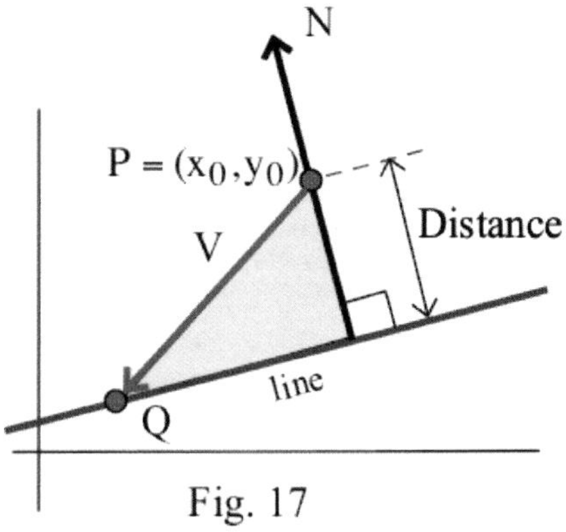

Fig. 17

Three dimensions: Distance from a **point** P to a **plane** with normal vector **N** (Fig. 18).

(1) Find a point Q on the plane and form the vector **V** from P to Q .

(2) Distance = | Projection of **V** onto **N** |

$$= \frac{|\mathbf{V} \bullet \mathbf{N}|}{|\mathbf{N}|} .$$

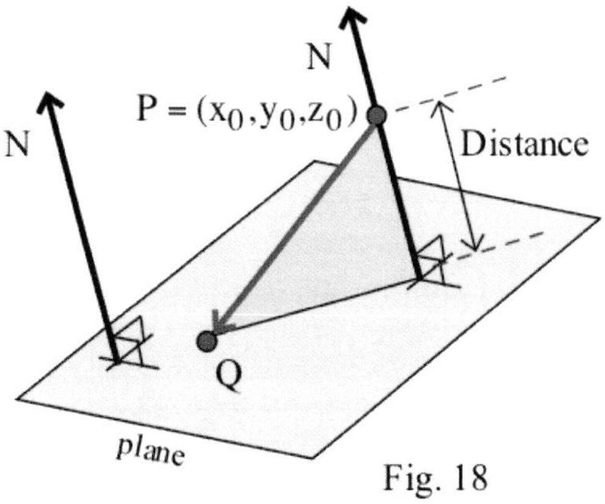

Fig. 18

Three dimensions: Distance from a **line** parallel to **U** to a **line** parallel to **V** (**U** not parallel to **V**) (Fig. 19).

(1) Find a point P on one line and a point Q on the other line .

(2) Form the vector **W** from P to Q .

(3) Calculate $\mathbf{N} = \mathbf{U} \times \mathbf{V}$.

(4) Distance = | projection of **W** onto **N** | $= \dfrac{|\mathbf{W} \bullet \mathbf{N}|}{|\mathbf{N}|}$.

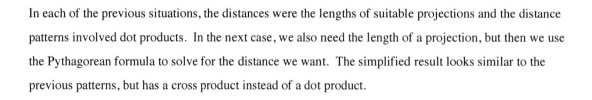

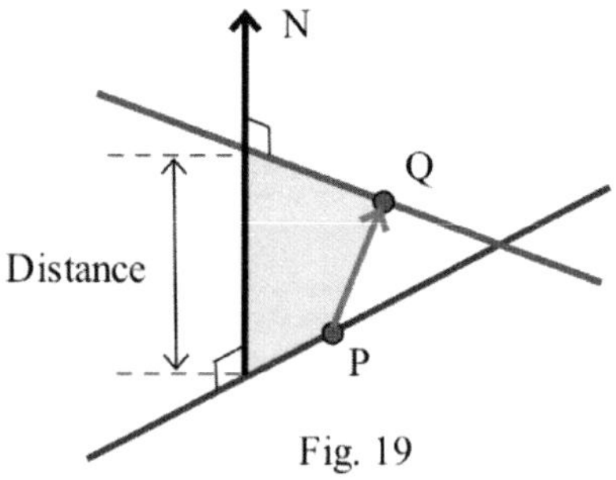

Fig. 19

(If **U** and **V** are parallel, pick a point on one line and then use the "point to line" formula.

In each of the previous situations, the distances were the lengths of suitable projections and the distance patterns involved dot products. In the next case, we also need the length of a projection, but then we use the Pythagorean formula to solve for the distance we want. The simplified result looks similar to the previous patterns, but has a cross product instead of a dot product.

Three dimensions: Distance from a **point** P to a **line** parallel to **U** (Fig. 20).

(1) Find a point Q on the line and form the vector **V** from P to Q .

(2) Put **W** = | projection of **V** onto **U** | = $\dfrac{|\mathbf{V}\cdot\mathbf{U}|}{|\mathbf{U}|} = \dfrac{|\mathbf{V}|\,|\mathbf{U}|\,|\cos(\theta)|}{|\mathbf{U}|} = |\mathbf{V}|\,|\cos(\theta)|$.

(3) Then $(\text{distance})^2 = |\mathbf{V}|^2 - |\mathbf{W}|^2$

$= |\mathbf{V}|^2 - |\mathbf{V}|^2\cos^2(\theta)$

$= |\mathbf{V}|^2(1-\cos^2(\theta))$

$= |\mathbf{V}|^2\sin^2(\theta)$

$= \dfrac{|\mathbf{V}|^2\,|\mathbf{U}|^2\sin^2(\theta)}{|\mathbf{U}|^2} = \left(\dfrac{|\mathbf{V}\times\mathbf{U}|}{|\mathbf{U}|}\right)^2$.

(4) Distance = $\dfrac{|\mathbf{V}\times\mathbf{U}|}{|\mathbf{U}|}$.

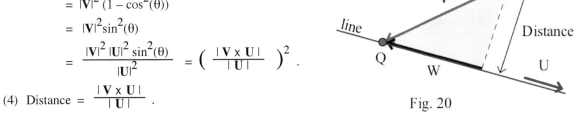

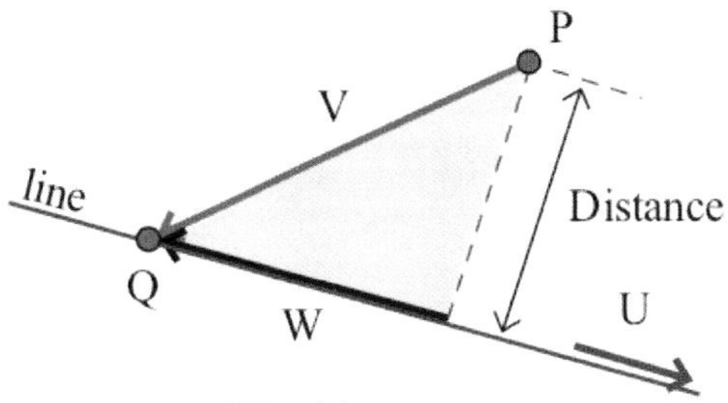

Fig. 20

Example 7: A power line runs in a straight line from the point A = (200, 0, 100) to the point B = (100, 700, 200). How close does the line come to the corner of a school at P = (300, 400, 0)?

Solution: This problem uses the pattern for the distance from a point to a line in three dimensions.

P = (300, 400, 0) and **U** = the vector from A to B = $\langle -100, 700, 100\rangle = 100\langle -1, 7, 1\rangle$:

$|\mathbf{U}| = 100\sqrt{51}$.

Since A is on the line, let **V** = vector from P to A = $\langle -100, -400, 100\rangle = 100\langle -1, -4, 1\rangle$.

$$\mathbf{V}\times\mathbf{U} = (100)(100)\begin{vmatrix} \mathbf{i} & \mathbf{j} & \mathbf{k} \\ -1 & -4 & 1 \\ -1 & 7 & 1 \end{vmatrix} = (100)^2(-11\mathbf{i} - 0\mathbf{j} - 11\mathbf{k}) . \quad |\mathbf{V}\times\mathbf{U}| = 100^2\sqrt{242} .$$

Finally, distance from power line to school = $\dfrac{|\mathbf{V}\times\mathbf{U}|}{|\mathbf{U}|} = \dfrac{100^2\sqrt{242}}{100\sqrt{51}} \approx 217.8$ feet.

PROBLEMS

In problems 1 – 8, find parametric equations for the lines.

1. The line through the point (2, –3, 1) and parallel to the vector $\langle 3, 4, 2\rangle$.
2. The line through the point (0, 5, –2) and parallel to the vector $\langle -3, 1, -4\rangle$.
3. The line through the point (–2, 1, 4) and parallel to the vector $\langle 5, 0, -3\rangle$.
4. The line through the origin and parallel to the vector $\langle 1, 2, 3\rangle$.
5. The line through the points (2, –1, 3) and (3, 4, –2).
6. The line through the points (7, 3, –4) and (5, 0, –2).
7. The line through the points (3, –2, 1) and (3, 4, –1).
8. The line through the origin and (3, 4, –2).

In problems 9 – 12, the equations for a pair of lines are given. Determine whether the lines intersect, and if they do intersect, find the point and angle of intersection.

9. Line L: $x = 2 + t, y = -1 + t, z = 3 + 2t$. Line K: $x = 2 - t, y = -1 + 2t, z = 3 + 4t$.

10. Line L: $x = 6 + 2t, y = 3 - 2t, z = -2 + 3t$. Line K: $x = 6 + t, y = 3 + 5t, z = -2 - 3t$.

11. Line L: $x = 1 + 3t, y = 5 - t, z = -2 + 2t$. Line K: $x = 9 + 4t, y = 5, z = 4 + 3t$.

12. Line L: $x = 1 + 2t, y = 1 + 3t, z = 3 + 5t$. Line K: $x = 2 + t, y = 3 + t, z = 1 + t$.

In problems 13 – 26, find equations for the planes.

13. The plane through the point $(2, 3, 1)$ and perpendicular to the vector $\langle 5, -2, 4 \rangle$.

14. The plane through the point $(4, 0, -2)$ and perpendicular to the vector $\langle 3, 1, -5 \rangle$.

15. The plane through the point $(-3, 5, 6)$ and perpendicular to the vector $\langle 0, 3, 0 \rangle$.

16. The plane through the origin and perpendicular to the vector $\langle 2, -2, 1 \rangle$.

17. The plane through the points $(1, 2, 3)$, $(5, 2, 1)$, and $(4, -1, 3)$.

18. The plane through the points $(3, 5, 3)$, $(-2, 5, 4)$, and $(1, 5, 6)$.

19. The plane through the points $(-4, 2, 5)$, $(1, 2, 1)$, and $(3, -3, 5)$.

20. The plane through the origin, $(1, 2, 3)$, and $(4, 5, 6)$.

21. The plane through the point $(2, -5, 7)$ and parallel to the xy–plane.

22. The plane through the point $(2, 4, 1)$ and parallel to the yz–plane.

23. The plane through the point $(4, 2, 3)$ and parallel to the plane $3x - 2y + 5z = 15$.

24. The plane through the origin and parallel to the plane $2x + 3y - z = 12$.

25. The plane through the point $(4, 1, 3)$ and perpendicular to the line $x = 2 + 5t, y = 1 - 3t, z = 2t$.

26. The plane through the point $(2, 7, 4)$ and perpendicular to the y–axis.

Try to answer problems 27 – 32 without doing any algebra — just think visually.

27. Where does the plane $x = 0$ intersect the plane $z = 0$?

28. Where does the plane $y = 2$ intersect the plane $z = 3$?

29. Where does the plane $3x + 2y + z = 30$ intersect the x–axis?

30. Where does the plane $3x + 2y + z = 30$ intersect the y–axis?

31. Where do the three planes $x = 4$, $y = 2$, and $z = 1$ intersect?

32. Where does the y–axis intersect the plane $z = 3$?

In problems 33 – 36, two planes are given. Represent their line of intersection using parametric equations and find the angle between the planes.

33. $4x - 2y + 2z = 10$ and $3x - 2y + 3z = 36$.

34. $x + y + 3z = 9$ and $2x + y - 3z = 18$.

35. $5y - z = 10$ and $x + 2y + 4z = 16$.

36. $x = 4$ and $3x - 5y + z = 20$.

In problems 37 – 40, find where the given line intersects the plane and find the angle the line makes with the plane.

37. Line L: $x = 7 + 2t, y = 5 + t, z = 2 + 4t$ and the plane $5x + y - 2z = 20$.

38. Line L: $x = -2 + t, y = 12 - 3t, z = -5 - t$ and the plane $5x + y - 2z = 20$.

39. Line L: $x = 4 + 2t, y = 2 - 3t, z = 7 + t$, and the plane $z = 5$.

40. The x–axis and the plane $4x - 2y + 5z = 12$.

41. Is it possible for three distinct planes to intersect at a single point?

42. Is it possible for three distinct planes to intersect along a line?

43. A bird is flying in a straight line from the feeder (Fig. 21) to the corner of the house. Write an equation for its line of flight.

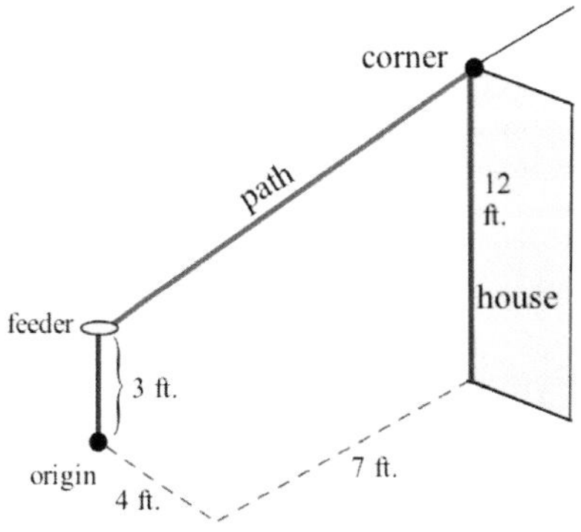

Fig. 21

44. What angle does the plane $3x + 2y + z = 10$ make with
(a) the xy–plane? (b) the xz–plane? (c) the yz–plane?

45. What angle does the plane $ax + by + cz = d$ make with each coordinate plane?

46. The four corners of a mirror are located at A(2,2,0), B(2,6,0), C(0,2,3), and D(0,6,3) (Fig. 22). A laser at L(5,0,0) directs a beam of light along the line $x = 5 - t$, $y = t$, $z = 0.5t$.

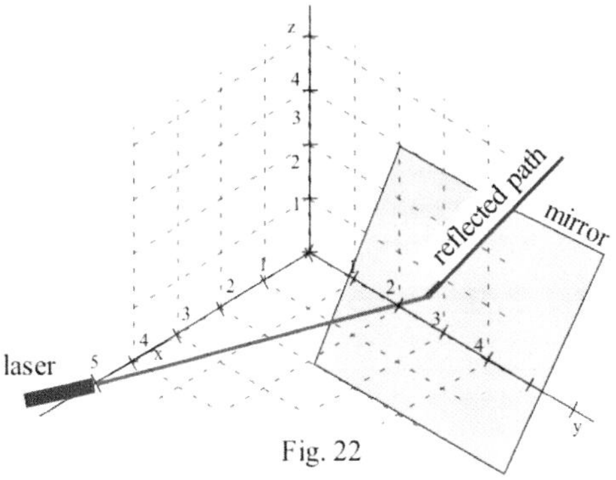

Fig. 22

(a) Write an equation for the plane of the mirror.
(b) Where does the light beam hit the mirror?
(c) What is the angle of incidence of the light with the mirror?

Many of the following applications can be solved in several ways including the methods of vectors and dot products from section 11.4.

47. When it was built, the Great Pyramid at Giza, Egypt was 481 feet tall and had a square base with length 756 feet on each side.
(a) Find the angle each side of the pyramid makes with the base.
(b) Find the angle each side makes with an adjacent side.
(c) Find the angle each edge makes with the base.

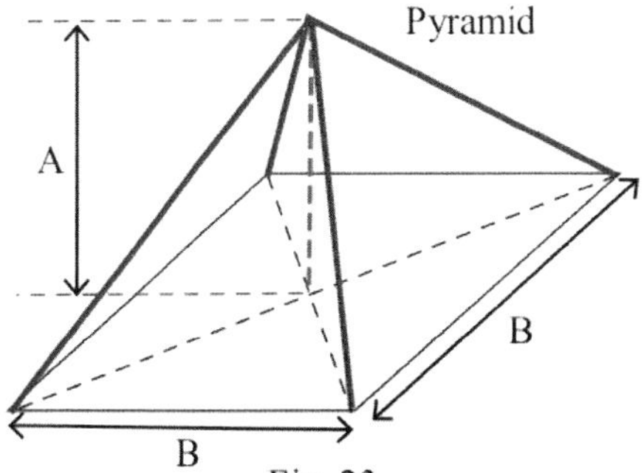

Fig. 23

48. In order to write a program to analyze a general pyramid (Fig. 23) with square base of length B and height A you need to determine the following:

(a) the angle each side makes with the base,

(b) the area of each side,

(c) the angle each edge makes with the base,

(d) the angle each side makes with an adjacent side, and

(e) the volume of the pyramid.

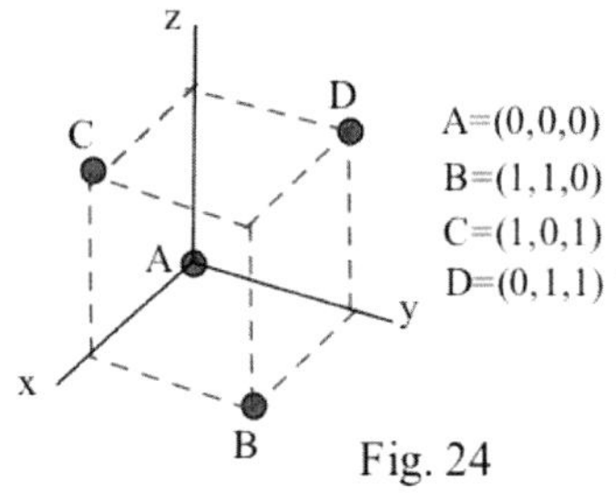

Fig. 24

49. Four molecules are located at the corners of a cube as shown in Fig. 24. The center of the cube is the point O = (0.5, 0.5, 0.5) Show that the angles AOB, AOC, and BOC are equal.

50. The four points (1,0,0), (–0.5,0.866,0), (–0.5,–0.866,0), and (0,0,1.414) are the vertices of an equilateral tetrahedron with center O = (0, 0, 0.354) (Fig. 25), and a molecule is located at each vertex.

(a) Show that the tetrahedron is really equilateral.

(b) Show that the angles AOB, AOC, and BOC are equal.

(c) Find the angle between the plane determined by AOB and the plane determined by AOC.

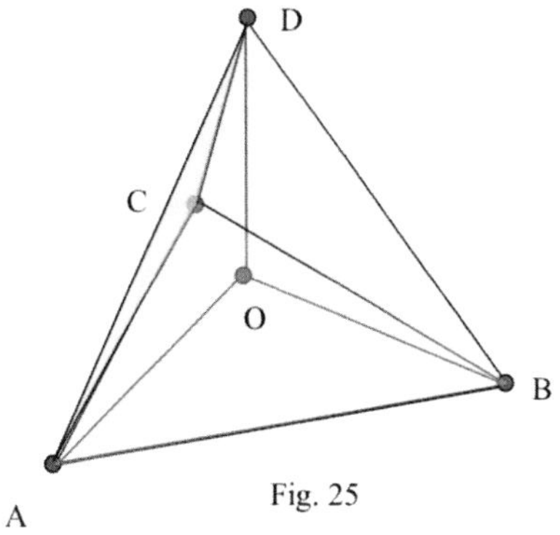

Fig. 25

Distances in Three Dimensions

51. Find the distance from the point (4, –1, 2) to the line x = 2 + t, y = 3 + 2t, z = 4t.

52. Find the distance from the point (4, –1, 2) to the line x = 1 – 3t, y = 2 + t, z = 3.

53. Find the distance from the point (2, 3, 1) to the plane 4x + 3y – z = 10.

54. Find the distance from the point (4, –3, 0) to the plane –2x + 5y + 3z = 15.

55. Find the distance (shortest distance) between the line x = 1 + 3t, y = 2 + 5t, z = 1 – t and the line x = 3 – 2t, y = 5 + t, z = 2 + 2t.

56. Find the distance (shortest distance) between the line x = 5 + 2t, y = 2 + t, z = 2 – 3t and the line x = 3 – 2t, y = 1 – 3t, z = 1.

In problems 57 – 62, the parameterized straight–line paths of two objects are given.

(a) Do the objects "crash" (so they are at the same location at the same time)? If so, at what time?

(b) Do the paths of the objects intersect (so the objects are at the same point but at different times)? If so, how close do the objects get to each other?

(c) Do the objects and their paths miss each other? If so, how close do the objects get to each other and how close do their paths get to each other?

57. Object A is at $x = 9 + t, y = 18 + t, z = 25 - 2t$ and object B is at $x = 3 + 2t, y = 30 - t$, and $z = 7 + t$.

58. Object A is at $x = -1 + t, y = -3 + 2t, z = 1 + t$ and object B is at $x = 2, y = t$, and $z = -2 + 2t$.

59. Object A is at $x = 5 - 5t, y = t, z = 5t$ and object B is at $x = 6 - 3t, y = 5 - 2t$, and $z = -3 + 4t$.

60. Object A is at $x = -4 + 5t, y = 3 + 2t, z = 16 - 3t$ and object B is at $x = 12 - t, y = 6 + 3t$, and $z = 3 + 4t$.

61. Object A is at $x = 5 - 2t, y = 0, z = 1$ and object B is at $x = 0, y = -1 + t$, and $z = 0$.

62. Object A is at $x = 1 + 3t, y = 2 + 2t, z = 3 + t$ and object B is at $x = 7 - t, y = 5 + 2t$, and $z = 3 + 3t$.

63. The bearing of an airplane is due north, and it passes directly above the origin at an altitude of 30,000 feet. How close does the airplane come to a balloon 5,000 feet directly above the point $x = 2{,}000$ and $y = 4{,}000$? (Assume the earth is "almost flat" in this region.)

64. At time t, airplane A is at $(-3 + t, 0, 1)$ and car B is at $(0, -5 + 2t, 0)$.

(a) How close do they come to each other?

(b) How close do their paths come to each other?

65. At time t, car A is at $(-3 + t, 2 + 2t, 0)$ and airplane B is at $(t, -5 + 2t, t)$.

(a) How close do they come to each other?

(b) How close do their come to each other?

66. Create your own problem, like problems 59 – 64, so the objects on different paths crash at the point $(3, 4, 5)$ at $t = 2$.

67. Create your own problem, like problems 59 – 64, with two objects on different paths so object A goes through the point $(3, 4, 5)$ at $t = 2$ and object B goes through the point $(3, 4, 5)$ at $t = 3$. (Then the paths intersect, but the objects do not crash.)

Practice Answers

Practice 1: (a) $x(t) = 3 + 2t, y(t) = -1 - 4t$.

(b) $x(t) = 3 + t, y(t) = -1 + 5t$.

The graphs are shown in Fig. 26.

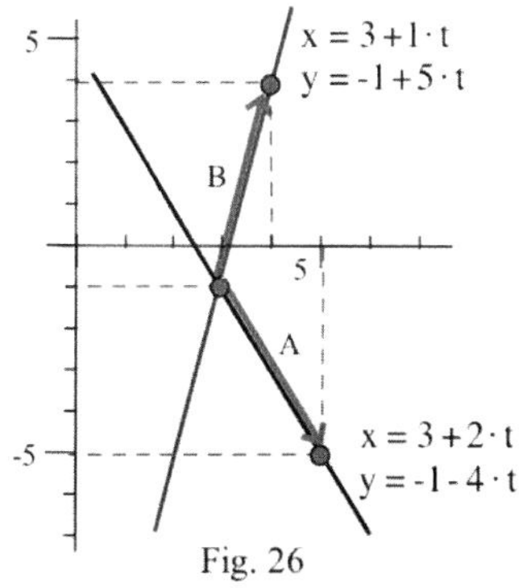

Fig. 26

Practice 2: (a) $x(t) = 2 + 3t, y(t) = -1 - 4t, z(t) = 0 + t$.

(b) The line is parallel to the vector $\mathbf{B} = \langle -1, 5, 2 \rangle$ from P to Q. Using P as the starting point, a parametric equation of the line is $x(t) = 3 - 1t$, $y(t) = 0 + 5t, z(t) = 2 + 2t$. (Using Q as the starting point, $x(t) = 2 - t$, $y(t) = 5 + 5t$, $z(t) = 4 + 2t$.)

Practice 3: Replacing the t parameter with s for line L and then setting the components equal to each other, we get: x: $1 + t = 8 + 4s$, y: $1 - 2t = -4 + s$, and z : $-3 + 2t = -5 - 8s$.

Solving the first two equations for t and s, $t = 3$ and $s = -1$. These values for t and s also satisfy the third equation so $x = 1 + (3) = 4, y = 1 - 2(3) = -5$, and $z = -3 + 2(3) = 3$.

The point $(4, -5, 3)$ lies on both lines.

The lines are parallel to $\mathbf{A} = \langle 1, -2, 2 \rangle$ and $\mathbf{B} = \langle 4, 1, -8 \rangle$ so

$$\cos(\theta) = \frac{\mathbf{A} \bullet \mathbf{B}}{|\mathbf{A}||\mathbf{B}|} = \frac{-14}{(3)(9)} \quad \text{and } \theta \approx 2.12 \text{ (about } 121.2°\text{)}.$$

(b) Replacing the t parameter with s for line L and then setting the components equal to each other, we get: x: $1 + t = 8 + 4s$, y: $1 - 2t = -4 + s$, and z : $2 + 2t = 3 + s$.

Solving the first two equations (they are the same as in part (a)) for t and s, $t = 3$ and $s = -1$. But these values do not satisfy the third equation, $2 + 2(3) \neq 3 + (-1)$, so the lines do not intersect.

(c) Every point in the yz–plane has x coordinate 0, so set $8 + 4t = 0$ to get $t = -2$. Then $y = -4 + (-2) = -6$ and $z = 3 + (-2) = 1$ so the line intersects the yz–plane at $(0, -6, 1)$.

Practice 4: The parametric equations for the line of travel of the arrow are $x = 1 + 4t, y = 2 + 5t$, $z = 3 + t$. The arrow reaches the wall when $y = 20$ so $2 + 5t = 20$ and $t = 18/5$. At that time, $t = 18/5$, the height of the arrow is $z = 3 + t = 3 + (18/5) = 33/5 = 6.6$ feet so the arrow does not go over the wall.

Practice 5: (a) The point (x,y,z) is on the plane if and only if $3(x-4) - 2(y-1) + 6(z-0) = 0$, or, equivalently, $3x - 2y + 6z = 10$.

(b) Taking the directions of the lines K and L from their parametric equations, we know line K is parallel to $\mathbf{U} = \langle 1, 2, -1 \rangle$ and line L is parallel to $\mathbf{V} = \langle -3, 1, 4 \rangle$. Then

$$\mathbf{N} = \mathbf{U} \times \mathbf{V} = \begin{vmatrix} \mathbf{i} & \mathbf{j} & \mathbf{k} \\ 1 & 2 & -1 \\ -3 & 1 & 4 \end{vmatrix} = \mathbf{i}\begin{vmatrix} 2 & -1 \\ 1 & 4 \end{vmatrix} - \mathbf{j}\begin{vmatrix} 1 & -1 \\ -3 & 4 \end{vmatrix} + \mathbf{k}\begin{vmatrix} 1 & 2 \\ -3 & 1 \end{vmatrix} = 9\mathbf{i} - 1\mathbf{j} + 7\mathbf{k} .$$

The point (2,2,4) is on both lines so it is on the plane. Using the point (2,2,4) and the normal vector $\mathbf{N} = \langle 9, -1, 7 \rangle$, the equation of the plane is

$$9(x-2) - 1(y-2) + 7(z-4) = 0 \text{ or } 9x - y + 7z = 44.$$

Practice 6: (a) $\mathbf{N} = \langle 3, 10, -4 \rangle$ and all nonzero scalar multiples of $\mathbf{N}$ are normal to the plane.

(b) The plane crosses the x–axis at (10,0,0), the y–axis at (0,3,0), and the z–axis at (0,0,–7.5) .

(c) Solving $3(4 + t) + 10(-2 + 2t) - 4(4 - t) = 30$ for t, we get

$12 + 3t - 20 + 20t - 16 + 4t = 30$ so $-24 + 27t = 30$ and $t = 2$. Then $x = 6, y = 2$, and $z = 2$. The point (6,2,2) is on the line and on the plane.

Practice 7: (a) We can eliminate y and solve for x in terms of z: (equation for R) + 3(equation for S) is $8x + 8z = 16$ so $x = 2 - z$. Then we can eliminate x and solve for y in terms of z: (equation for R) – (equation for S) is $4y - 4z = 12$ so $y = 3 + z$. Treating the variable z as our parameter "t" we have the line **x = 2 – t, y = 3 + t, and z = t** .

As a check, when $t = 0$ then $x = 2, y = 3$, and $z = 0$, and the point (2,3,0) lies on both planes. When $t = 1$, then $x = 1, y = 4$, and $z = 1$, and the point (1,4,1) also lies on both planes.

(b) The angle between the planes is the angle between the normal vectors of the planes.

$\mathbf{N}_R = \langle 2, 3, -1 \rangle, \mathbf{N}_S = \langle 2, -1, 3 \rangle$ so

$$\cos(\theta) = \frac{\mathbf{N}_R \bullet \mathbf{N}_S}{|\mathbf{N}_R||\mathbf{N}_S|} = \frac{-2}{\sqrt{14}\sqrt{14}} = \frac{-1}{7} \approx -0.143 \text{ so } \theta \approx 1.71 \text{ (about } 98.2°) .$$

11.7 Vector Reflections

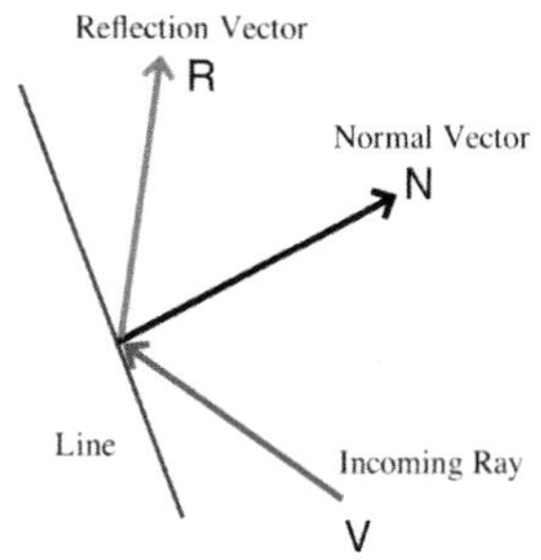

Figure 1

Vectors were originally used in physics and engineering (forces, work, ...), but a very common use now is in computer graphics, and a single frame in an animated movie such as "Cars" may require working with hundreds of millions of vectors. Realistic lighting and surfaces require calculating what happens when a ray is reflected off of a mirror or shiny surface. We will start examining that situation in 2D (Fig. 1), a ray reflected off of a line or curve, and later extend it to 3D, a ray reflecting off of a surface. The Reflection Theorem illustrates a modern use of the projection vector.

Reflection Theorem:

If an incident vector ray **V** is reflected off of a line L that has normal vector **N**,

then the reflection vector is $\mathbf{R} = \mathbf{V} - 2 \cdot \text{Proj}_\mathbf{N}\mathbf{V}$. (Fig. 1)

Proof: The basis of the proof is very visual and geometric (Fig. 2). Put (i) $\mathbf{A} = \text{Proj}_\mathbf{N}\mathbf{V}$, (ii) $\mathbf{B} = -2 \cdot \text{Proj}_\mathbf{N}\mathbf{V}$, and (iii) $\mathbf{C} = \mathbf{V} - 2 \cdot \text{Proj}_\mathbf{N}\mathbf{V}$. Then we can use geometry to justify that the vector **C** in (iii) is the reflection vector we want.

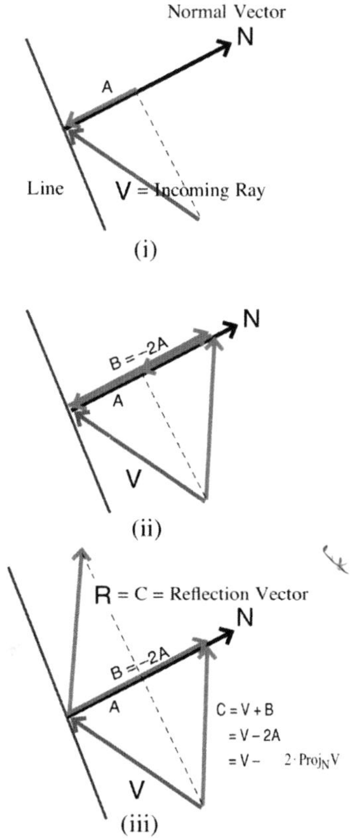

Fig. 2

Example 1: Find the reflection vector R when an incoming vector $\mathbf{V} = \langle -3, 1 \rangle$ is reflected by the line 3x + 2y = 12 (Fig. 3). Then give parametric equations for the reflected line when the reflection occurs at the point (2, 3).

Solution: The line has normal vector $\mathbf{N} = \langle 3, 2 \rangle$ so

$$\text{Proj}_\mathbf{N}\mathbf{V} = \frac{\mathbf{V} \bullet \mathbf{N}}{\mathbf{N} \bullet \mathbf{N}}\mathbf{N} = \frac{-7}{13}\langle 3, 2 \rangle = \left\langle \frac{-21}{13}, \frac{-14}{13} \right\rangle$$

and the reflection vector is

$$\mathbf{R} = \mathbf{V} - 2 \cdot \text{Proj}_\mathbf{N}\mathbf{V} = \langle -3, 1 \rangle - 2\left\langle \frac{-21}{13}, \frac{-14}{13} \right\rangle$$

$$= \left\langle -3 + \frac{42}{13}, 1 + \frac{28}{13} \right\rangle = \left\langle \frac{3}{13}, \frac{41}{13} \right\rangle$$

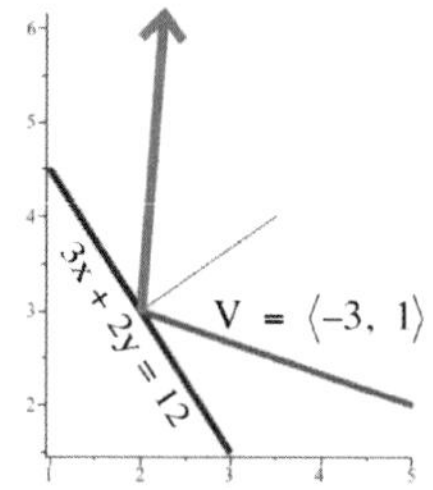

Fig. 3

One representation of the reflected line is $x(t) = 2 + \frac{3}{13}t, \; y(t) = 3 + \frac{41}{13}t$.

Since this operation is done millions of times for each frame, it is important that the operation be simple and quick.

Practice 1: Find the reflection vector R when an incoming vector $\mathbf{V} = \langle -1, -2 \rangle$ is reflected by the line –x + 2y = 2 (Fig. 4). Give parametric equations for the reflected line when the reflection occurs at the point (1.5, 1.75).

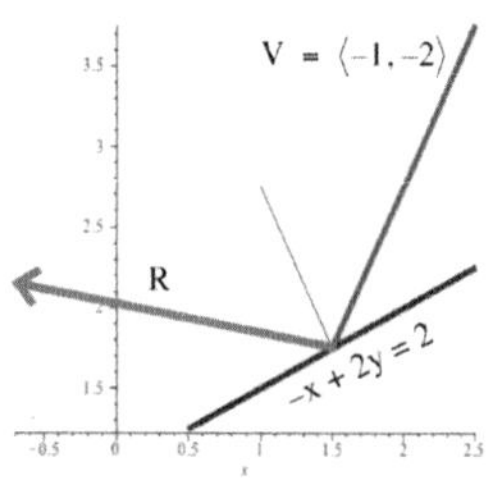

Fig. 4

We can also reflect off a curve at a point – just reflect off the tangent line to the curve at the point.

Example 2: Find the reflection vector R when an incoming vector $\mathbf{V} = \langle 3, -1 \rangle$ is reflected by the ellipse (2cos(t), sin(t)) at the point where t = 1.2 on the ellipse (Fig. 5).

Solution: $x'(t) = -2\sin(t)$ and $y'(t) = \cos(t)$ so a tangent vector to the ellipse when t = 1.2 is $\mathbf{T} = \langle -2\sin(1.2), \cos(1.2) \rangle \approx \langle -1.864, 0.362 \rangle$ and then a normal vector is $\mathbf{N} \approx \langle 0.362, 1.864 \rangle$. Finally,

$$\text{Proj}_{\mathbf{N}}\mathbf{V} = \frac{\mathbf{V} \bullet \mathbf{N}}{\mathbf{N} \bullet \mathbf{N}}\mathbf{N} \approx \frac{-0.778}{3.606}\langle 0.362, 1.864 \rangle = \langle -0.078, -0.402 \rangle$$

so $\mathbf{R} = \mathbf{V} - 2 \cdot \text{Proj}_{\mathbf{N}}\mathbf{V} = \langle 3, -1 \rangle - 2\langle -0.078, -0.402 \rangle = \langle 3.156, -0.197 \rangle$

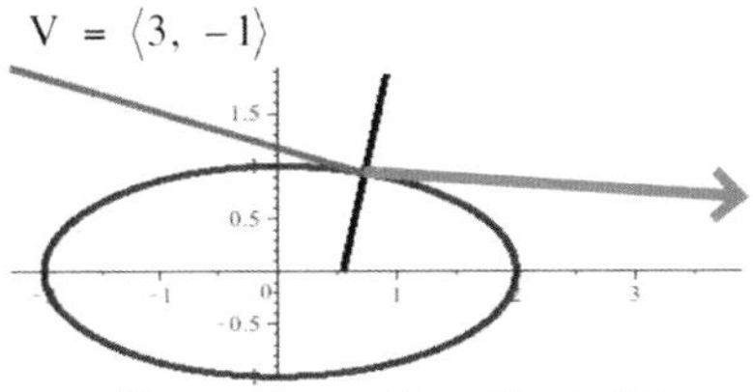

Fig. 5

Practice 2: Find the reflection vector **R** when an incoming vector $\mathbf{V} = \langle -1, 2 \rangle$ is reflected by the parabola (t, t^2) at the point where t = 1.2 on the parabola.

These ideas and computations extend very nicely to reflection vectors in 3D, and those reflections use the same formula: $\mathbf{R} = \mathbf{V} - 2 \cdot \text{Proj}_{\mathbf{N}}\mathbf{V}$. In the case of the reflection of a vector by a plane, we already know how to quickly find a normal vector to the plane so the reflection calculation is straightforward. To reflect off some other surface we will need a normal vector, and we will see how to find such a normal vector in Section 13.4.

Practice 3: Find the reflection vector **R** when an incoming vector $\mathbf{V} = \langle 2, -3, -1 \rangle$ is reflected by the plane $2x - 3y + z = 10$ at the point where (5, 1, 3) on the plane.

Reflection Problems

For Problems 1 – 6, (a) determine the reflection vector **R** when an incoming vector **V** is reflected by the given line, and (b) determine parametric equations for the reflected line when the reflection occurs at the given point.

1. The incoming vector $\mathbf{V} = \langle 2, -1 \rangle$ is reflected by the line $3x + y = 6$ at the point (1, 3).
2. The incoming vector $\mathbf{V} = \langle -1, 1 \rangle$ is reflected by the line $3x + y = 6$ at the point (1, 3).
3. The incoming vector $\mathbf{V} = \langle 2, 3 \rangle$ is reflected by the line $5x - 2y = 7$ at the point (3, 4).
4. The incoming vector $\mathbf{V} = \langle 0, -2 \rangle$ is reflected by the line $5x - 2y = 7$ at the point (3, 4).

Problems 5 and 6 can be easily done without the Reflection Theorem, but, of course, the theorem works.

5. The incoming vector $\mathbf{V} = \langle -3, 2 \rangle$ is reflected by the y-axis at the point (0, 3).
6. The incoming vector $\mathbf{V} = \langle 3, -1 \rangle$ is reflected by the x-axis at the point (2, 0).

For Problems 7 – 12, (a) determine the reflection vector **R** when an incoming vector **V** is reflected by the given curve, and (b) determine parametric equations for the reflected line when the reflection occurs at the given point.

7. The incoming vector $\mathbf{V} = \langle -3, 1 \rangle$ is reflected by the ellipse (2cos(t), sin(t)) at the point where t = 0.9 on the ellipse.

8. The incoming vector $\mathbf{V} = \langle 2, 1 \rangle$ is reflected by the ellipse (2cos(t), sin(t)) at the point where t = 0.9 on the ellipse.

9. The incoming vector $\mathbf{V} = \langle 2, 1 \rangle$ is reflected by the curve $\left(t^2, t^3 \right)$ at the point where t = 2.

10. The incoming vector $\mathbf{V} = \langle -1, 1 \rangle$ is reflected by the curve $\left(t^2, t^3 \right)$ at the point where t = 2.

11. The incoming vector $\mathbf{V} = \langle -1, 1 \rangle$ is reflected by the curve $y = x^2$ at the point (2, 4).

12. The incoming vector $\mathbf{V} = \langle 2, 1 \rangle$ is reflected by the curve $y = x^2$ at the point (2, 4).

For Problems 13 – 16, (a) determine the reflection vector R when an incoming vector V is reflected by the given plane, and (b) determine parametric equations for the reflected line when the reflection occurs at the given point.

13. The incoming vector $V = \langle 2, 6, 3 \rangle$ is reflected by the plane x + 2y + 3z = 13 at the point (2, 4, 1).

14. The incoming vector $V = \langle 4, 1, 3 \rangle$ is reflected by the plane 3x – 2y + 4z = 5 at the point (1, 3, 2).

15. The incoming vector $V = \langle 3, 2, 1 \rangle$ is reflected by the plane x = 0 at the point (0, 4, 2).

16. The incoming vector $V = \langle 2, -3, -1 \rangle$ is reflected by the plane z = 0 at the point (3, 4, 0).

Practice Answers

Practice 1: The line has normal vector $\mathbf{N} = \langle -1, 2 \rangle$ so $\text{Proj}_{\mathbf{N}}\mathbf{V} = \dfrac{\mathbf{V} \bullet \mathbf{N}}{\mathbf{N} \bullet \mathbf{N}}\mathbf{N} \approx \dfrac{-3}{5}\langle -1, 2 \rangle = \left\langle \dfrac{3}{5}, \dfrac{-6}{5} \right\rangle$

and the reflection vector is $\mathbf{R} = \mathbf{V} - 2 \cdot \text{Proj}_{\mathbf{N}}\mathbf{V} = \langle -1, -2 \rangle - 2\left\langle \dfrac{3}{5}, \dfrac{-6}{5} \right\rangle = \left\langle \dfrac{-11}{5}, \dfrac{2}{5} \right\rangle$.

The reflected line is $x(t) = 1.5 - \dfrac{11}{5}t,\ y(t) = 1.75 + \dfrac{2}{5}t$.

Practice 2: $x'(t) = 1$ and $y'(t) = 2t$ so a tangent vector to the parabola when $t = 1.2$ is $\mathbf{T} = \langle 1, 2.4 \rangle$ and then a normal vector is $\mathbf{N} \approx \langle -2.4, 1 \rangle$. Finally, $\text{Proj}_\mathbf{N}\mathbf{V} = \dfrac{\mathbf{V} \bullet \mathbf{N}}{\mathbf{N} \bullet \mathbf{N}}\mathbf{N} \approx \dfrac{4.4}{6.76}\langle -2.4, 1 \rangle = \langle -1.56, 0.65 \rangle$

so $\mathbf{R} = \mathbf{V} - 2 \cdot \text{Proj}_\mathbf{N}\mathbf{V} = \langle -1, 2 \rangle - 2\langle -1.56, 0.65 \rangle = \langle 2.12, 0.7 \rangle$

$y(t) = 0.783 - 2.099t$.

Related web sites:

Ray tracing in Wikipedia – a nice overview

http://en.wikipedia.org/wiki/Ray_tracing_(graphics)

Just look at the pictures in this one (about movie "Cars" from Pixar)

http://graphics.pixar.com/library/RayTracingCars/paper.pdf

"Rendering this image used **111 million diffuse rays, 37 million specular rays, and 26 million shadow rays**. The rays cause **1.2 billion ray-triangle intersection tests**. With multiresolution geometry caching, the render time is 106 minutes."

Appendix: Sketching Planes and Conics in the XYZ Coordinate System

Some mathematicians draw horrible sketches of 3–dimensional objects and they still lead productive, happy lives. Some mathematics students are terrible 3D artists and still get A grades. But it really takes very little time and practice to do decent sketches of planes and conic sections in 3D, and it can be satisfying to have the curve you know is an ellipse actually look like an ellipse. This Appendix illustrates some step–by–step ways to draw the basic building blocks for many 3D shapes: the axis system, planes and conic sections.

First Step: Invest in a pencil and a small, inexpensive, clear plastic $30-60-90^o$ triangle with a centimeter scale (Fig. 1). They make it much easier to create good 3D sketches. Many of the directions in this Appendix suggest drawing "help lines" to assist you in putting the various objects in the appropriate positions and orientations. The final drawings often look better if these "help lines" are removed when the sketch is finished, so a pencil is better than a pen. Many of the directions require you to draw lines parallel to the axes and to draw parallelograms, and a clear plastic straightedge (or a triangle) is very useful for this. Finally, the centimeter scale produces small sketches appropriate for personal work.

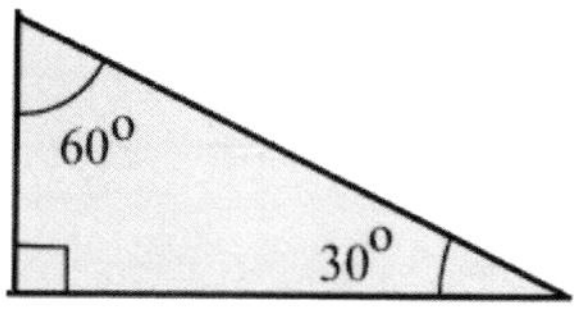

Fig. 1

The following directions are for "by hand" sketches, but they can also help you when using CAD (computer–aided design) and Draw computer programs.

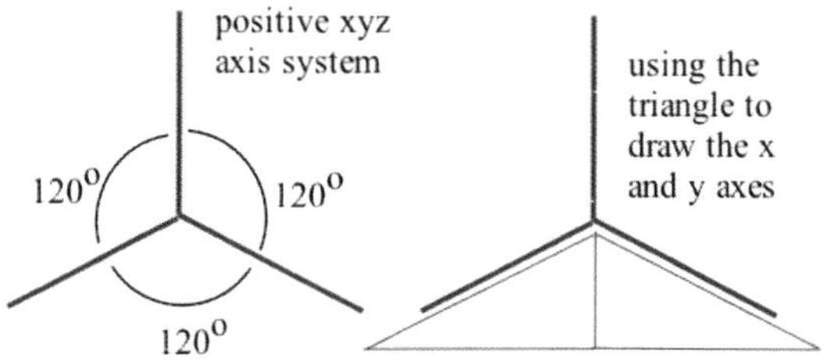

Fig. 2

A . The XYZ axis system — everything starts here!

Including an axis system in a 3D sketch gives the viewer an orientation and a perspective for the location of an object.

(1) Draw a vertical line segment.

(2) Sketch 2 other line segments making 120° angles with the vertical segment. The 30–60–90° triangle is useful here (Fig. 2).

(3) If any of the variables take negative values, you can extend the appropriate axis to the negative values using dashed lines.

(4) If a scale is needed, put "tick" marks along the axes at uniform intervals. (Fig. 3)

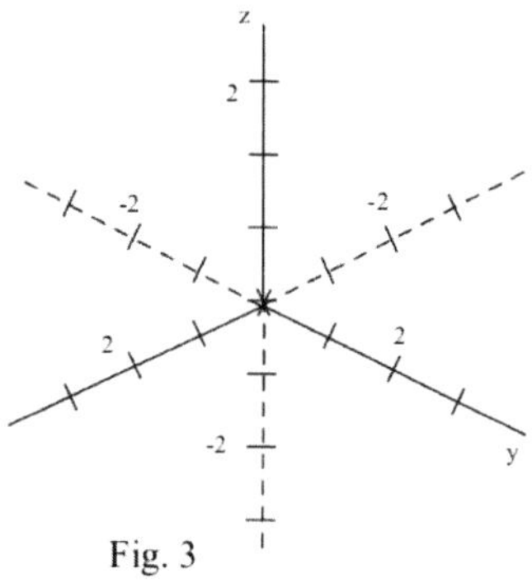

Fig. 3

Since the xyz–coordinate system is needed in so many sketches, you should eventually get used to sketching it without the aid of a ruler. ("Trick:" To get the y-axis, start at the origin, go right 2 units and down one unit, plot a point, and draw the line through this point and the origin. This line is the y-axis. The x-axis is similar: go left two and down one.)

B . The XY, XZ, and YZ coordinate planes. (Actually, rectangular pieces of the coordinate planes.)

The following rectangular pieces of the coordinate planes are useful by themselves, and they are very important aids for drawing curves and conics later.

Keys: Start with an xyz–axis system. Use lines parallel to each axis. The result should be a parallelogram.

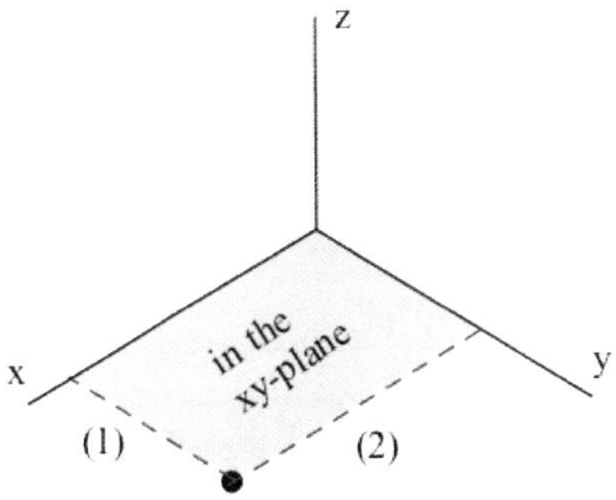

Fig. 4

Piece of the XY coordinate plane: (Fig. 4)

(1) Pick a point on the x–axis and draw a line segment to the left of and parallel to the y–axis.

(2) From the end of the segment in step (1), draw a line to the y–axis that is parallel to the x–axis.

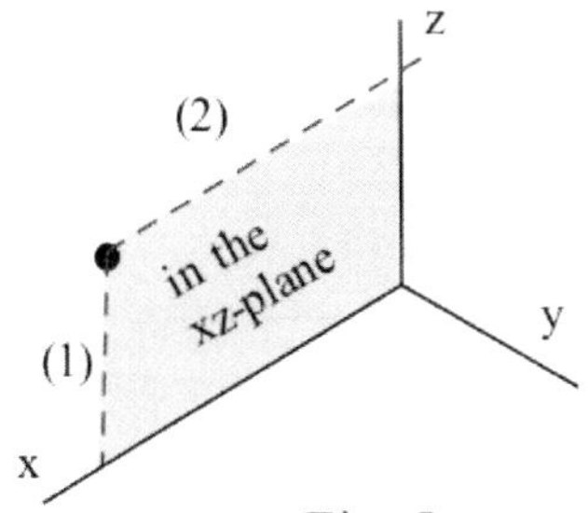

Fig. 5

Piece of the XZ coordinate plane: (Fig. 5)

(1) Pick a point on the x–axis and draw a vertical line segment (up and parallel to the z–axis).

(2) From the end of the segment in step (1), draw a line to the z–axis that is parallel to the x–axis.

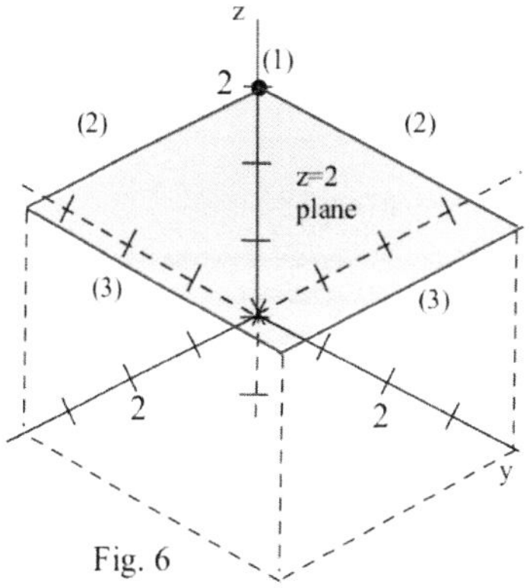

Fig. 6

Plane (piece) parallel to the xy–plane: (Fig. 6)

(1) Locate and label the appropriate point on the z–axis, for example, z = 2.

(2) From the point in step (1), draw line segments parallel to the x–axis and the y–axis.

(3) From the ends of the segments in step (2), draw additional lines parallel to the y–axis and the x–axis to complete the parallelogram.

Plane (piece) parallel to the xz–plane: (Fig. 7)

(1) Locate and label the appropriate point on the y–axis, for example, y = 2.

(2) From the point in step (1), draw line segments parallel to the x–axis and the z–axis.

(3) From the ends of the segments in step (2), draw additional lines parallel to the z–axis and the x–axis to complete the parallelogram.

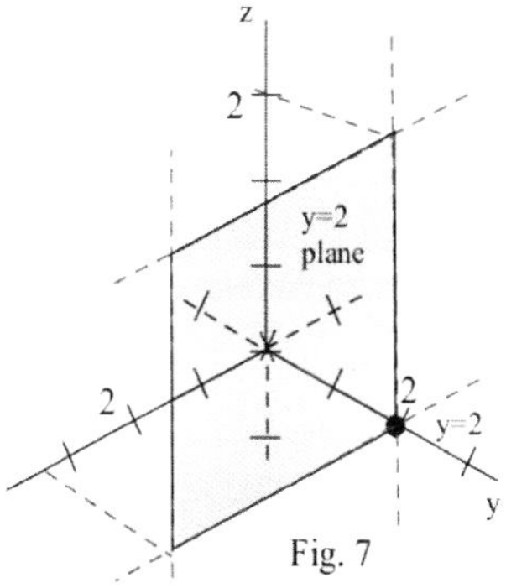

Fig. 7

Practice 1: Sketch the planes z = 1 and x = 2 in Fig. 8.

Practice 2: Sketch the planes z = –1 and y = 1 on an XYZ system.

Practice 3: Sketch the rectangular box with opposite corners at

(0, 0, 0) and (4, 3, 2).

Challenge 1: Sketch the plane that contains the points

(0, 0, 3), (4, 0, 3), and (0, 2, 3).

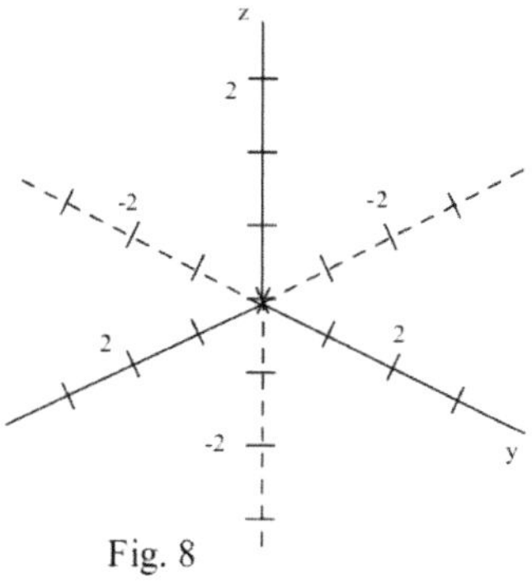

Fig. 8

C . Ellipses

Keys: Plot the vertices and make a parallelogram "frame" for the ellipse. Sketch short tangent segments at the midpoints of the sides of the parallelogram.

Example 1: An ellipse in the xy–plane with vertices at (3,0,**0**), (–3,0,**0**), (0,2,**0**), and (0,–2,**0**). (Fig. 9)

(1) Plot the vertices in the xy–plane (z=0).

(2) Sketch a parallelogram "frame" by drawing lines through (3,0) and (–3,0) that are parallel to the y–axis, and lines through (0,2) and (0,–2) that are parallel to the x–axis.

(3) At each vertex sketch a short "tangent segment" that lies on the side of the parallelogram frame.

(4) Finish the sketch by smoothly connecting the "tangent segments."

If your sketch requires several conics, it is useful to erase all or most of the parallelogram frame.

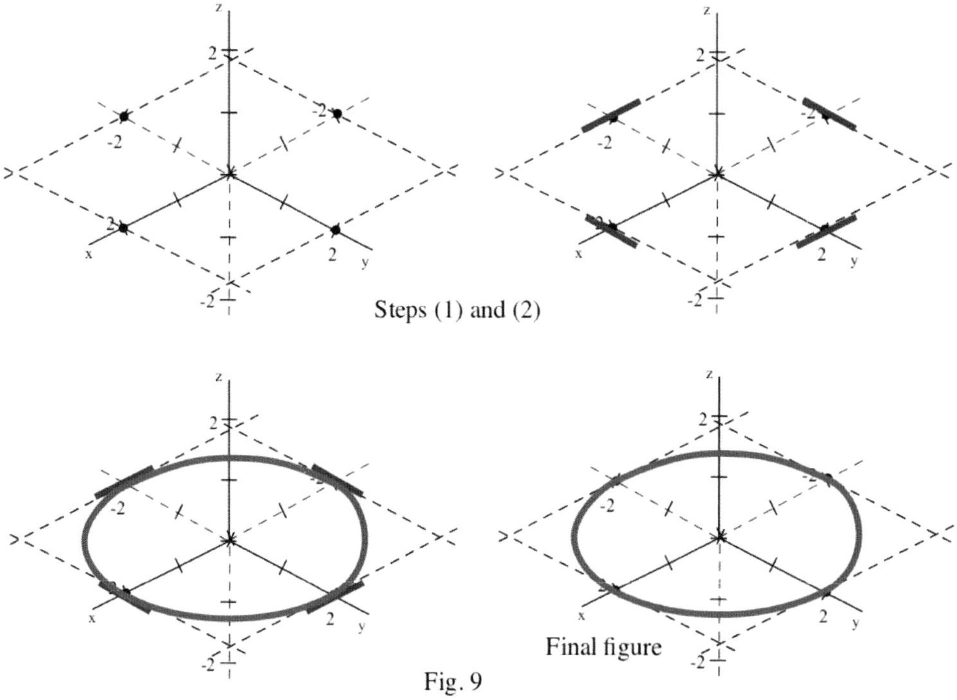

Fig. 9

Example 2: An ellipse in the xz–plane with vertices at (3,**0**,0), (–3,**0**,0), (0,**0**,2), and (0,**0**,–2). Fig. 10 shows the intermediate steps to construct this ellipse as well as the final result.

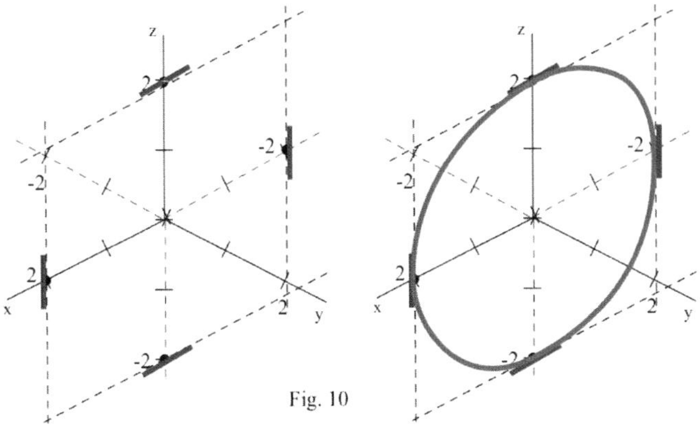
Fig. 10

Practice 4: Sketch the ellipse with vertices at (1, **0**, 0), (0, **0**, 0), (0, **0**, 3), and (0, **0**, –3) on the XYZ system in Fig. 11.

Practice 5: Sketch the ellipse with vertices at (3, 0, 2), (–3, 0, 2), (0, 1, 2), and (0, –1, 2) .

Practice 6: Sketch the elliptical cylinder $\frac{y^2}{9} + z^2 = 1$.

Challenge 2: Sketch the ellipse with vertices at (4,2,0), (2,0,3), (0,2,6), and (2,4,3).

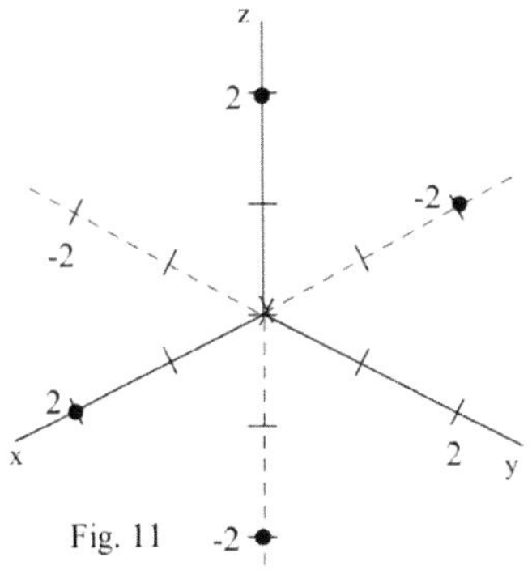
Fig. 11

D . Parabolas

Keys: Plot the vertex and one other point on the parabola. Make a parallelogram "frame" for the parabola and plot a symmetric point. Sketch short "tangent segments" at the plotted points.

Example 3: Sketch the parabola lying in the xy-plane that satisfies $y = x^2 - 4x + 6$. (Fig. 12)

(1) Plot the vertex (2,1,0). When x = 1, then y = (1)2 – 4(1) + 6 = 3. Plot (1,3,0).

(2) Sketch a parallelogram "frame" by drawing lines through (2,1,0) parallel to the x–axis and the y–axis, and through (1,3,0) parallel to the x–axis and the y–axis.

(3) Plot the "symmetric point" (3,3,0) and draw a line through it parallel to the x–axis and the y–axis.

(4) Sketch a short "tangent segment" at the vertex (2,1,0) (this "tangent segment" lies on the side of the parallelogram). Add "tangent segments" at the other two plotted points (1,3,0) and (3,3,0) (these "tangent segments are **not** parallel to the sides of the parallelogram).

(5) Finish the sketch by smoothly connecting the "tangent segments."

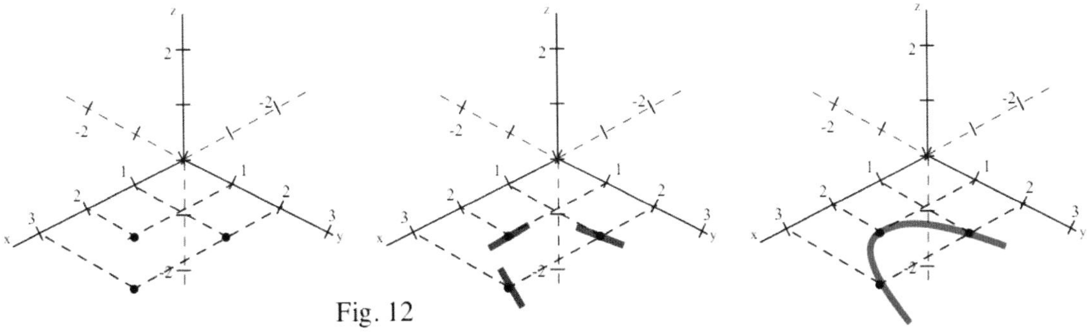
Fig. 12

Example 4: Sketch the parabola lying in the xz-plane that satisfies $z = x^2 + 1$.

Fig. 13 shows a beginning step and the final result.

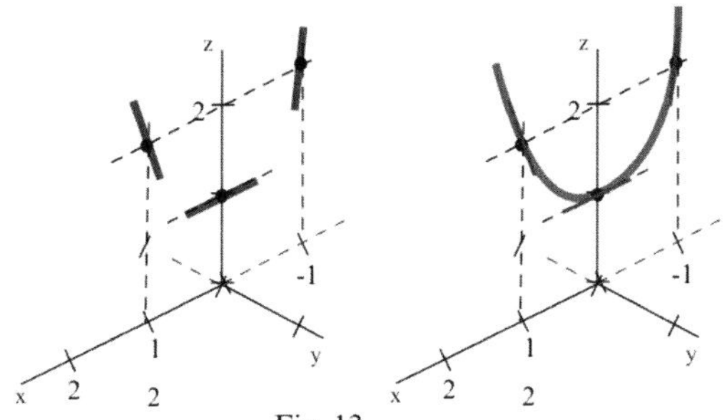

Fig. 13

Practice 7: Sketch the parabola with vertex at (0,2,0) that contains the point (0,1,2) on the XYZ system in Fig. 14.

Practice 8: Sketch the parabola with vertex at (2,0,2) that contains the point (0,0,0) .

Practice 9: Sketch the parabola $x = (y-1)^2$ on the plane z = 1.

Challenge 3: Sketch the parabola with vertex (4,2,0) that contains the symmetric points (0,1,3) and (0,3,3).

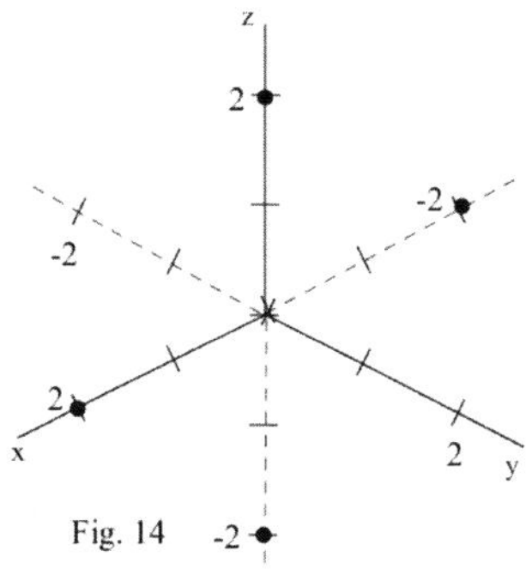

Fig. 14

E . General curves in coordinate planes

Keys: Start with a regular rectangular coordinate graph of the function and a few "tangent segments." Plot a domain–range parallelogram in 3D along with the points and the "tangent segments."

Example 5: Sketch the curve y = sin(x), $0 \le x \le 3\pi$, in the xy–plane. (Fig. 15)

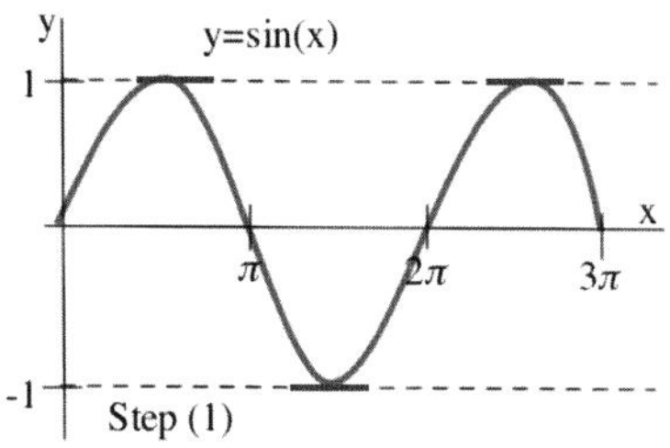

(1) Graph y = sin(x), $0 \le x \le 3\pi$, in the 2D rectangular coordinate system, label a few points and add "tangent segments" at those points.

(2) Sketch the domain–range parallelogram, $0 \le x \le 3\pi$ and $-1 \le y \le 1$.

(3) Sketch the points and "tangent segments" in the parallelogram in step (2).

(4) Finish the sketch by smoothly connecting the "tangent segments."

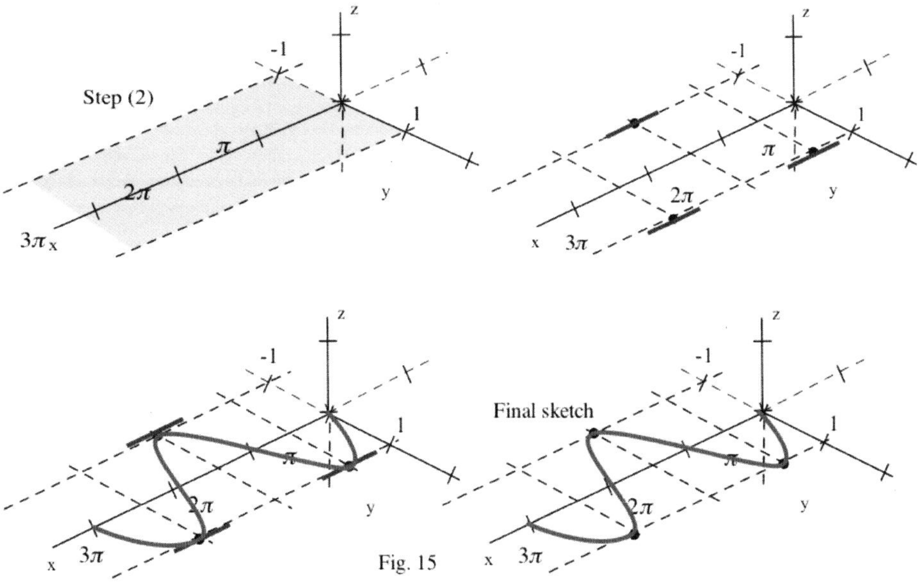

Fig. 15

Practice 10: On the XYZ system in Fig. 16 sketch a curve

(a) in the xz-plane (y=0) that satisfies $z = 1 + \cos(x), -\pi \le x \le 3\pi$,

(b) in the y = 2 plane that satisfies $z = 1 + \cos(x), -\pi \le x \le 3\pi$,.

Practice 11: On an XYZ system sketch the curve (a) in the yz-plane (x=0) that satisfies $z = y$ for $0 \le y \le 4$, (b) in the x = 2 plane that satisfies $z = y$ for $0 \le y \le 4$.

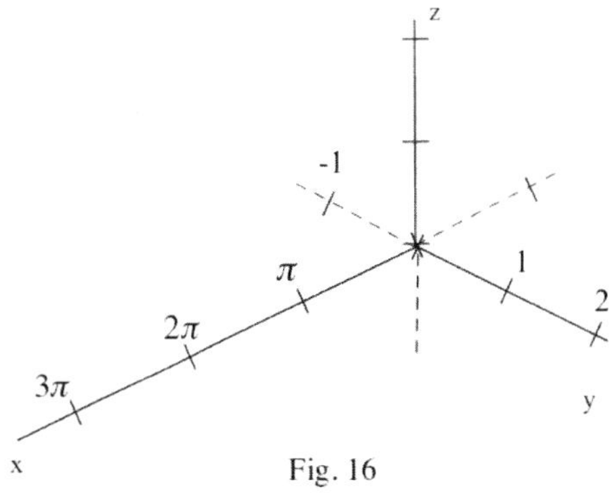

Fig. 16

Practice 12: Sketch the curve $y = 4 - x^2$, $-2 \le x \le 3$ (a) in the xy–plane and (b) in the z = 3 plane.

Challenge 4: Sketch the graph of $z = |x| + |y|$ on the plane $\{y = -x$, no restrictions on $z\}$.

Final Note

Sketching 3D curves on a 2D piece of paper presents challenges for most of us, but with a bit of practice (and a systematic approach) most of us can learn to do decent sketches of a some simple 3D curves. It certainly helps if the curve lies in a plane.

11.1 Selected Answers

5. (b) $|\mathbf{U}| = \sqrt{17}$, $|\mathbf{V}| = \sqrt{13}$

direction of **U** is $\langle 1/\sqrt{17}\ ,\ 4/\sqrt{17} \rangle$, direction of **V** is $\langle 3/\sqrt{13}\ ,\ 2/\sqrt{13} \rangle$

(c) slope of **U** is 4/1 = 4, slope of **V** is 2/3

angle of **U** with x-axis is $\theta = \arctan(4) = 1.326\ (\approx 76^o)$

angle of **V** with x-axis is $\theta = \arctan(2/3) = 0.588\ (\approx 33.7^o)$

7. (b) $|\mathbf{U}| = \sqrt{29}$, $|\mathbf{V}| = \sqrt{58}$

direction of **U** is $\langle -2/\sqrt{29}\ ,\ 5/\sqrt{29} \rangle$, direction of **V** is $\langle 3/\sqrt{58}\ ,\ -7/\sqrt{58} \rangle$

(c) slope of **U** is -5/2 , slope of **V** is -7/3

angle of **U** with x-axis is $\theta = \arctan(-5/2) = -1.190\ (\approx -68.8^o)$

angle of **V** with x-axis is $\theta = \arctan(-7/3) = -1.166\ (\approx -66.8^o)$

9. (b) $|\mathbf{U}| = \sqrt{25} = 5$, $|\mathbf{V}| = \sqrt{25} = 5$

direction of **U** is $\langle -4/5\ ,\ -3/5 \rangle$, direction of **V** is $\langle 3/5\ ,\ -4/5 \rangle$

(c) slope of **U** is 3/4 , slope of **V** is -4/3

angle of **U** with x-axis is $\theta = \arctan(3/4) = 0.643\ (\approx 36.9^o)$

angle of **V** with x-axis is $\theta = \arctan(-4/) = -0.927\ (\approx -53.1^o)$

13. $\mathbf{U} = \mathbf{A} + \mathbf{B} - \mathbf{C} = \langle -1, 3 \rangle$, $\mathbf{V} = \mathbf{A} - \mathbf{B} + \mathbf{C} = \langle 3, 5 \rangle$

15. $\mathbf{V} = 3\langle 0.6, 0.8 \rangle = \langle 1.8, 2.4 \rangle$

17. $\mathbf{V} = \langle 4.10, 2.87 \rangle$

19. $\mathbf{V} = \langle 7/\sqrt{10}, 21/\sqrt{10} \rangle$ or $\langle -7/\sqrt{10}, -21/\sqrt{10} \rangle$

21. $\mathbf{V} = \langle 1/\sqrt{26}, 5/\sqrt{26} \rangle$ or $\langle -1/\sqrt{26}, -5/\sqrt{26} \rangle$

23. $\mathbf{V} = \langle 1, 0 \rangle$ or $\langle -1, 0 \rangle$

25. $\mathbf{V} = \langle 1, 0 \rangle$ or $\langle -1, 0 \rangle$

31. shadow on x-axis $1\mathbf{i} + 0\mathbf{j}$, on y-axis is $0\mathbf{i} + 4\mathbf{j}$

33. shadow on x-axis $5\mathbf{i} + 0\mathbf{j}$, on y-axis is $0\mathbf{i} - 2\mathbf{j}$

37. $\mathbf{C} = \langle -4, -6 \rangle$

38. $\mathbf{C} = \langle 3, 7 \rangle$

39. $\langle 25.36, 0 \rangle$ and $\langle 0, 54.37 \rangle$

40. $\langle 96.59, 0 \rangle$ and $\langle 0, 25.88 \rangle$

41. magnitude = 119.5 pounds, angle $\approx 39.2^o$

42. magnitude = 139 pounds, angle $\approx 21.4^o$

43. magnitude = 268.45 pounds, angle $\approx 23.9^o$

44. (a) path = $\langle 230, 40 \rangle$ (b) aim 11.5^o south of east

46. (a) You are 19.6 miles from home (b) You should hike in the direction 34.6^o west of south

47. (a) The tension in each rope is 90.45 pounds.

(b) The tension in the short rope is 97.1 pounds, The tension in the long rope is 59.2 pounds.

11.2 Selected Answers

13. dist(A,B)= $\sqrt{5}$, dist(A,C)= $\sqrt{11}$, dist(A,d)= 5 , dist(B,C)= $\sqrt{6}$, dist(B,D)= $\sqrt{8}$, dist(C,D)= $\sqrt{18}$
    No three of these points are colinear.

15. dist(A,B)= 6 , dist(A,C)= 3 , dist(A,d)= $\sqrt{6}$, dist(B,C)= 9 , dist(B,D)= $\sqrt{50}$, dist(C,D)= $\sqrt{11}$
    The points A, B, and C are colinear.

17. corners: (1,2,3), (4,2,3), (4,4,3), (4,4,1), (1,4,1). volume = (3)(2)(2) = 12

19. corners: (1,4,0), (1,5,0), (4,4,0), (4,4,3), (4,5,3). volume = (3)(1)(3) = 9

25. $(x-4)^2+(y-3)^2+(z-5)^2=9$

26. $x^2+(y-3)^2+(z-6)^2=4$

27. $(x-5)^2+(y-1)^2+z^2=25$

29. center (3, -4, 1), radius = 4

30. center (-2, 0, 4), radius = 5

31. center (2, 3, 4), radius = 10

33. empty set (no intersection), a point, a line

34. empty set (no intersection), a line, a plane

35. empty set (no intersection), a point, a circle

36. empty set (no intersection), a point, a circle, a sphere

45. (a) $\frac{16\pi}{3}$ (for half sphere) (b) $\frac{8\pi}{3}$ (for quarter sphere) (c) $\frac{4\pi}{3}$ (for 1/8 sphere)

46. (a) 18π (b) 9π (c) $\frac{9\pi}{2}$

S1. (1, 2, 0) on xy-plane, (1, 0, 3) on xz-plane, (0, 2, 3) on yz-plane

S2. (4, 1, 0) on xy-plane, (4, 0, 2) on xz-plane, (0, 1, 2) on yz-plane

S3. (a, b, 0) on xy-plane, (a, 0, c) on xz-plane, (0, b, c) on yz-plane

S4. (4, 2, 0) to (1, 3, 0) on xy-plane, (4, 0, 1) to (1, 0, 3) on xz-plane, (0, 2, 1) to (0, 3, 3) on yz-plane

S7. xy-plane: line segment from (0,0,0) to (4,0,0). yz-plane: line segment from (0,0,0) to (0,0,3),
    xz-plane: triangle with vertices (0,0,0), (4,0,3), and (4,0,2)

S8. xy-plane: triangle with vertices (1,2,0), (4,3,0), and (2,3,0)
    xz-plane: triangle with vertices (1,0,3), (4,0,1), and (2,0,4)
    yz-plane: triangle with vertices (0,2,3), (0,3,1), and (0,3,4)

S10. (a) 0 (b) 10

S11. (a) 0 (b) 12

11.3 Selected Answers

5. $\mathbf{W} = \langle 6, -3, 18 \rangle$, $|\mathbf{U}| = 7$, $|\mathbf{V}| = 11$, $|\mathbf{W}| = \sqrt{369} \approx 19.21$

{ dir. of **U**} = $\mathbf{U}/|\mathbf{U}| = \langle 2/7, 3/7, 6/7 \rangle$, { dir. of **V**} = $\mathbf{V}/|\mathbf{V}| = \langle 2/11, -9/11, 6/11 \rangle$

{ dir. of **W**} = $\mathbf{W}/|\mathbf{W}| = \langle 6/\sqrt{369}, -3/\sqrt{369}, 18/\sqrt{369} \rangle \approx \langle 0.31, -0.16, 0.94 \rangle$

7. $\mathbf{W} = \langle 14, -3, 32 \rangle$, $|\mathbf{U}| = 15$, $|\mathbf{V}| = 9$, $|\mathbf{W}| = \sqrt{1229} \approx 35.06$

{ dir. of **U**} = $\mathbf{U}/|\mathbf{U}| = \langle 5/15, 2/15, 14/15 \rangle$, { dir. of **V**} = $\mathbf{V}/|\mathbf{V}| = \langle 4/9, -7/9, 4/9 \rangle$

{ dir. of **W**} = $\mathbf{W}/|\mathbf{W}| = \langle 14/\sqrt{1229}, -3/\sqrt{1229}, 32/\sqrt{1229} \rangle \approx \langle 0.40, -0.09, 0.91 \rangle$

9. $\mathbf{W} = \langle 21, 18, -2 \rangle$, $|\mathbf{U}| = 11$, $|\mathbf{V}| = 9$, $|\mathbf{W}| = \sqrt{769} \approx 27.73$

{ dir. of **U**} = $\mathbf{U}/|\mathbf{U}| = \langle 9/11, 6/11, 2/11 \rangle$, { dir. of **V**} = $\mathbf{V}/|\mathbf{V}| = \langle 1/3, 2/3, -2/3 \rangle$

{ dir. of **W**} = $\mathbf{W}/|\mathbf{W}| = \langle 21/\sqrt{769}, 18/\sqrt{769}, -2/\sqrt{769} \rangle \approx \langle 0.76, 0.65, -0.07 \rangle$

11. $\mathbf{W} = \langle 26, 25, -2 \rangle$, $|\mathbf{U}| = 15$, $|\mathbf{V}| = 9$, $|\mathbf{W}| = \sqrt{1305} \approx 36.12$

{ dir. of **U**} = $\mathbf{U}/|\mathbf{U}| = \langle 10/15, 11/15, 2/15 \rangle$, { dir. of **V**} = $\mathbf{V}/|\mathbf{V}| = \langle 2/3, 1/3, -2/3 \rangle$

{ dir. of **W**} = $\mathbf{W}/|\mathbf{W}| = \langle 26/\sqrt{1305}, 25/\sqrt{1305}, -2/\sqrt{1305} \rangle \approx \langle 0.72, 0.69, -0.06 \rangle$

13. $\mathbf{C} = \langle -8, -6, 6 \rangle$ 14. $\mathbf{C} = \langle 4, 1, -4 \rangle$ 15. $\mathbf{C} = \langle -3-e, -9-\pi, -1 \rangle$

17. smallest magnitude is **C**, largest magnitude is **D**

18. smallest magnitude is **D**, largest magnitude is **C**

23. $\langle 0, 1, 0 \rangle$, $\langle 0, 0, 1 \rangle$, $\langle 0, 3, -4 \rangle$ are all perpendicular to **A** as is ever non-zero vector with x-coordinate equal to 0 (= vectors that lie in the yz-plane). There are an infinite number of nonparallel vectors that are perpendicular to **A**.

24. $\langle 1, 0, 0 \rangle$, $\langle 0, 3, 0 \rangle$, $\langle 5, -4, 0 \rangle$ are all perpendicular to **B** as is ever non-zero vector with z-coordinate equal to 0 (= vectors that lie in the xy-plane). There are an infinite number of nonparallel vectors that are perpendicular to **B**.

25. $\langle 0, 0, 2 \rangle$, $\langle -2, 1, 0 \rangle$, $\langle 2, -1, 7 \rangle$ are all perpendicular to **C.** There are an infinite number of nonparallel vectors that are perpendicular to **C**.

31. $x(t) = 3 + 4t$, $y(t) = 5 - t$, $z(t) = 1$

32. $x(t) = 1 + 4t$, $y(t) = 2 - 2t$, $z(t) = 3 + 2t$

33. $x(t) = 2 + 3t$, $y(t) = 3$, $z(t) = 6 - 5t$

11.4 Selected Answers

1. $\mathbf{A} \bullet \mathbf{B} = 3$, $\mathbf{B} \bullet \mathbf{A} = 3$, $\mathbf{A} \bullet \mathbf{A} = 14$, $\mathbf{A} \bullet (\mathbf{B}+\mathbf{A}) = 17$, and $(2\mathbf{A}+3\mathbf{B}) \bullet (\mathbf{A} - 2\mathbf{B}) = -101$

2. $\mathbf{A} \bullet \mathbf{B} = 2$, $\mathbf{B} \bullet \mathbf{A} = 2$, $\mathbf{A} \bullet \mathbf{A} = 41$, $\mathbf{A} \bullet (\mathbf{B}+\mathbf{A}) = 43$, and $(2\mathbf{A}+3\mathbf{B}) \bullet (\mathbf{A} - 2\mathbf{B}) = -94$

3. $\mathbf{U} \bullet \mathbf{V} = 2$, $\mathbf{U} \bullet \mathbf{U} = 41$, $\mathbf{U} \bullet \mathbf{i} = 6$, $\mathbf{U} \bullet \mathbf{j} = -1$, $\mathbf{U} \bullet \mathbf{k} = 2$, and $(\mathbf{V} + \mathbf{k}) \bullet \mathbf{U} = 8$

4. $\mathbf{U} \bullet \mathbf{V} = 0$, $\mathbf{U} \bullet \mathbf{U} = 22$, $\mathbf{V} \bullet \mathbf{i} = 2$, $\mathbf{V} \bullet \mathbf{j} = 4$, $\mathbf{V} \bullet \mathbf{k} = -3$, and $(\mathbf{V} + \mathbf{k}) \bullet \mathbf{U} = 2$

5. $\mathbf{S} \bullet \mathbf{T} = -3$, $\mathbf{T} \bullet \mathbf{U} = -4$, $\mathbf{T} \bullet \mathbf{T} = 35$, $(\mathbf{S} + \mathbf{T}) \bullet (\mathbf{S} - \mathbf{T}) = -14$, and $(\mathbf{S} \bullet \mathbf{T})\mathbf{U} = \langle -3, -9, -6 \rangle$

7. angle between **A** and **B** is 1.39 ($\approx 79.9^o$), angle between **A** and x-axis is 1.30 ($\approx 74.5^o$)
angle between **A** and y-axis is 1.01 ($\approx 57.7^o$), angle between **A** and z-axis is 0.641 ($\approx 36.7^o$)

9. angle between **U** and **V** is 1.51 ($\approx 86.7^o$), angle between **U** and x-axis is 0.36 ($\approx 20.4^o$)
angle between **U** and y-axis is 1.73 ($\approx 98.98^o$), angle between **U** and z-axis is 1.25 ($\approx 71.8^o$)

11. angle between **S** and **T** is 1.68 ($\approx 96.3^o$), angle between **S** and x-axis is 1.12 ($\approx 64.1^o$)
angle between **S** and y-axis is 2.63 ($\approx 150.5^o$), angle between **S** and z-axis is 1.35 ($\approx 77.4^o$)

13. angle between **A** and **B** is 1.57 ($\approx 90^o$), angle between **A** and x-axis is 1.24 ($\approx 71.1^o$)
angle between **A** and y-axis is 2.08 ($\approx 119.1^o$), angle between **A** and z-axis is 0.624 ($\approx 35.8^o$)

15. angle between **A** and **B** is 2.59 ($\approx 148.7^o$), angle between **A** and x-axis is 0.38 ($\approx 21.8^o$)
angle between **A** and y-axis is 1.95 ($\approx 111.8^o$), angle between **A** and z-axis is 1.57 ($\approx 90^o$)

17. angle between **U** and **V** is 1.43 ($\approx 81.9^o$), angle between **U** and x-axis is 1.25 ($\approx 71.6^o$)
angle between **U** and y-axis is 1.57 ($\approx 90^o$), angle between **U** and z-axis is 0.322 ($\approx 18.4^o$)

19. 0.124 ($\approx 7.13^o$)
20. 2.48 ($\approx 142.1^o$)
21. 0.785 ($\approx 45^o$)
22. 1.23 ($\approx 70.5^o$)
23. 0.897 ($\approx 51.4^o$)
24. 0 ($= 0^o$)
25. 0 ($= 0^o$)
26. 1.57 ($\approx 90^o$)
27. 0.32 ($\approx 18.4^o$)

29. $\mathbf{N} = \langle -2, -1, 0 \rangle$ is one correct answer.
30. $\mathbf{N} = \langle 3, 0, 5 \rangle$ is one correct answer.
31. $\mathbf{N} = \langle 3, -7 \rangle$ is one correct answer.
32. $\mathbf{N} = \langle 3, 7 \rangle$ is one correct answer.
33. $\mathbf{N} = \langle -1, 1, 1 \rangle$ is one correct answer.
34. $\mathbf{N} = \langle -5, 0, 1 \rangle$ is one correct answer.
36. $\mathbf{N} = \langle 0, 2, -3 \rangle$ is one correct answer.
37. $\mathbf{N} = \langle 1, 1, 2 \rangle$ is one correct answer.
38. $\mathbf{N} = \langle 1, 0, 6 \rangle$ is one correct answer.
39. $\mathbf{N} = \langle 3, 2 \rangle$ is one correct answer.
40. $\mathbf{N} = \langle -2, 3 \rangle$ is one correct answer.
41. $\mathbf{N} = \langle 2, 3 \rangle$ is one correct answer.

43. $\mathbf{N} = \langle 3, 2 \rangle$ is one correct answer.

45. $\mathbf{N} = \langle 1, -4 \rangle$ is one correct answer.

46. $\mathbf{N} = \langle 5, 1 \rangle$ is one correct answer.

47. $\mathbf{N} = \langle 0, 3 \rangle$ is one correct answer.

57. $\mathbf{Proj_B\ A} = \langle 25/34, 0, -15/34 \rangle$, $\mathbf{Proj_A\ B} = \langle -1, 2, 0 \rangle$

59. $\mathbf{Proj_B\ A} = \langle 0,0,0 \rangle$, $\mathbf{Proj_A\ B} = \langle 0,0,0 \rangle$. **A** and **B** are perpendicular.

61. $\mathbf{Proj_B\ A} = \langle 0,0,0 \rangle$, $\mathbf{Proj_A\ B} = \langle 0,0,0 \rangle$. **A** and **B** are perpendicular.

63. $\mathbf{Proj_B\ A} = \langle 0, -2, 0 \rangle$, $\mathbf{Proj_A\ B} = \langle -1/7, 2/7, -3/7 \rangle$

65. $|\mathbf{Proj_B\ A}| = \left| \frac{\mathbf{A \cdot B}}{|\mathbf{B}|} \right|$, $\mathbf{Proj_A\ B} = \left| \frac{\mathbf{A \cdot B}}{|\mathbf{A}|} \right| = \left| \frac{\mathbf{A \cdot B}}{3|\mathbf{B}|} \right| = \frac{1}{3} \left| \frac{\mathbf{A \cdot B}}{|\mathbf{B}|} \right|$. $|\mathbf{Proj_B\ A}|$ is larger.

67. distance is $3/\sqrt{2} \approx 2.12$

68. distance is 3.88

69. distance is 3.06

70. distance is 0.95

71. distance is 1.40

73. distance is 15.4 feet

74. distance is 27.76 inches

75. 9.18 m

77. work = $\frac{5000}{\sqrt{141}} \approx 421$ foot-pounds

78. work = 842 foot-pounds

79. work = 480 foot-pounds

80. work = 240 foot-pounds

81. work = 1212.1 foot-pounds

82. A work = 2349 ft-lbs, B work = 2441 ft-lbs

83. $\mathbf{A \bullet B} = 15$, angle ≈ 1.05 ($= 60^\circ$)

84. $\mathbf{A \bullet B} = 16$, angle ≈ 1.21 ($\approx 69.3^\circ$)

85. angle $\approx 112^\circ$, "different"

86. angle $\approx 19.9^\circ$, "very alike"

87. angle $\approx 9.45^\circ$, "very alike"

88. angle $= 90^\circ$, "different"

11.5 Selected Answers

1. 7 2. 7 3. $2x - 5y$ 4. $15 - ab$

5. 1 6. -1 7. -6 8. -42

9. $x + 7y - 5z$ 10. $4a + 10b - 6c$ 11. -77 12. $3x - 3x^2$

13. (a) $\langle -10, -5, 10\rangle$ (b) 0 (c) 0 (d) 15

15. (a) $\langle -24, -8, 0\rangle$ (b) 0 (c) 0 (d) $\sqrt{640}$

17. (a) $\langle 3, -11, -5\rangle$ (b) 0 (c) 0 (d) ≈ 12.45

19. scalar 20. not defined 21. not defined 22. not defined

25. The angle between a vector **A** and itself is 0 so **AxA** = |**A**| |**A**| sin(0) = 0. Alternately, | **AxA** | = the area of the parallelogram determined by **A** and **A**, and that area is 0.

26. | **AxB** | = |**A**| |**B**| | sin(θ) | which is maximum when | sin(θ) | = 1, and | sin(θ) | = 1 when $\theta = \pm\pi/2$.

31. (a) torque = **AxB** = $\langle 12\cos(-30^o),\ 12\sin(-30^o),\ 0\rangle \mathrm{x} \langle 0,\ 70,\ 0\rangle$

$= \{840\cos(-30^o)\}\mathbf{k} \approx 727.46\mathbf{k}$ inch-pounds

(b) torque = **AxB** = $\langle 8\cos(-30^o),\ 8\sin(-30^o),\ 0\rangle \mathrm{x} \langle 20\cos(40^o),\ 20\sin(40^o),\ 0\rangle$

$= \{160\cos(-30^o)\sin(40^o) - 160\sin(-30^o)\cos(40^o)\}\mathbf{k} \approx 150.35\mathbf{k}$ inch-pounds

33. Yes. {torque on **A** by **B**} + {torque on **A** by **C**} = **AxB** + **AxC** = **Ax**(**B**+**C**) = {torque on **A** by (**B**+**C**) }

34. parallelogram area = | **AxB** | = 18 , triangle area = 9

35. parallelogram area = | **AxB** | = 6 , triangle area = 3

37. triangle area = $\dfrac{\sqrt{180}}{2} \approx 6.71$ 38. triangle area = $\dfrac{\sqrt{(bc)^2 + (ac)^2 + (ab)^2}}{2}$

39. parallelpiped volume = | (**AxB**) • **C** | = 17 40. parallelpiped volume = | (**AxB**) • **C** | = 78

41. parallelpiped volume = | (**AxB**) • **C** | = | a | · | b | · | c | = | abc | cubic units

42. tetrahedron volume = $\dfrac{1}{6}| 36\mathbf{k} | = 6$ cubic units 43. tetrahedron volume = $\dfrac{15}{6}$ cubic units

44. tetrahedron volume = 8 cubic units 45. tetrahedron volume = $\dfrac{| abc |}{6}$ cubic units

47. $A_{xy} = 4$, $A_{xz} = 4$, $A_{yz} = 8$, and $A_{xyz} = \sqrt{96}$ 48. $A_{xy} = 4$, $A_{xz} = 6$, $A_{yz} = 12$, and $A_{xyz} = \sqrt{196} = 14$

11.6 Selected Answers

1. x(t) = 2 + 3t , y(t) = –3 + 4t , z(t) = 1 + 2t

3. x(t) = –2 + 5t , y(t) = 1 , z(t) = 4 – 3t

5. x(t) = 2 + t , y(t) = –1 + 5t , z(t) = 3 – 5t

7. x(t) = 3 , y(t) = –2 + 6t , z(t) = 1 – 2t

9. Lines intersect at the point (2, –1, 3) when t = 0, $\theta = \arccos\left(\frac{9}{\sqrt{6}\sqrt{21}}\right) \approx \arccos(0.802) \approx 0.604\ (\approx 36.7^o)$

11. L(0) = (1, 5, –2) = K(–2_. The lines intersect at the point (1, 5, –2).
$\theta = \arccos\left(\frac{18}{5\sqrt{14}}\right) \approx \arccos(0.962) \approx 0.277\ (\approx 15.8^o)$

13. 5(x–2) + (–2)(y–3) + 4(z–1) = 0 or 5x – 2y + 4z =8

14. 3x + y – 5z = 22

15. 0(x + 3) + 3(y – 5) + 0(z–6) = 0 or 3y = 15

16. 2x – 2y + z = 0

17. (–6)(x–1) + (–6)(y–2) + (–12)(z–3) = 0 or x + y + 2z = 9

18. y = 5

19. 20x + 28y + 25z = 101

20. –x + 2y – z = 0

21. z = 7

22. x = 2

23. 3x – 2y + 5z = 23

24. 2x + 3y – z = 0

25. 5x – 3y + 2z = 23

26. y = 7

27. They intersect along the y-axis

29. Plane intersects the x-axis at x=10

31. They intersect at the point (4, 2, 1)

32. y-axis never intersects the plane z = 3

33. x(t) = –26 + t, y(t) –57 + 3t, z(t) = t. $\mathbf{N_1} = \langle 4,-2,2\rangle$, $\mathbf{N_2} = \langle 3,-2,3\rangle$
$\theta = \arccos\left(\frac{22}{\sqrt{24}\sqrt{22}}\right) \approx \arccos(0.957) \approx 0.294\ (\approx 16.8^o)$

34. x(t) = 9 + 6t, y*(t) = –9t, z(t) = t. $\theta \approx 1.066\ (\approx 61.1^o)$

35. $x(t) = 12 - \frac{22}{5}t,\ y(t) = 2 + \frac{1}{5}t,\ z(t) = t.$ $\mathbf{N_1} = \langle 0,\ 5,\ -1\rangle$, $\mathbf{N_2} = \langle 1,\ 2, 4\rangle$
$\theta = \arccos\left(\frac{6}{\sqrt{26}\sqrt{21}}\right) \approx 1.311\ (\approx 75.1^o)$

37. They intersect at the point (–11/3, –1/3, –58/3). $\arccos\left(\frac{3}{\sqrt{30}\sqrt{21}}\right) \approx 1.451\ (\approx 83.1^o)$ so the angle of intersection is $\theta = \pi/2 - 1.451 \approx 0.120\ \ (\approx 6.9^o)$

38. They intersect at the point (0, 6, –7). Angle of intersection is approximately 0.222 $(\approx 12.7^o)$

39. They intersect at the point (0, 8, 5). Angle of intersection is approximately 0.271 $(\approx 15.5^o)$

41. Yes

42. Yes

43. x(t) = –7t, y(t) = 4t, z(t) = 3 + 9t

44. (a) $\approx 1.30\ (\approx 74.5^o)$ (b) $\approx 1.01\ (\approx 57.7^o)$ (c) $\approx 0.64\ (\approx 36.7^o)$

45. $\theta = \arccos\left(\frac{a}{\sqrt{a^2+b^2+c^2}}\right)$ with the xy-plane. $\theta = \arccos\left(\frac{b}{\sqrt{a^2+b^2+c^2}}\right)$ with the xz-plane.

$\theta = \arccos\left(\frac{c}{\sqrt{a^2+b^2+c^2}}\right)$ with the yz-plane.

47. (a) $\theta = \arctan(481/378) \approx 0.905$ ($\approx 51.8^\circ$) (b) $\cos(\varphi) \approx 0.382$ so $\varphi \approx 1.179$ ($\approx 84.8^\circ$)

(c) $\alpha = \arctan(481/534.6) \approx 0.733$ ($\approx 42.0^\circ$)

51. distance $\approx$ 4.879 52. distance $\approx$ 2.145 53. distance $\approx$ 1.18 54. distance $\approx$ 6.164

55. distance $\approx$ 1.32 56. distance $\approx$ 1.35

57. (a) The objects "crash" at the point (15, 24, 13) when t=6. (b) Paths intersect fo {min. dist.} = 0

(c) No, the objects crash, and their paths intersect.

58. (a) The objects "crash" at the point (2, 3, 4) when t=3.

59. (a) The objects do not crash. They are never at the same point at the same time.

(b) The paths of the objects intersect: object A is at (0,1,5) when t=1 and B is at (0,1,5) when t=2.

(c) {minimum distance between objects} ≈ 0.85 , {min. distance between paths} = 0 since paths intersect.

61. (a) The objects do not crash. They are never at the same point at the same time.

(b) The paths of the objects do not intersect.

(c) {minimum distance between objects} ≈ 2.79 , {min. distance between paths} = 1

63. (a) Shortest distance between the airplane and the car is $\sqrt{58} \approx 7.62$.

(b) Shortest distance between paths is $13/\sqrt{5} \approx 5.81$.

11.7 Selected Answers

1. $\mathbf{V} = \langle 2, -1\rangle$, $\mathbf{N} = \langle 3, 1\rangle$, $\text{Proj}_{\mathbf{N}}\mathbf{V} = \left\langle \frac{3}{2}, \frac{1}{2}\right\rangle$, $\mathbf{R} = \mathbf{V} - 2\cdot\text{Proj}_{\mathbf{N}}\mathbf{V} = \langle -1, -2\rangle$.

Point is (1, 3) and the line is $x(t) = 1 - t$, $y(t) = 3 - 2t$.

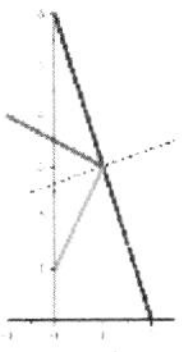
Prob. 1

2. $\mathbf{V} = \langle -1, 1\rangle$, $\mathbf{N} = \langle 3, 1\rangle$, $\text{Proj}_{\mathbf{N}}\mathbf{V} = \left\langle -\frac{3}{5}, -\frac{1}{5}\right\rangle$, $\mathbf{R} = \left\langle \frac{1}{5}, \frac{7}{5}\right\rangle$.

Point is (1, 3) and the line is $x(t) = 1 + \frac{1}{5}t$, $y(t) = 3 + \frac{7}{5}t$.

3. $\mathbf{V} = \langle 2, 3 \rangle$, $\mathbf{N} = \langle 5, -2 \rangle$, $\text{Proj}_{\mathbf{N}}\mathbf{V} = \left\langle \frac{20}{29}, -\frac{8}{29} \right\rangle$, $\mathbf{R} = \mathbf{V} - 2 \cdot \text{Proj}_{\mathbf{N}}\mathbf{V} = \left\langle \frac{18}{29}, \frac{103}{29} \right\rangle$.

Point is (3, 4) and the line is $x(t) = 3 + \frac{18}{29}t$, $y(t) = 4 + \frac{103}{29}t$.

4. $\mathbf{V} = \langle 0, -2 \rangle$, $\mathbf{N} = \langle 5, -2 \rangle$, $\text{Proj}_{\mathbf{N}}\mathbf{V} = \left\langle \frac{20}{29}, -\frac{8}{29} \right\rangle$, $\mathbf{R} = \left\langle -\frac{40}{29}, -\frac{42}{29} \right\rangle$.

Point is (3, 4) and the line is $x(t) = 3 - \frac{40}{29}t$, $y(t) = 4 - \frac{42}{29}t$.

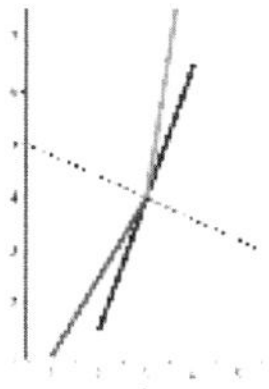
Prob. 3

5. $\mathbf{V} = \langle -3, 2 \rangle$, $\mathbf{N} = \langle 1, 0 \rangle$, $\text{Proj}_{\mathbf{N}}\mathbf{V} = \langle -3, 0 \rangle$, $\mathbf{R} = \mathbf{V} - 2 \cdot \text{Proj}_{\mathbf{N}}\mathbf{V} = \langle 3, 2 \rangle$.

Point is (0, 3) and the line is $x(t) = 3t$, $y(t) = 3 + 2t$.

6. $\mathbf{V} = \langle 3, -1 \rangle$, $\mathbf{N} = \langle 0, 1 \rangle$, $\text{Proj}_{\mathbf{N}}\mathbf{V} = \langle 0, -1 \rangle$, $\mathbf{R} = \langle 3, 2 \rangle$.

Point is (2, 0) and the line is $x(t) = 2 + 3t$, $y(t) = 2t$.

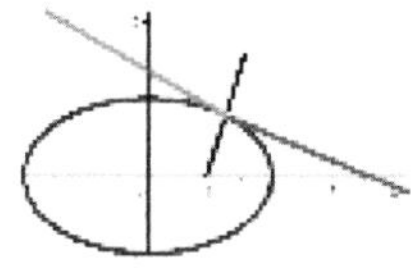
Prob. 7

7. $\mathbf{V} = \langle -3, 1 \rangle$, $\mathbf{N} = \langle -0.622, -1.567 \rangle$, $\text{Proj}_{\mathbf{N}}\mathbf{V} = \langle -0.065, -0.164 \rangle$,

$\mathbf{R} = \mathbf{V} - 2 \cdot \text{Proj}_{\mathbf{N}}\mathbf{V} = \langle -2.869, 1.329 \rangle$. Point is (1.243, 0.783) and the line is $x(t) = 1.243 - 2.869t$, $y(t) = 0.783 + 1.329t$.

8. $\mathbf{V} = \langle 2, 1 \rangle$, $\mathbf{N} =$, $\text{Proj}_{\mathbf{N}}\mathbf{V} = \langle 0.165, 1.550 \rangle$, $\mathbf{R} = \langle 0.770, -2.099 \rangle$.

Point is (1.243, 0.783) and the line is $x(t) = 1.243 + 0.770t$,

$y(t) = 0.783 - 2.099t$.

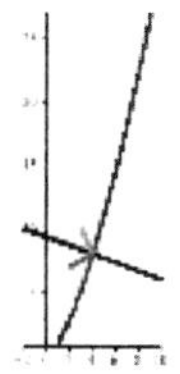
Prob. 9

9. $\mathbf{V} = \langle 2, 1 \rangle$, $\mathbf{N} = \langle -12, 4 \rangle$, $\text{Proj}_{\mathbf{N}}\mathbf{V} = \langle 1.5, -0.5 \rangle$,

$\mathbf{R} = \mathbf{V} - 2 \cdot \text{Proj}_{\mathbf{N}}\mathbf{V} = \langle -1, 2 \rangle$. Point is (4, 8) and the line is $x(t) = 4 - t$, $y(t) = 8 + 2t$.

10. $\mathbf{V} = \langle -1, 1 \rangle$, $\mathbf{N} = \langle -12, 4 \rangle$, $\text{Proj}_{\mathbf{N}}\mathbf{V} = \langle -1.2, 0.4 \rangle$, $\mathbf{R} = \langle 1.4, 0.2 \rangle$.

Point is (4, 8) and the line is $x(t) = 4 + 1.4t$, $y(t) = 8 + 0.2t$.

11. $\mathbf{V} = \langle -1, 1 \rangle$, $\mathbf{N} = \langle -4, 1 \rangle$, $\text{Proj}_{\mathbf{N}}\mathbf{V} = \langle -1.176, 0.294 \rangle$, $\mathbf{R} = \mathbf{V} - 2 \cdot \text{Proj}_{\mathbf{N}}\mathbf{V} = \langle 1.353, 0.412 \rangle$.

Point is (2, 4) and the line is $x(t) = 2 + 1.353t$, $y(t) = 4 + 0.412t$.

12. $\mathbf{V} = \langle 2, 1 \rangle$, $\mathbf{N} = \langle -4, 1 \rangle$, $\text{Proj}_{\mathbf{N}}\mathbf{V} = \langle 1.647, -0.412 \rangle$, $\mathbf{R} = \langle -1.294, 1.824 \rangle$.

Point is (2, 4) and the line is $x(t) = 2 - 1.294t$, $y(t) = 4 + 1.824t$.

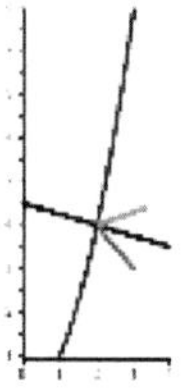
Prob. 11

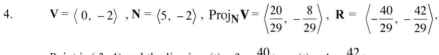

13. $\mathbf{V} = \langle\ 2,\ 6,\ 3\rangle$, $\mathbf{N} = \langle\ 1,\ 2,\ 3\rangle$, $\text{Proj}_{\mathbf{N}}\mathbf{V} = \left\langle \frac{23}{14},\ \frac{23}{7},\ \frac{69}{14} \right\rangle$,

$\mathbf{R} = \mathbf{V} - 2\cdot\text{Proj}_{\mathbf{N}}\mathbf{V} = \left\langle -\frac{9}{7},\ -\frac{4}{7},\ -\frac{48}{7} \right\rangle$. Point is (2, 4, 1) and the

line is $x(t) = 2 - \frac{9}{7}t$, $y(t) = 4 - \frac{4}{7}t$, $z(t) = 1 - \frac{48}{7}t$.

Prob. 13

14. $\mathbf{V} = \langle\ 4,\ 1,\ 3\rangle$, $\mathbf{N} = \langle\ 3,\ -2,\ 4\rangle$, $\text{Proj}_{\mathbf{N}}\mathbf{V} = \left\langle \frac{66}{29},\ -\frac{44}{29},\ \frac{88}{29} \right\rangle$,

$\mathbf{R} = \left\langle -\frac{16}{29},\ \frac{117}{29},\ -\frac{89}{29} \right\rangle$.

Point is (1, 3, 2) and the line is $x(t) = 1 - \frac{16}{29}t$, $y(t) = 3 + \frac{117}{29}t$, $z(t) = 2 - \frac{89}{29}t$.

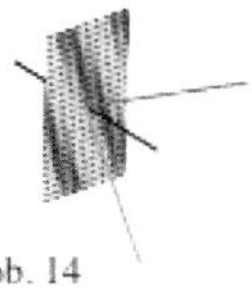
Prob. 14

15. $\mathbf{V} = \langle\ 3,\ 2,\ 1\rangle$, $\mathbf{N} = \langle\ 1,\ 0,\ 0\rangle$, $\text{Proj}_{\mathbf{N}}\mathbf{V} = \langle\ 3,\ 0,\ 0\rangle$,

$\mathbf{R} = \mathbf{V} - 2\cdot\text{Proj}_{\mathbf{N}}\mathbf{V} = \langle\ -3,\ 2,\ 1\rangle$.

Point is (0, 4, 2) and line is $x(t) = -3t$, $y(t) = 4 + 2t$, $z(t) = 2 + t$

16. $\mathbf{V} = \langle\ 2,\ -3,\ -1\rangle$, $\mathbf{N} = \langle\ 0,\ 0,\ 1\rangle$, $\text{Proj}_{\mathbf{N}}\mathbf{V} = \langle\ 0,\ 0,\ -1\rangle$, $\mathbf{R} = \langle\ 2,\ -3,\ 1\rangle$.

Point is (2, –3, –1) and line is $x(t) = 2 + 2t$, $y(t) = -3 - 3t$, $z(t) = -1 + t$.

12.0 INTRODUCTION TO VECTOR–VALUED FUNCTIONS

So far, our excursion into 3–dimensional space has been rather static — we examined points, lines, planes, and vectors, but they did not move (except for points along lines). Those ideas and techniques are important for representing the positions of objects, but objects change position, and calculus is the study of "change." This chapter begins our extension of the ideas of calculus beyond two dimensions, and that extension is the focus for most of the rest of the book.

In earlier chapters, the functions we worked with generally had the form $y = f(x)$ — a single input value x resulted in a single output value y. As we move to higher dimensions, we expand the notion of function keeping the idea that each input produces a single output: f(input) = output. Now, however, we expand the types of objects that can be valid outputs, the range of f, and we expand the types of objects that can be valid inputs, the domain of f.

$r(t) = \langle \cos(t), \sin(t), t \rangle$

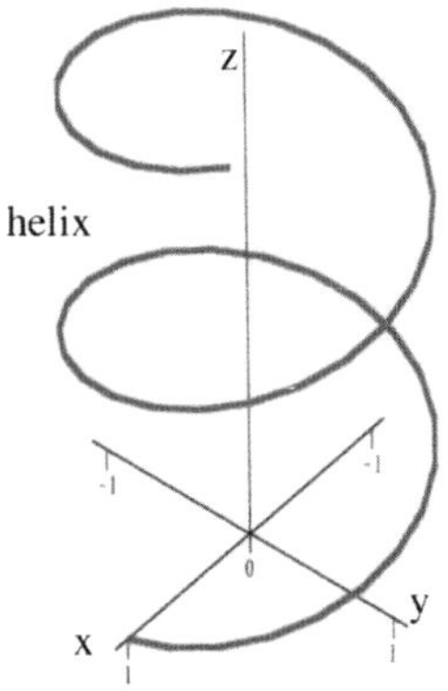

Fig. 1

Chapter 12 focuses on functions whose domains consist of numbers, but whose ranges consist of vectors: functions of the form f(number) = vector such as $\mathbf{r}(t) = \langle \cos(t), \sin(t), t \rangle$. These are called vector–valued functions, and typically their graphs are curves in space (Fig. 1).

Chapters 13 and 14 focus on functions whose domains consist of more than one variable and whose ranges consist of numbers: functions of the type f(number, number) = number or $f(x,y) = z$. These are called functions of several variables, and typically their graphs are surfaces in space (Fig. 2).

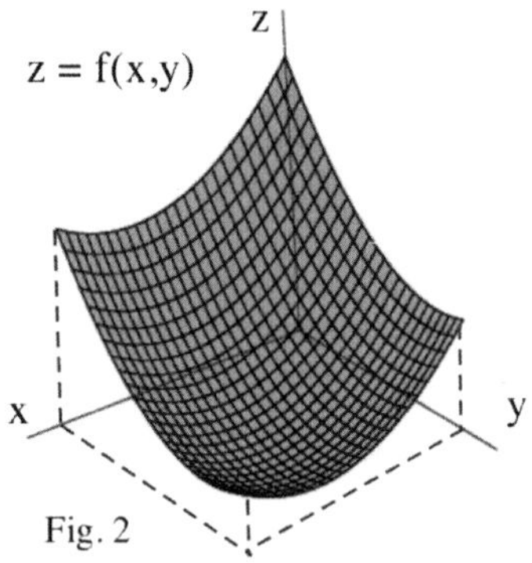

Fig. 2

In this chapter we examine vector–valued functions, and the discussion of vector–valued functions is similar in outline to our discussion of functions $y = f(x)$ in the early chapters of this book. First we discuss the meaning of vector–valued functions and their graphs. Then we look at the calculus ideas of limit, derivative and integral as they apply to vector–valued functions and examine some applications of these calculus ideas. As before, the meaning of these topics is inherently geometric, and you need to be able to work "visually" as well as "analytically."

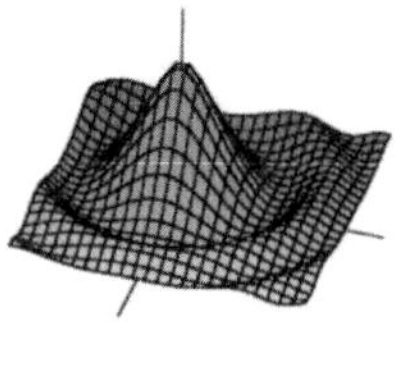

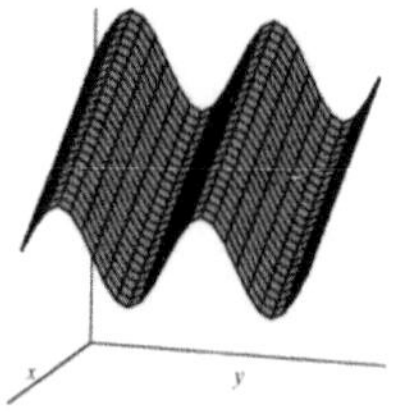

12.1 VECTOR–VALUED FUNCTIONS AND CURVES IN SPACE

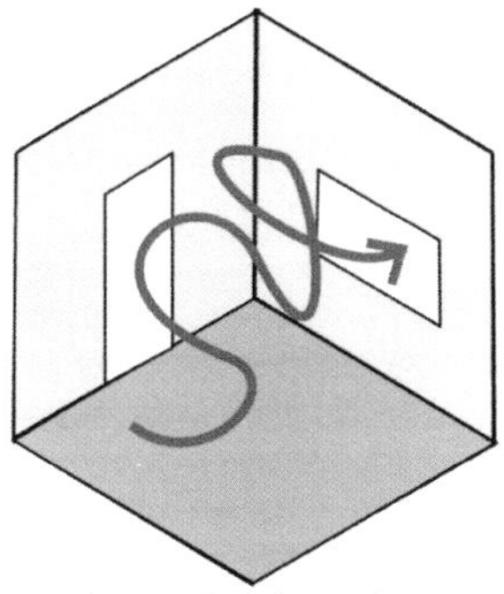
Fig. 1: Bug's path

As a bug buzzes around a room (Fig. 1) or a submarine explores the oceans or a planet orbits a moving star, we could describe the location of the bug or submarine or planet at any particular time as a point (x, y, z) in 3–dimensional space. But to describe the object's movement we need the locations at many times, and that leads very naturally to the idea of representing the path of the object as a set of points P(t) = (x(t), y(t), z(t)) given by parametric equations x = x(t), y = y(t), and z = z(t) where the variable t represents the parameter time. Fig. 2 shows the paths of P(t) = (1 + t, 0 + 3t, 1 + 2t) and Q(t) = (cos(t), sin(t), t). Such parametric equations and their graphs are fundamental to our study of motion in 3–dimensional space, but for many uses it is more effective to work with vectors rather than points, and that leads to **vector–valued** functions.

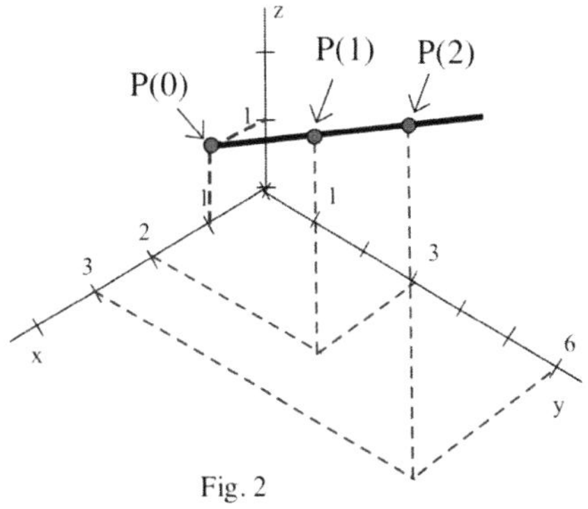

Fig. 2

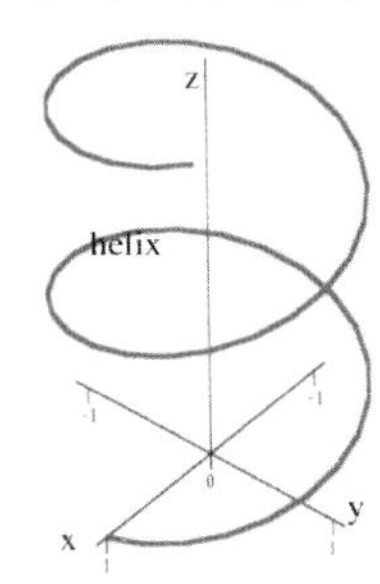

> Definition: A **vector–valued function** is a rule that assigns a vector to each input number.
> Typically, a vector–valued function has the form
> $$\mathbf{r}(t) = x(t)\mathbf{i} + y(t)\mathbf{j} + z(t)\mathbf{k} = \langle x(t), y(t), z(t) \rangle$$
> where x, y, and z are scalar–valued functions.
>
> The domain of a vector–valued function $\mathbf{r}(t)$ is a set of real numbers: the domain of $\mathbf{r}$ consists of those t in the domains of x, y, and z.
>
> The range of a vector–valued function is a collection of vectors.

Vector–valued functions offer two advantages that should become clearer as you work with them. First, vector–valued functions are often notationally simpler than parametric equations — it is easier to write $\mathbf{r}(t)$ than (x(t), y(t), z(t)) . This is not a big advantage; P(t) is also easy to write, but the $\mathbf{r}(t)$ notation does make some ideas and computations easier. The second advantage of vector–valued functions is that they allow us to use the powerful machinery we have already developed for working with vectors. This vector machinery is particularly useful when we consider **tangent vectors** to curves and **angles** between curves.

Our discussion of vector–valued functions is similar to the discussion of functions of the form $y = f(x)$ in the early chapters. We begin by considering their graphs and the ideas of limits and continuity for vector–valued functions.

Graphs of Vector–Valued Functions — Space Curves

If an object is located at the point $P(t) = (x(t), y(t), z(t))$ at time t, then we say that

$$\mathbf{r}(t) = x(t)\mathbf{i} + y(t)\mathbf{j} + z(t)\mathbf{k} = \langle x(t), y(t), z(t) \rangle ,$$

a vector from the origin to the point $P(t)$, is the object's **position vector** . The functions $x(t), y(t)$, and $z(t)$ are called the **components** of $\mathbf{r}(t)$ or the **component functions** of $\mathbf{r}(t)$. Fig. 3 shows a point $P(t)$ and the position vector $\mathbf{r}(t)$ for that point. It is difficult to draw and difficult to interpret a collection of vectors, so typically when we work with the graphs of vector–valued functions we draw only the path of the endpoints of the vectors: the graph of $\mathbf{r}(t) = \langle x(t), y(t), z(t) \rangle$ is the collection of points $P(t) = (x(t), y(t), z(t))$.

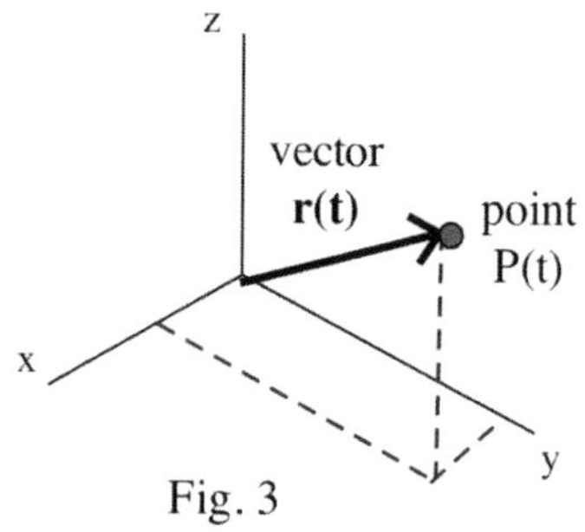

Fig. 3

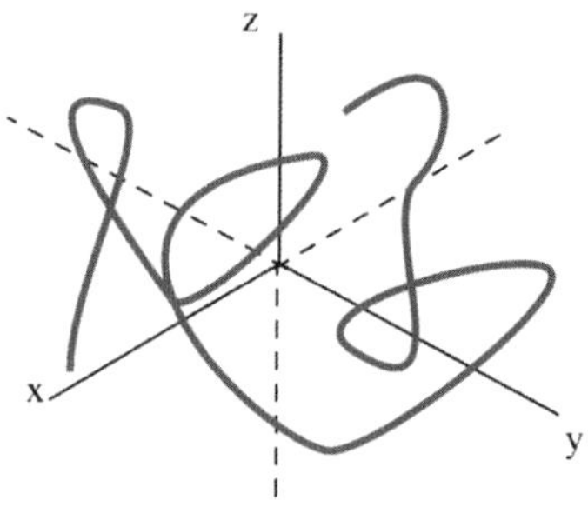

Fig. 4

The graphs of vector–valued functions can be very complex and difficult to sketch, for example Fig. 4, but many are manageable "by hand" and you should practice sketching some of their graphs.

Example 1: Sketch the graphs of $\mathbf{r}(t) = \langle 3 - t, 2t, 4 - 2t \rangle$ for $0 \le t \le 3$, and $\mathbf{s}(t) = \langle t, 0, \sin(t) \rangle$ for $0 \le t \le 2\pi$.

Solution: The graphs are shown in Fig. 5.

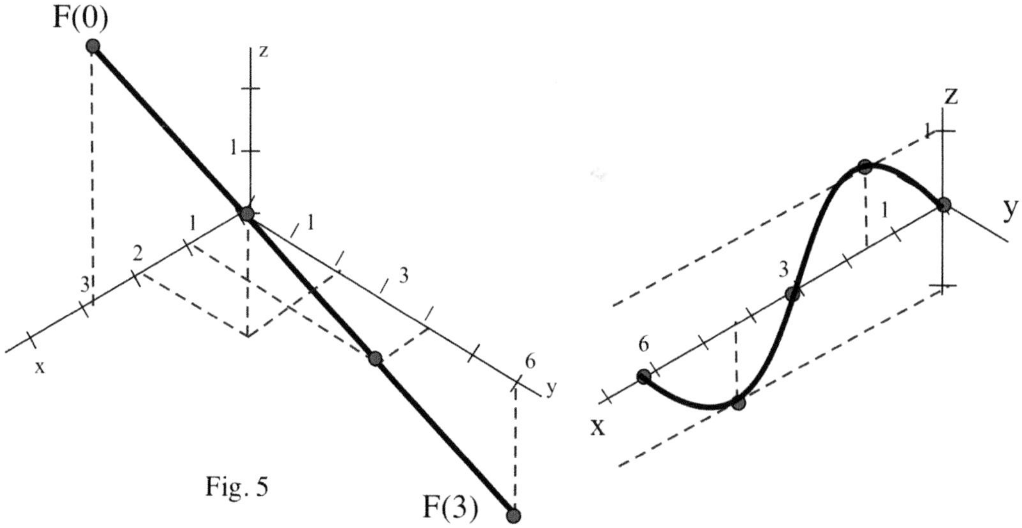

Fig. 5

Practice 1: Sketch the graphs of $\mathbf{r}(t) = \langle t, 4-2t, 2-t \rangle$ for $0 \le t \le 2$, and $\mathbf{s}(t) = \langle t, t^2, 1 \rangle$ for $0 \le t \le 2$.

Sometimes the component functions $x(t), y(t)$, and $z(t)$ may only be given as graphic information, but we can still sketch the graph of the vector–valued function they define.

Example 2: The graphs of $x(t), y(t)$, and $z(t)$ are shown in Fig. 6. Use this information about the component functions of $\mathbf{r}(t) = x(t)\mathbf{i} + y(t)\mathbf{j} + z(t)\mathbf{k}$
$= \langle x(t), y(t), z(t) \rangle$
to sketch a graph of $\mathbf{r}(t)$.

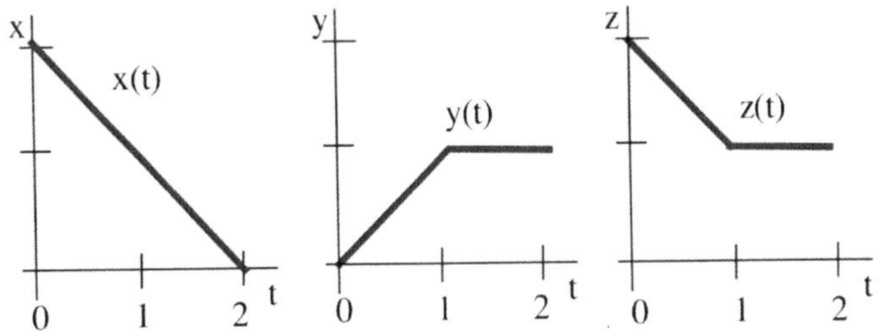

Fig. 6

Solution: The graph of $\mathbf{r}(t)$ is shown in Fig. 7.

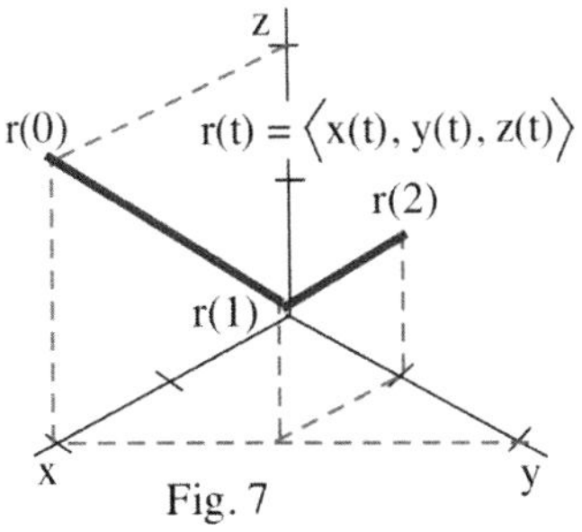

Fig. 7

Practice 2: The graphs of $x(t), y(t)$, and $z(t)$ are shown in Fig. 8. Use this information about the component functions of $\mathbf{r}(t) = x(t)\mathbf{i} + y(t)\mathbf{j} + z(t)\mathbf{k} = \langle x(t), y(t), z(t) \rangle$ to sketch a graph of $\mathbf{r}(t)$.

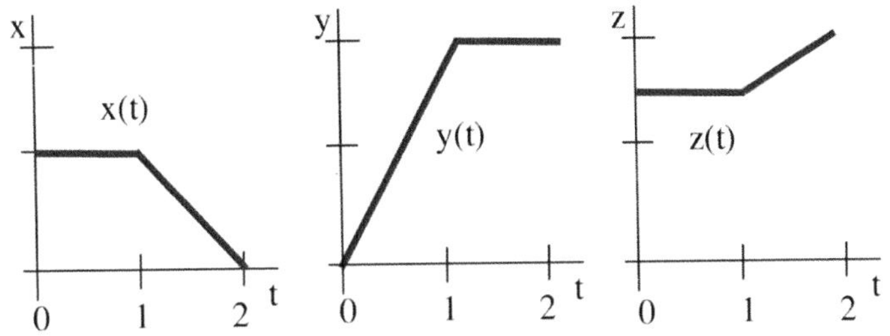

Fig. 8

Limits and Continuity of Vector–Valued Functions

Limits and continuity of vector–valued functions are defined in terms of the components of the function. If you understand the meaning of limits and continuity for functions $y = f(x)$, then these concepts are not difficult for vector–valued functions.

Definition: For $\mathbf{r}(t) = \langle x(t), y(t), z(t) \rangle$,

$$\lim_{t \to a} \mathbf{r}(t) = \left\langle \lim_{t \to a} x(t), \lim_{t \to a} y(t), \lim_{t \to a} z(t) \right\rangle$$

provided that each of the limits of the component functions exists.

To calculate the limit of a vector–valued function we simply need the limits of each of the component scalar–valued functions. If any of the limits of the component functions fail to exist, we say that the limit of the vector–valued function does not exist. The various properties of limits of vector–valued functions follow directly from the corresponding properties of scalar–valued functions as they are applied to each component separately.

Example 3: Determine $\lim_{t\to 0} \mathbf{r}(t)$ and $\lim_{t\to\infty} \mathbf{s}(t)$ for $\mathbf{r}(t) = \langle \cos(t), \sqrt{4+t}, 3e^{2t} \rangle$

and $\mathbf{s}(t) = \langle \frac{2t}{t+3}, \frac{5}{t}, 3 + \frac{\sin(t)}{2t} \rangle$

Solution: $\lim_{t\to 0} \mathbf{r}(t) = \langle \lim_{t\to 0} \cos(t), \lim_{t\to 0} \sqrt{4+t}, \lim_{t\to 0} 3e^{2t} \rangle = \langle 1, 2, 3 \rangle$ and

$$\lim_{t\to\infty} \mathbf{s}(t) = \langle \lim_{t\to\infty} \frac{2t}{t+3}, \lim_{t\to\infty} \frac{5}{t}, \lim_{t\to\infty} 3 + \frac{\sin(t)}{2t} \rangle = \langle 2, 0, 3 \rangle$$

Practice 3: Determine $\lim_{t\to\pi} \mathbf{r}(t)$ and $\lim_{t\to\infty} \mathbf{r}(t)$ for $\mathbf{r}(t) = \langle \cos(t), \frac{\sin(t)}{t}, 3 \rangle$

Continuity of vector–valued functions is stated in terms of limits, so the continuity of a vector–valued function depends on the continuity (and limits) of the component functions.

> Definition: A vector–valued function $\mathbf{r}(t) = \langle x(t), y(t), z(t) \rangle$ is **continuous at the point $t = t_0$**
>
> if $\lim_{t\to t_0} \mathbf{r}(t) = \mathbf{r}(t_0)$.

The following result about continuity follows directly from the definition and is usually easier to use.

> Component Continuity Theorem for Vector–valued Functions
>
> A vector–valued function $\mathbf{r}(t) = \langle x(t), y(t), z(t) \rangle$ is continuous at the point $t = t_0$ if and only if each of the component functions $x(t)$, $y(t)$, and $z(t)$ is continuous at $t = t_0$.

Proof: The proof follows directly from the definitions of continuity and limits of vector–valued functions. If $\mathbf{r}(t)$ is continuous at $t = t_0$, then

$$\lim_{t\to t_0} \mathbf{r}(t) = \mathbf{r}(t_0) = \langle x(t_0), y(t_0), z(t_0) \rangle.$$

But, from the definition of limit of $\mathbf{r}(t)$, $\lim_{t\to a} \mathbf{r}(t) = \langle \lim_{t\to a} x(t), \lim_{t\to a} y(t), \lim_{t\to a} z(t) \rangle$.

Two vectors are equal if and only if their respective components are equal, so

$$\lim_{t \to t_0} x(t) = x(t_0), \lim_{t \to t_0} y(t) = y(t_0), \text{ and } \lim_{t \to t_0} z(t) = z(t_0),$$

and we have shown that x, y, and z are continuous at $t = t_0$.

The proof that "if x(t), y(t), and z(t) are continuous at $t = t_0$, then **r**(t) is continuous at $t = t_0$" is similar but starts with the assumption that x, y, and z are continuous at $t = t_0$.

Example 4: Where are $\mathbf{r}(t) = \langle \cos(t), \sin(t), t^2 \rangle$ and $\mathbf{s}(t) = \langle 2 + t, \frac{1}{t-3}, \ln(t) \rangle$ continuous?

Solution: (a) $\mathbf{r}(t) = \langle \cos(t), \sin(t), t^2 \rangle$ is continuous everywhere (for all t) since all of the component functions $x(t) = \cos(t)$, $y(t) = \sin(t)$, and $z(t) = t^2$ are continuous for all values of t.

(b) $\mathbf{s}(t) = \langle 2 + t, \frac{1}{t-3}, \ln(t) \rangle$ is continuous for $0 < t < 3$ and $3 < t$ since $x(t) = 2 + t$ is continuous for all values of t; $y(t) = \frac{1}{t-3}$ is continuous for $t \neq 3$; and $z(t) = \ln(t)$ is continuous for $t > 0$.

Practice 4: Where is $\mathbf{r}(t) = \langle 1/t, e^t, \text{INT}(t) \rangle$ continuous?

Fig. 9 shows the graphs of continuous components x(t) and y(t) and a discontinuous component z(t). It also shows the discontinuous vector–valued function $\mathbf{r}(t) = \langle x(t), y(t), z(t) \rangle$.

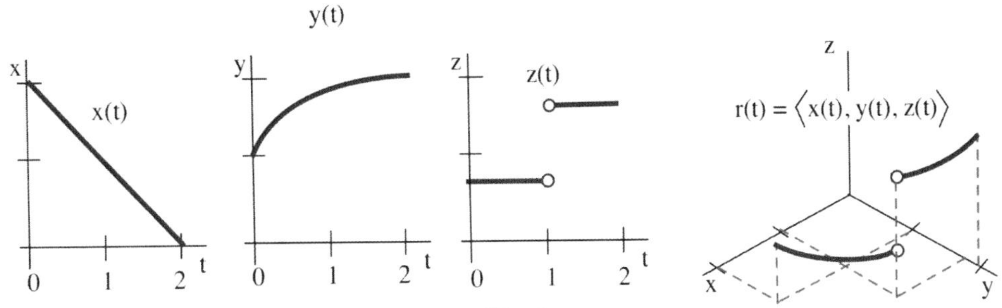

Fig. 9

Some Useful Space Curves: Lines, Helix, and Bezier Curves

By this point in your mathematical development, the graphs of some functions should be almost "automatic." You should be able to visualize the graphs of $y = x^2$, $y = 3 + 2\sin(x)$, $y = |x-2|$ and a variety of others with little effort. It is useful to **start** to develop similar skills with the graphs of a few vector–valued functions. Lines and helices are useful shapes to begin with, and Bezier curves in three dimensions have a number of useful properties.

Lines

If $\mathbf{r}(t) = \langle x(t), y(t), z(t) \rangle$, and each of the component functions $x(t)$, $y(t)$, and $z(t)$ is a linear function $(at + b)$, then the graph of $\mathbf{r}(t)$ is a straight line in space. In this case we need only evaluate and plot $\mathbf{r}(t)$ for a couple values of t before finishing the graph with a straightedge.

Example 5: Sketch the graphs of $\mathbf{r}(t) = \langle 2 + t, 3 - t, 1 + 2t \rangle$ and $\mathbf{s}(t) = \langle 2, t, 3 \rangle$.

Solution: $\mathbf{r}(0) = \langle 2, 3, 1 \rangle$ and $\mathbf{r}(1) = \langle 3, 2, 3 \rangle$. Fig. 10 shows these two points and a graph of $\mathbf{r}(t)$. $\mathbf{s}(0) = \langle 2, 0, 3 \rangle$ and $\mathbf{s}(1) = \langle 2, 1, 3 \rangle$. Fig. 11 shows these two points and a graph of $\mathbf{s}(t)$.

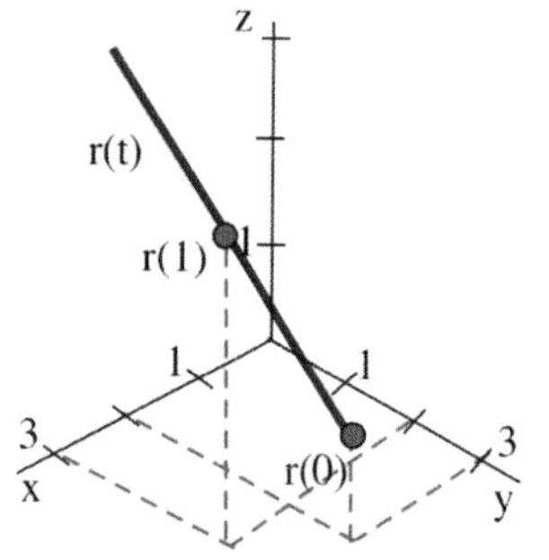

Fig. 10

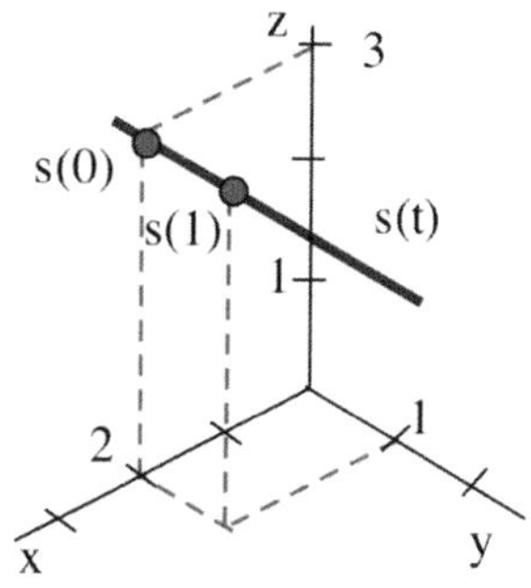

Fig. 11

Practice 5: Sketch the graphs of $\mathbf{r}(t) = \langle t - 1, 0, 1 + 2t \rangle$ and $\mathbf{s}(t) = \langle 2, 4, 3t \rangle$.

If $x(t)$, $y(t)$, and $z(t)$ are linear functions, then they are continuous for all values of t so $\mathbf{r}(t) = \langle x(t), y(t), z(t) \rangle$ is also continuous for all values of t (by the Component Continuity Theorem).

The Helix and Some Variations

The graph of $\mathbf{r}(t) = \langle \cos(t), \sin(t), t \rangle$ is shown in Fig. 12, and it is called a **helix**, or a **circular helix around the z–axis**. The parametric graph of $(\cos(t), \sin(t))$ is a circle in the xy–plane (Fig. 13), and $z(t) = t$ then "stretches" this circle into the z direction to create the spiral shape (Fig. 12). The helix is sometimes useful, and it provides a focus to investigate the effects on the shape of the graph of certain changes in the component functions.

$r(t) = \langle \cos(t), \sin(t), t \rangle$

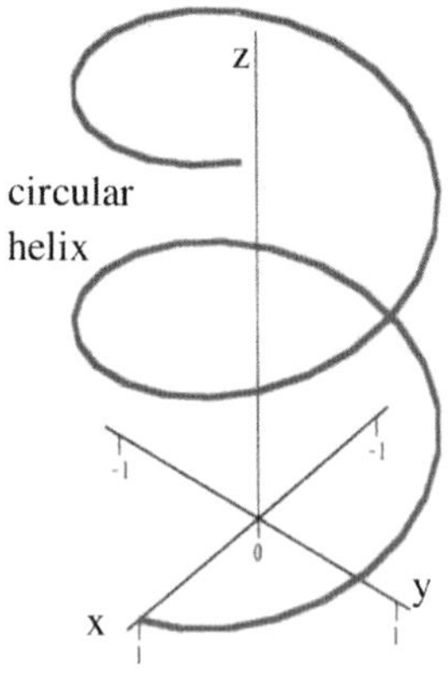

Fig. 12

$P(t) = (\cos(t), \sin(t))$
a circle in the xy-plane

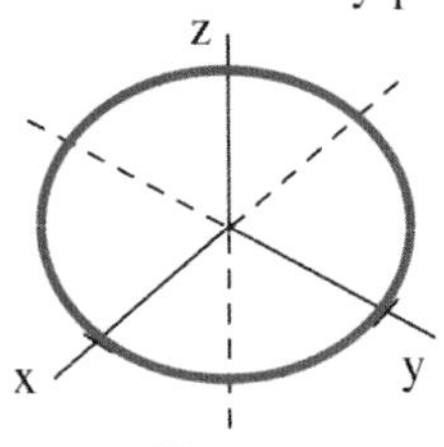

Fig. 13

Example 6: (a) Without plotting any points, describe the shapes of the graphs of $\mathbf{r}(t) = \langle \cos(t), t, \sin(t) \rangle$ and $\mathbf{S}(t) = \langle \cos(t), 2\cdot\sin(t), t \rangle$.

(b) Then sketch the graphs of $\mathbf{r}(t)$ and $\mathbf{S}(t)$.

Solution: (a) The graph of $\mathbf{r}(t) = \langle \sin(t), t, \cos(t) \rangle$ is a circular helix around the y–axis (Fig. 14)

(b) The parametric graph of $(\cos(t), 2\cdot\sin(t))$ in the xy–plane is an ellipse (Fig. 15a), so the graph of $\mathbf{S}(t) = \langle \cos(t), 2\cdot\sin(t), t \rangle$ is an elliptical helix around the z–axis (Fig. 15b)

r(t) = ⟨ sin(t), t, cos(t) ⟩

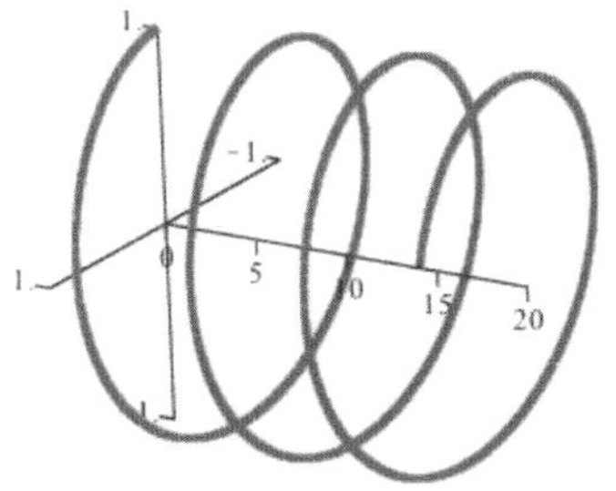

Fig. 14

P(t) = (cos(t), 2sin(t))

ellipse in the xy-plane

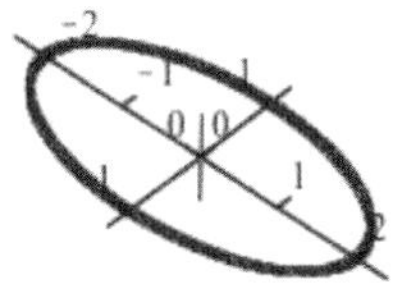

(a)

r(t) = ⟨ cos(t), 2sin(t), t ⟩

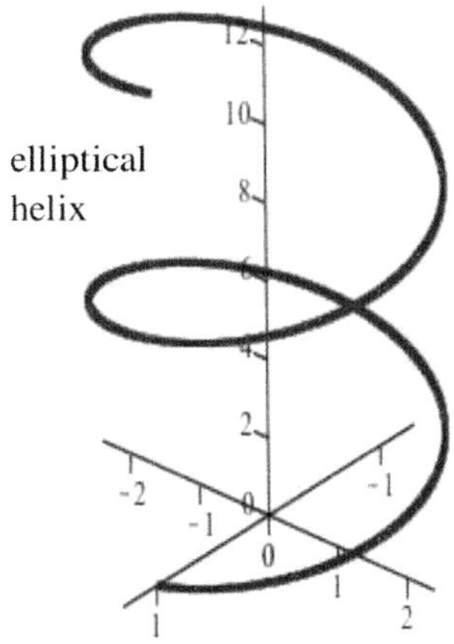

(b)

Fig. 15

Practice 6: (a) Without plotting any points, describe the shapes of the graphs of $\mathbf{r}(t) = \langle t, \cos(t), \sin(t) \rangle$ and $\mathbf{S}(t) = \langle 2\cdot\cos(t), t, 3\cdot\sin(t) \rangle$.

(b) Then sketch the graphs of $\mathbf{r}(t)$ and $\mathbf{S}(t)$.

Figs. 16 and 17 show the graphs of two more variations on the component functions of the original helix.

r(t) = ⟨ cos(t), sin(t), t^2 ⟩

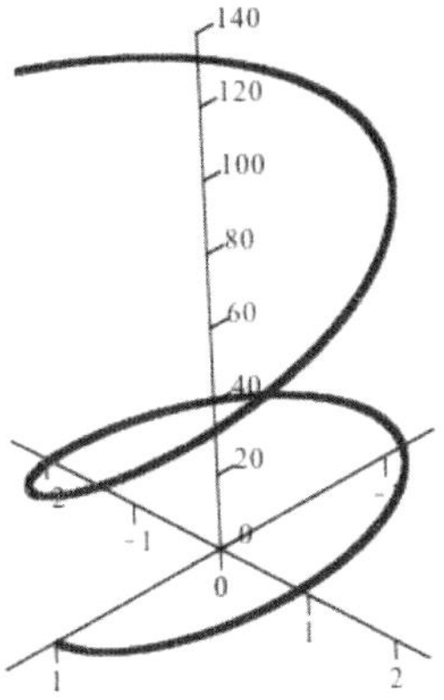

Fig. 16

r(t) = ⟨ tcos(t), tsin(t), t ⟩

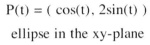

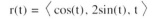

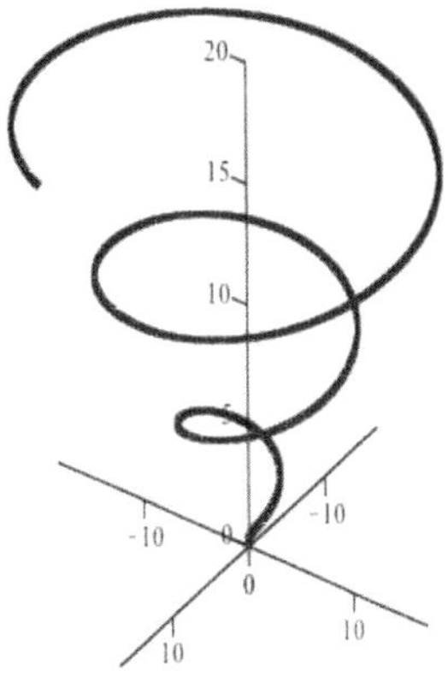

Fig. 17

Bezier Curves in Three Dimensions

One powerful aspect of the Bezier curves we saw in Section 9.4.5 is that the ideas and even the formulas for these curves extend very easily to Bezier curves in three dimensions. In Section 9.4.5 we saw that if we start with four points $P_0 = (x_0, y_0), P_1 = (x_1, y_1), P_2 = (x_2, y_2)$, and $P_3 = (x_3, y_3)$ then, for $0 \le t \le 1$, the curve given by

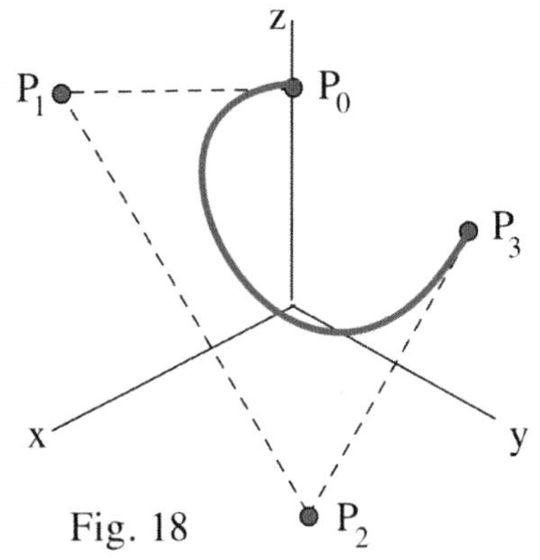

Fig. 18

$$x(t) = (1-t)^3 \cdot x_0 + 3(1-t)^2 t \cdot x_1 + 3(1-t)t^2 \cdot x_2 + t^3 \cdot x_3 \quad \text{and}$$
$$y(t) = (1-t)^3 \cdot y_0 + 3(1-t)^2 t \cdot y_1 + 3(1-t)t^2 \cdot y_2 + t^3 \cdot y_3$$

(or more simply as $B(t) = (1-t)^3 \cdot P_0 + 3(1-t)^2 t \cdot P_1 + 3(1-t)t^2 \cdot P_2 + t^3 \cdot P_3$)

has several useful properties:

(1) $B(0) = P_0$ and $B(1) = P_3$ so P_0 and P_3 are the "endpoints" of $B(t)$ for $0 \le t \le 1$.

(2) $B(t)$ is a cubic polynomial so it is continuous and differentiable.

(3) $B'(0)$ = slope of the line segment from P_0 to P_1:
$B'(1)$ = slope of the line segment from P_2 to P_3.

(4) For $0 \le t \le 1$, the graph of $B(t)$ is in the region whose corners are the control points.

A Bezier curve $B(t)$ for four given points P_0, P_1, P_2 and P_3 is shown in Fig. 18 .

This extends very simply to three dimensions.

Bezier Curve in Three Dimensions

If P_0, P_1, P_2 and P_3 are four points in 3–dimensional space, then the Bezier curve in three dimensions for those points is

$$B(t) = (1-t)^3 \cdot P_0 + 3(1-t)^2 t \cdot P_1 + 3(1-t)t^2 \cdot P_2 + t^3 \cdot P_3 \quad \text{for } 0 \le t \le 1:$$

$$x(t) = (1-t)^3 \cdot x_0 + 3(1-t)^2 t \cdot x_1 + 3(1-t)t^2 \cdot x_2 + t^3 \cdot x_3,$$
$$y(t) = (1-t)^3 \cdot y_0 + 3(1-t)^2 t \cdot y_1 + 3(1-t)t^2 \cdot y_2 + t^3 \cdot y_3, \text{ and}$$
$$z(t) = (1-t)^3 \cdot z_0 + 3(1-t)^2 t \cdot z_1 + 3(1-t)t^2 \cdot z_2 + t^3 \cdot z_3 .$$

This 3–dimensional Bezier curve has the same properties as those listed for the 2–dimensional Bezier curve with two modifications.

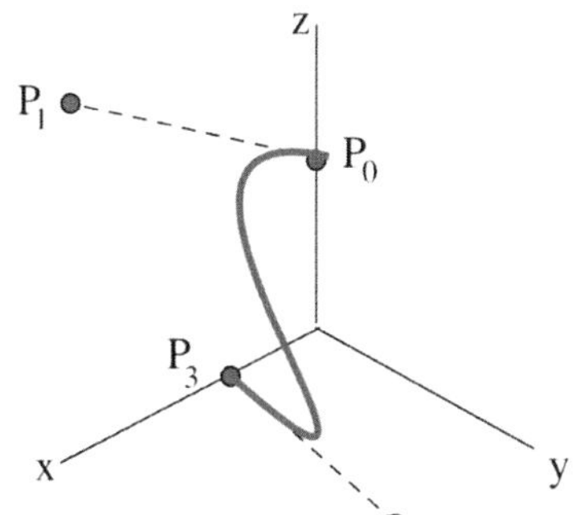

Fig. 19

(3') When t = 0 the direction of B(t) is the same as the direction of the line segment from P_0 to P_1 (Fig. 19): when t = 1 the direction of B(t) is the same as the direction of the line segment from P_2 to P_3 . (We will define the "direction" of a space curve in Section 12.2, but it is similar to "slope" in two dimensions.)

(4') For $0 \le t \le 1$, the graph of B(t) is "on the rubber sheet" whose corners are the control points (Fig. 20). (This is not very precise, but it should convey the idea without the need for more technical vocabulary.)

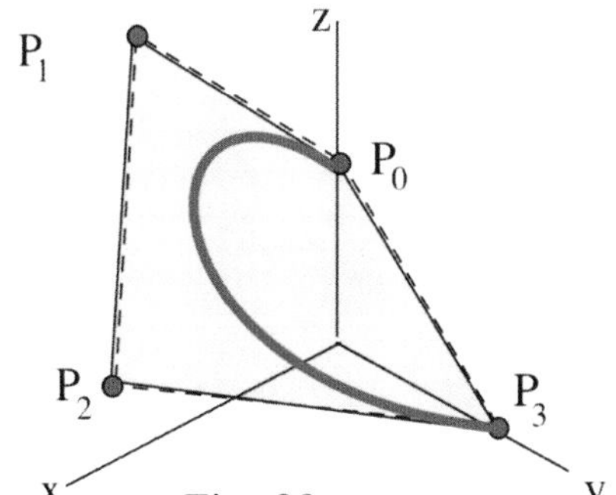

Fig. 20

PROBLEMS

In practice, most 3–dimensional graphs are created using computers. The point of many of the following problems is to help develop your 3–dimensional visualization skills.

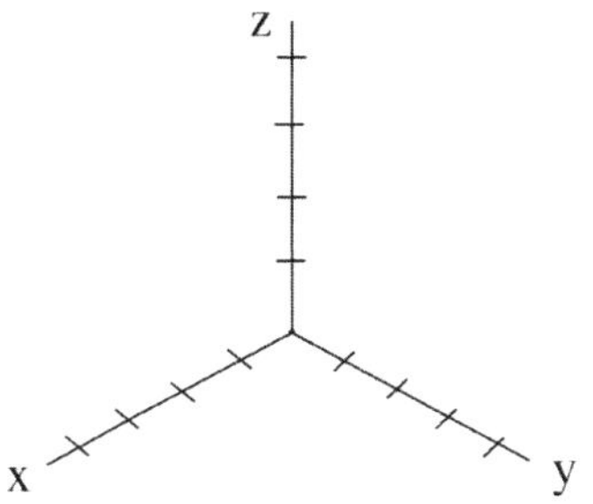

Fig. 21

In problems 1 – 10, sketch the graph of each vector–valued function. Many of these graphs lie on a coordinate plane so it is not difficult to sketch relatively good graphs by hand.

1. Sketch $\mathbf{r}(t) = \langle t, t^2, 0 \rangle$ on the axes system in Fig. 21 for $0 \le t \le 2$.

2. Sketch $\mathbf{s}(t) = \langle t, 0, t^2 \rangle$ on the axes system in Fig. 21 for $0 \le t \le 2$.

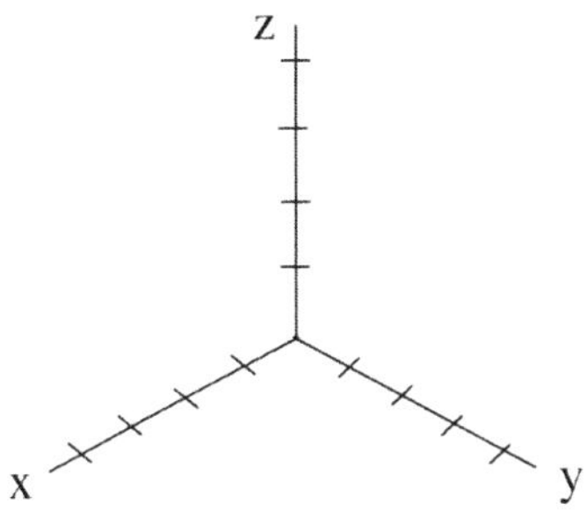

Fig. 22

3. Sketch $\mathbf{r}(t) = \langle 0, t^2, t \rangle$ on the axes system in Fig. 22 for $0 \le t \le 2$.

4. Sketch $\mathbf{s}(t) = \langle t, 2t, 0 \rangle$ on the axes system in Fig. 22 for $0 \le t \le 3$.

5. Sketch $\mathbf{r}(t) = \langle 3t, 0, t \rangle$ on the axes system in Fig. 23 for $0 \le t \le 3$.

6. Sketch $\mathbf{s}(t) = \langle 0, t, \sin(t) \rangle$ on the axes system in Fig. 23 for $0 \le t \le 2\pi$.

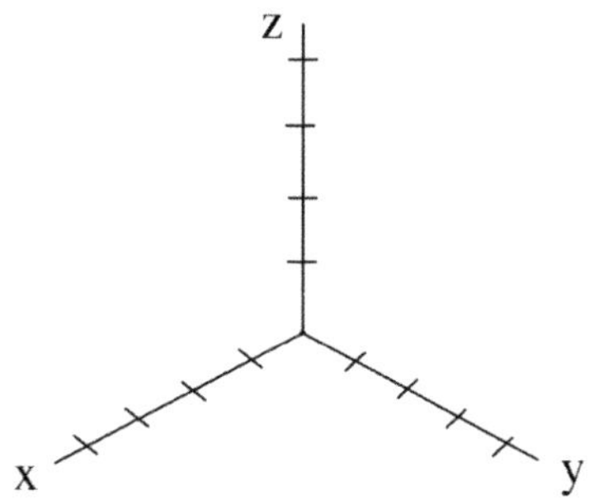

Fig. 23

7. Sketch $\mathbf{r}(t) = \langle 0, \sin(t), t \rangle$ on the axes system in Fig. 24 for $0 \le t \le 2\pi$.

8. Sketch $\mathbf{s}(t) = \langle \cos(t), \sin(t), 1 \rangle$ on the axes system in Fig. 24 for $0 \le t \le 2\pi$.

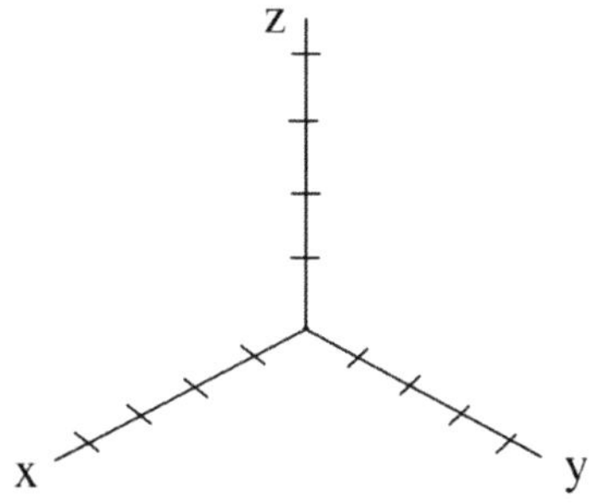

Fig. 24

9. Sketch $\mathbf{r}(t) = \langle 2, \sin(t), \cos(t) \rangle$ on the axes system in Fig. 25 for $0 \le t \le 2\pi$.

10. Sketch $\mathbf{s}(t) = \langle 0, \sin(t), \cos(t) \rangle$ on the axes system in Fig. 25 for $0 \le t \le 2\pi$.

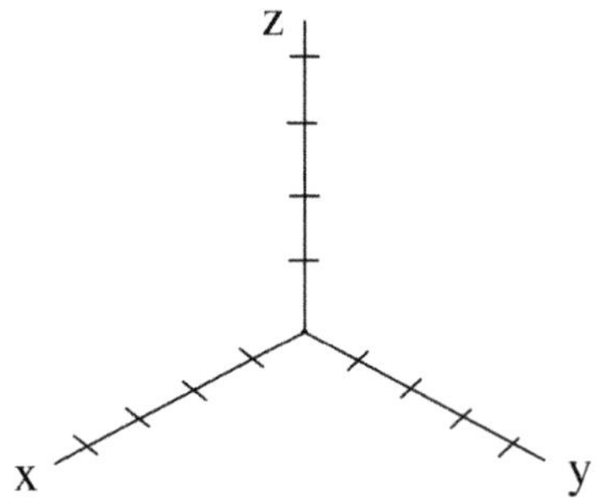

Fig. 25

In problems 11 – 16, calculate and carefully plot three points on each vector–valued linear function and then complete the graph of the line.

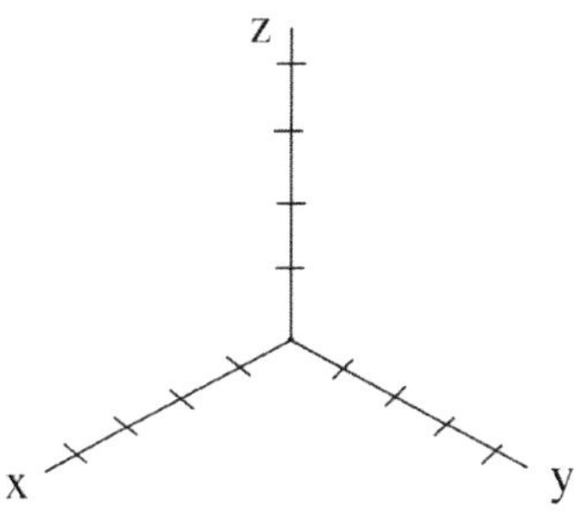

Fig. 26

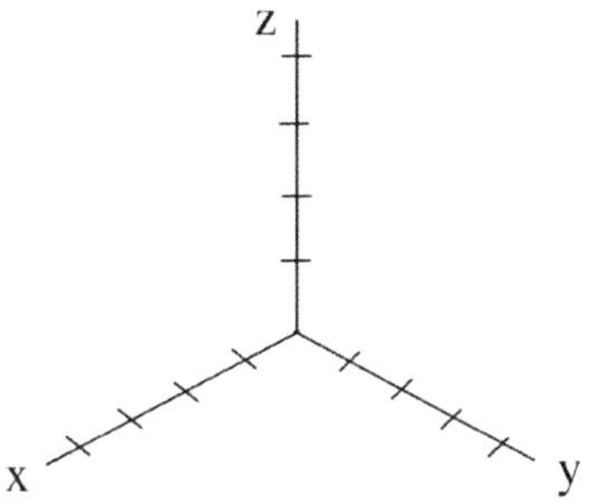

Fig. 27

11. $\mathbf{r}(t) = \langle 2 - t, t, 1 + 2t \rangle$ in Fig. 26.

12. $\mathbf{s}(t) = \langle 2, t, 1 + 2t \rangle$ in Fig. 26.

13. $\mathbf{r}(t) = \langle t, 2, 3 \rangle$ in Fig. 27.

14. $\mathbf{s}(t) = \langle 1, t, 3 \rangle$ in Fig. 27.

15. $\mathbf{r}(t) = \langle t + 1, t - 2, 2t \rangle$ in Fig. 28.

16. $\mathbf{s}(t) = \langle 1 + 2t, t, 3 \rangle$ in Fig. 28.

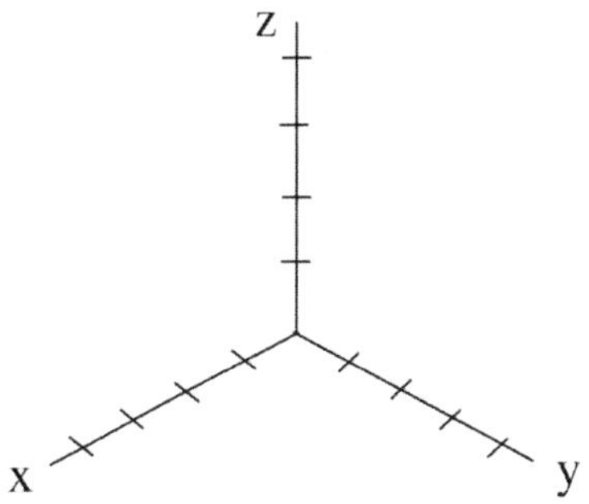

Fig. 28

In problems 17 – 20, graphs are given for x(t), y(t), and z(t). Use the information in these graphs to graph the vector–valued function $\mathbf{r}(t) = \langle x(t), y(t), z(t) \rangle$ on the given coordinate system.

17. The graphs of x(t), y(t), and z(t) are in Fig. 29.

18. The graphs of x(t), y(t), and z(t) are in Fig. 30.

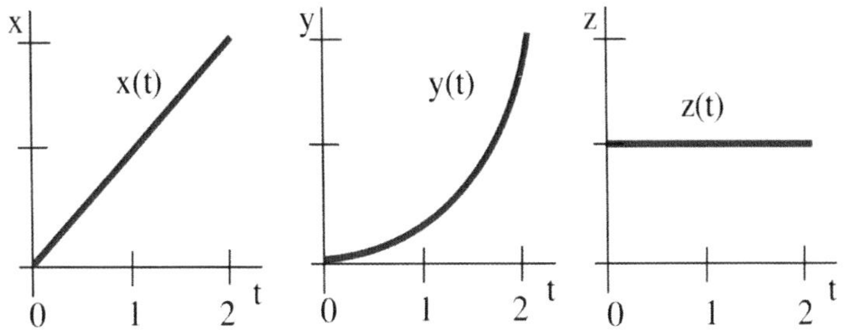

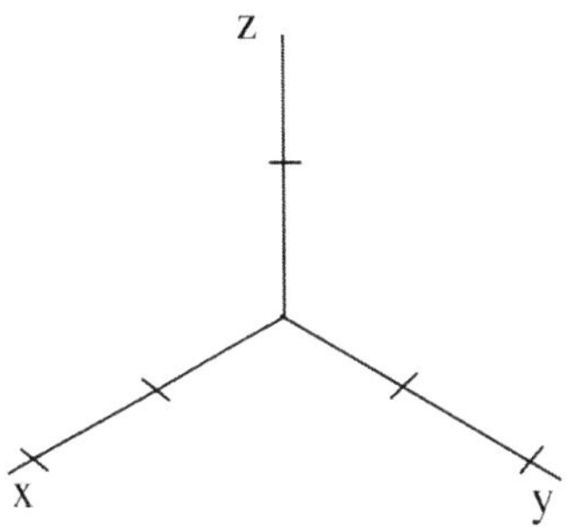

Fig. 29

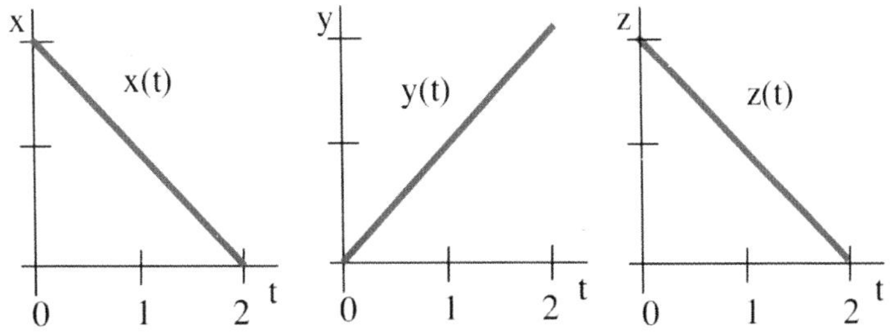

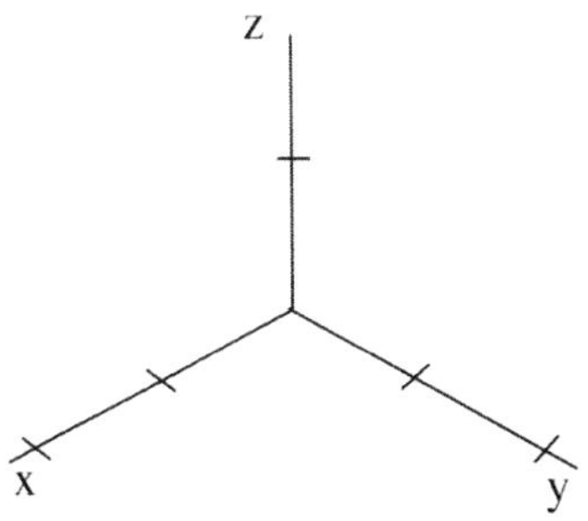

Fig. 30

19. The graphs of x(t), y(t), and z(t) are in Fig. 31.

20. The graphs of x(t), y(t), and z(t) are in Fig. 32.

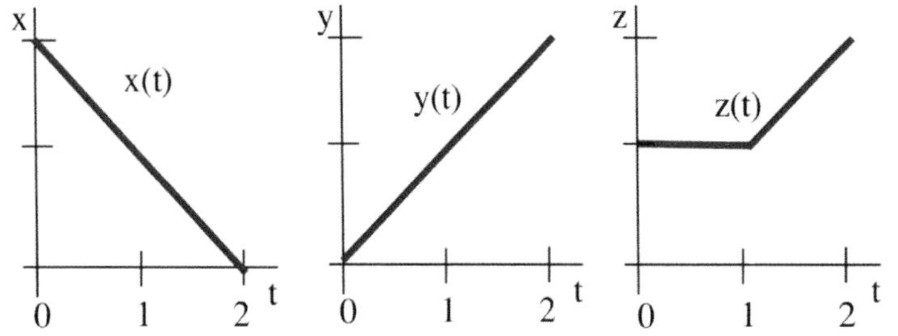

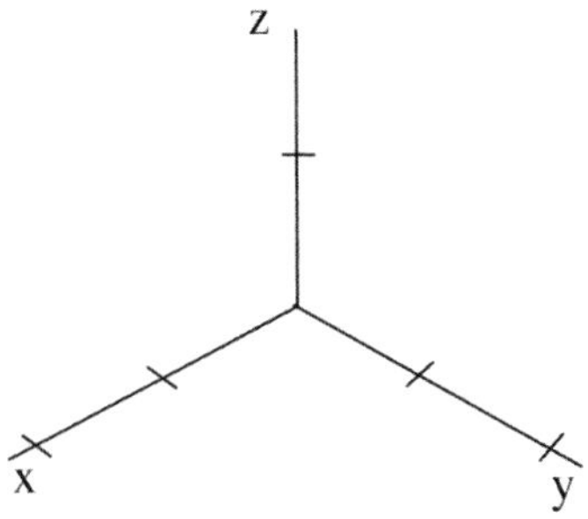

Fig. 31

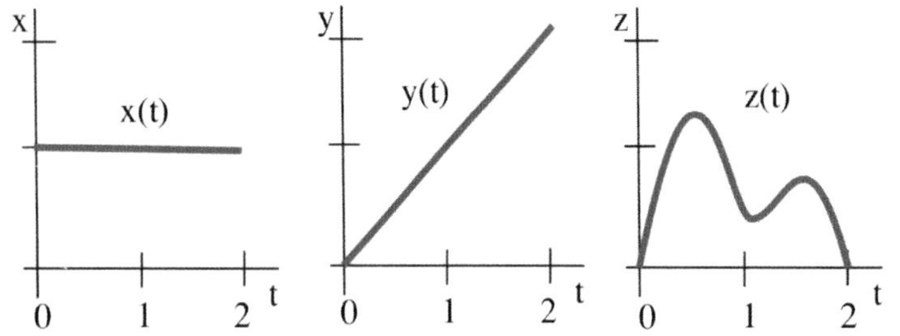

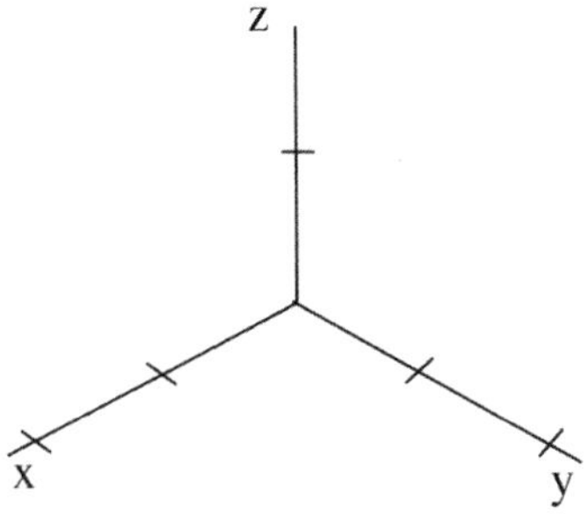

Fig. 32

In problems 21 – 28, calculate the limits.

21. $\lim_{t\to 3} \left\langle t, \frac{t^2}{2}, \frac{1}{t-1} \right\rangle$.

22. $\lim_{t\to 0} \left\langle \cos(t), \frac{t}{3-t}, t^2 \right\rangle$.

23. $\lim_{t\to 2} \left\langle 4, \frac{t}{t-2}, \sqrt{7+t} \right\rangle$.

24. $\lim_{t\to 2} \left\langle \sqrt{t-1}, t^3, 1 \right\rangle$.

25. $\lim_{t\to \infty} \left\langle \frac{4}{t+1}, \frac{3t+1}{t-2}, \frac{\sin(t)}{t} \right\rangle$.

26. $\lim_{t\to \infty} \left\langle \frac{5t^2+2t+1}{t^2+t-6}, \sin(t), \frac{1}{1+t^2} \right\rangle$.

27. $\lim_{t\to \infty} \left\langle \arctan(t), \frac{\ln(t)}{t}, \frac{3}{t^2} \right\rangle$.

28. $\lim_{t\to \infty} \left\langle \sin(\arctan(t)), 1, 5^{1/t} \right\rangle$.

In problems 29 – 36, determine where the given vector–valued functions are continuous.

29. $\mathbf{r}(t) = \left\langle t, \frac{t^2}{2}, \frac{1}{t-1} \right\rangle$.

30. $\mathbf{s}(t) = \left\langle \cos(t), \frac{t}{3-t}, t^2 \right\rangle$.

31. $\mathbf{u}(t) = \left\langle 4, \frac{t}{t-2}, \sqrt{7+t} \right\rangle$.

32. $\mathbf{v}(t) = \left\langle \sqrt{t-1}, t^3, 1 \right\rangle$.

33. $\mathbf{w}(t) = \left\langle \frac{4}{t+1}, \frac{3t+1}{t-2}, \frac{\sin(t)}{t} \right\rangle$.

34. $\mathbf{r}(t) = \left\langle \frac{5t^2+2t+1}{t^2+t-6}, \sin(t), \frac{1}{1+t^2} \right\rangle$.

35. $\mathbf{r}(t) = \left\langle \arctan(t), \frac{\ln(t)}{t}, \frac{3}{t^2} \right\rangle$.

36. $\mathbf{s}(t) = \left\langle \sin(\arctan(t)), 1, 5^{1/t} \right\rangle$.

In problems 37 – 40, determine the Bezier curve B(t) for the given control points P_0, P_1, P_2 and P_3 . If you have access to a computer with the appropriate software, graph B(t).

37. $P_0 = (0,0,1), P_1 = (1,0,1), P_2 = (1,2,0)$, and $P_3 = (0,2,0)$.

38. $P_0 = (0,0,2), P_1 = (1,0,1), P_2 = (0,2,0)$, and $P_3 = (1,3,0)$.

39. $P_0 = (0,1,2), P_1 = (0,0,2), P_2 = (2,1,0)$, and $P_3 = (2,0,0)$.

40. $P_0 = (0,1,2), P_1 = (0,0,2), P_2 = (2,0,0)$, and $P_3 = (1,1,0)$.

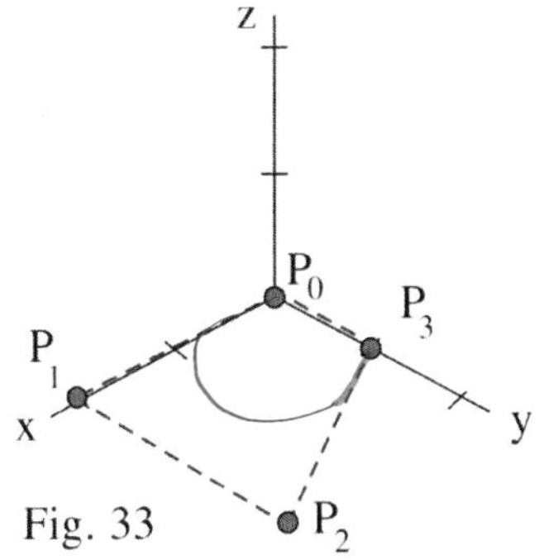

Fig. 33

In problems 41 – 44, control points P_0, P_1, P_2 and P_3 are shown on a graph. For the given control points sketch a curve with the properties of the Bezier curve. (In some graphs additional lines are included with the control points to help with your sketch.)

41. P_0, P_1, P_2 and P_3 are given in Fig. 33.

42. P_0, P_1, P_2 and P_3 are given in Fig. 34.

43. P_0, P_1, P_2 and P_3 are given in Fig. 35.

44. P_0, P_1, P_2 and P_3 are given in Fig. 36.

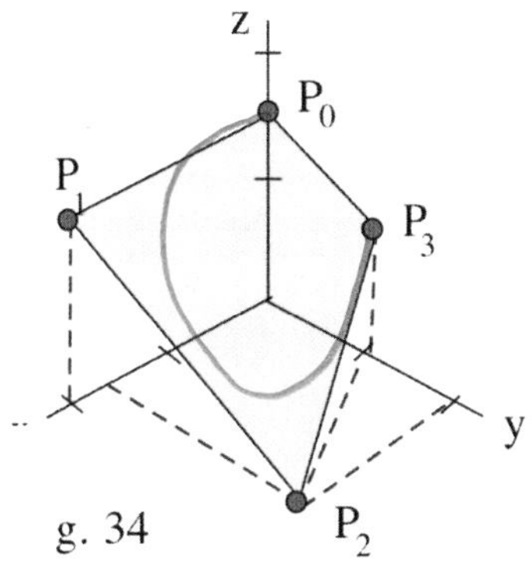

g. 34

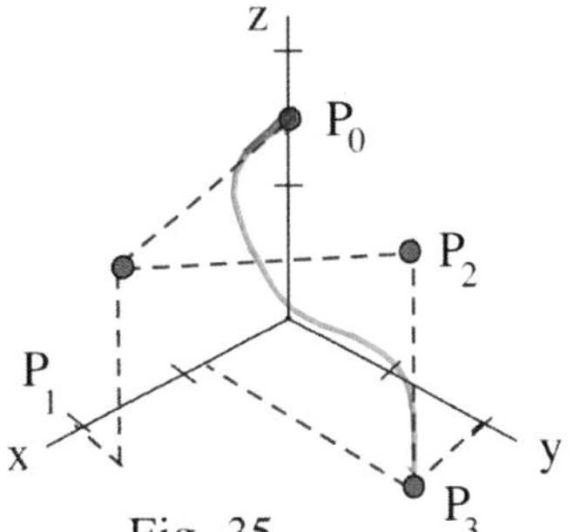

Fig. 35

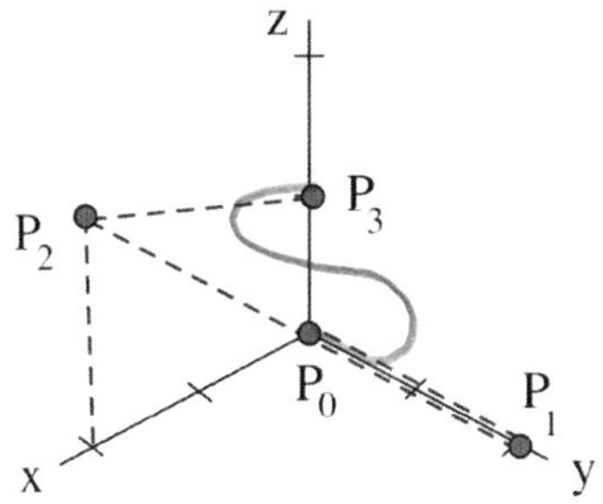

Fig. 36

In problems 45 – 48, graphs of vector–valued functions are given. At the points labeled A, B, and C on each curve determine whether each variable function x(t), y(t), and z(t) is increasing (I) or decreasing (D) and fill in the table for each function.

45. $\mathbf{r}(t) = \langle x(t), y(t), z(t) \rangle$ is given in Fig. 37.

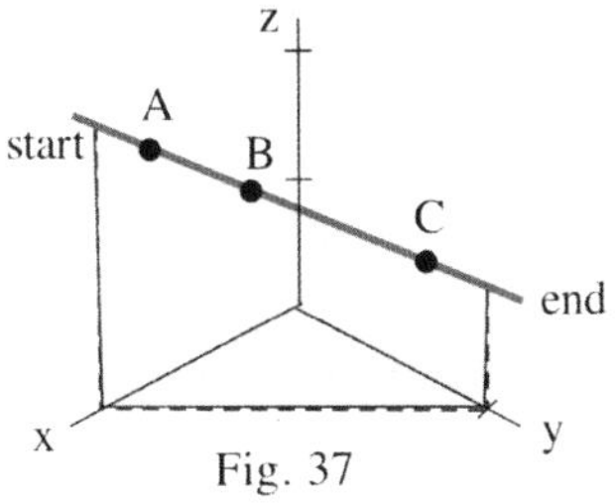

Fig. 37

	A	B	C
x			
y			
z			

46. $\mathbf{s}(t) = \langle x(t), y(t), z(t) \rangle$ is given in Fig. 38.

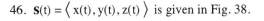

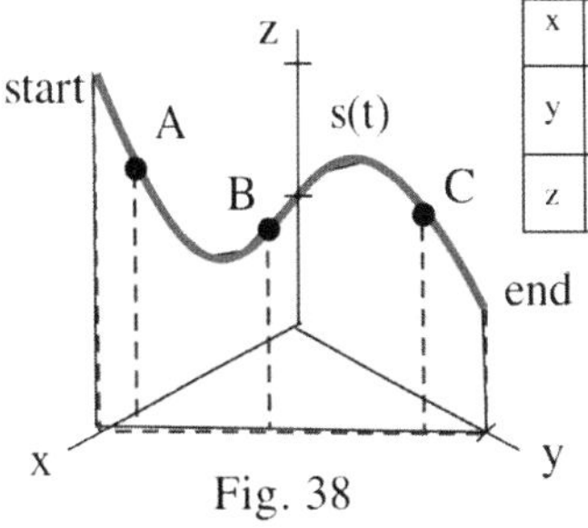

Fig. 38

	A	B	C
x			
y			
z			

47. $\mathbf{u}(t) = \langle x(t), y(t), z(t) \rangle$ is given in Fig. 39.

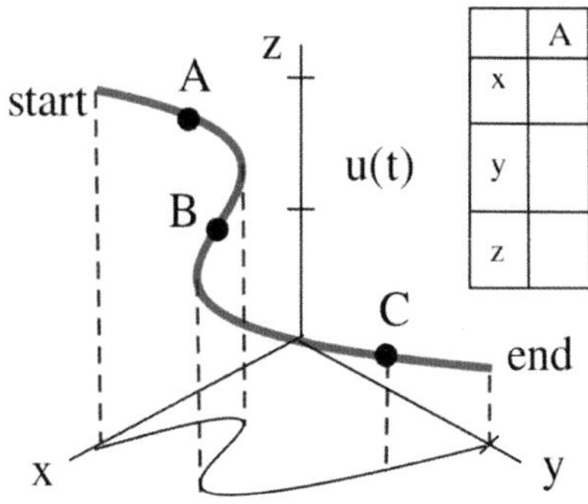

	A	B	C
x			
y			
z			

Fig. 39

48. $\mathbf{v}(t) = \langle x(t), y(t), z(t) \rangle$ is given in Fig. 40.

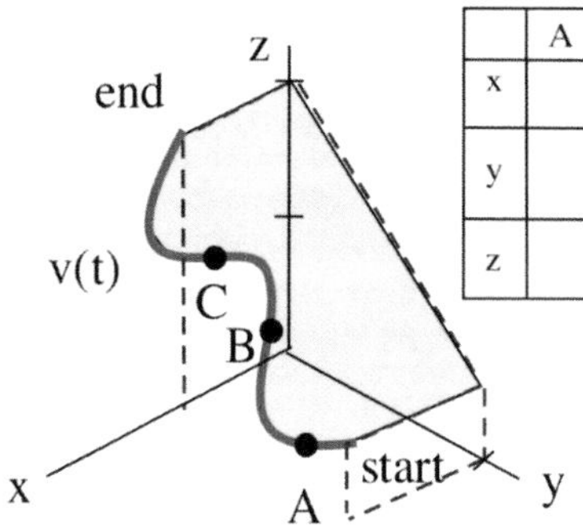

	A	B	C
x			
y			
z			

Fig. 40

Practice Answers

Practice 1: The graphs of $\mathbf{r}(t) = \langle t, 4-2t, 2-t \rangle$ and $\mathbf{s}(t) = \langle t, t^2, 1 \rangle$ for $0 \le t \le 2$ are shown in Fig. 41.

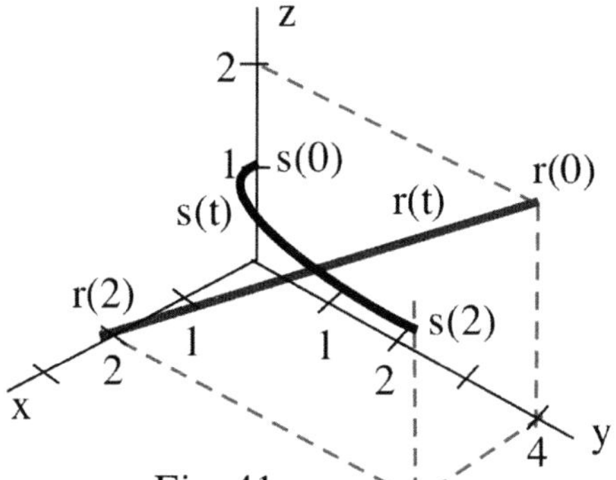

Fig. 41

Practice 2: The graphs of $x(t), y(t)$, and $z(t)$ are shown in Fig. 42a, 42b, and 42c. The graph of $\mathbf{r}(t) = x(t)\mathbf{i} + y(t)\mathbf{j} + z(t)\mathbf{k} = \langle x(t), y(t), z(t) \rangle$ is shown in Fig. 42d .

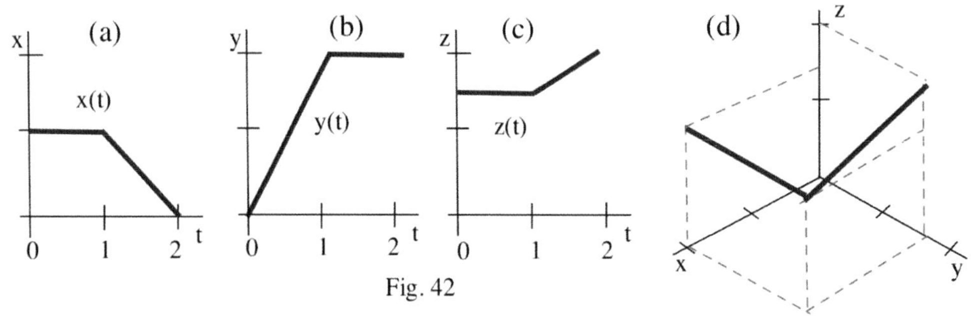

Fig. 42

Practice 3: For $\mathbf{r}(t) = \left\langle \cos(t), \frac{\sin(t)}{t}, 3 \right\rangle$, $\lim_{t \to \pi} \mathbf{r}(t) = \langle -1, 0, 3 \rangle$, and $\lim_{t\to\infty} \mathbf{r}(t)$ does not exist because $\lim_{t\to\infty} \cos(t)$ does not exist.

Practice 4: $\mathbf{r}(t) = \langle 1/t, e^t, \text{INT}(t) \rangle$ is continuous everywhere except where t is an integer since INT(t) is not continuous where t is an integer. 1/t is not continuous when $t = 0$, but we have already excluded $t = 0$ because 0 is an integer.

Practice 5: The graphs of $\mathbf{r}(t) = \langle t-1, 0, 1+2t \rangle$ and $\mathbf{s}(t) = \langle 2, 4, 3t \rangle$ are shown in Fig. 43 .

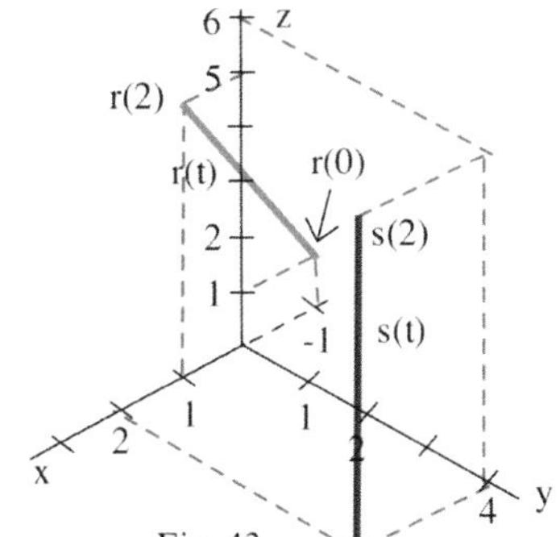

Fig. 43

Practice 6: The graph of $\mathbf{r}(t) = \langle t, \cos(t), \sin(t) \rangle$ is a circular helix around the x–axis. This graph is shown in Fig. 44.
The graph of $\mathbf{s}(t) = \langle 2\cdot\cos(t), t, 3\cdot\sin(t) \rangle$ is an elliptical helix around the y–axis. This graph is shown in Fig. 45.

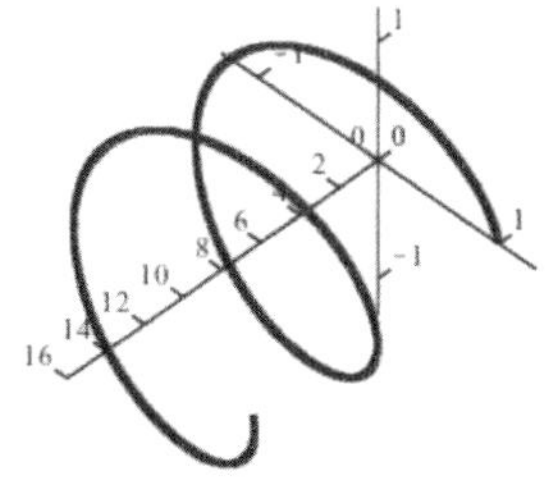
circular helix around the x-axis
Fig. 44

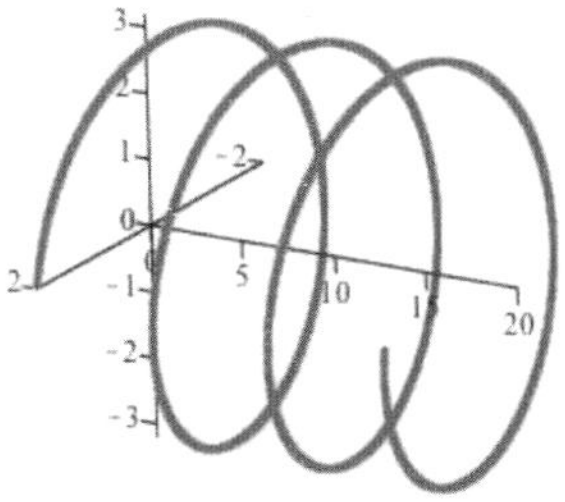
elliptical helix around the y-axis
Fig. 45

Appendix: MAPLE and graphs of vector–valued functions — spacecurve( )

The computer language MAPLE has a number of commands for creating graphs of 3–dimensional objects including vector–valued functions. In order to access these commands, we first need to load the "plots" package:

with(plots); *(then press the <enter> key)* This loads the "plots" package and lists the new commands available for us to use.

The command to create graphs of vector–valued functions is

spacecurve([x(t), y(t), z(t)] , t = a..b, *options*);

where x(t), y(t), and z(t) are formulas for the x, y, and z coordinates,

a and b are the starting and stopping values for the variable t

and *options* includes commands for the type of axes, the color, the thickness, and the number of points to be plotted.

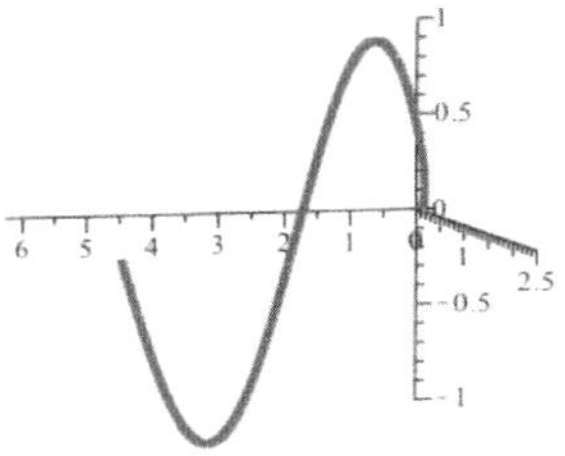
Fig. M1

For example, the command

spacecurve([t, sqrt(t), sin(t)], t=0..2*Pi, axes=NORMAL, numpoints=200, color=red, thickness=3);

creates the graph in Fig. M1.

By positioning the cursor on the graph, a "hand" appears and, while holding down the mouse button, we can rotate the graph by slowing moving the mouse and then releasing the button. Fig. M2 shows another view of the graph.

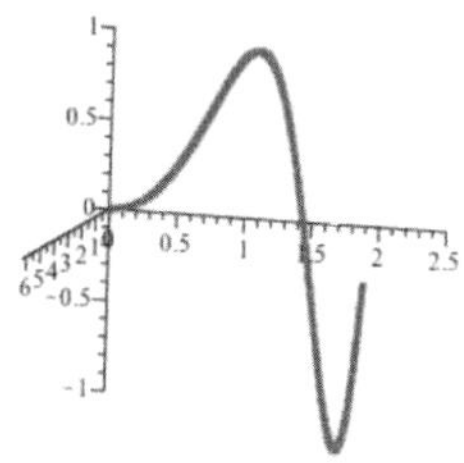

Fig M2

Fig. M3 shows three views created using the command

spacecurve([sin(t)*cos(25*t), t, sin(t)*sin(25*t)], t=0..Pi, axes=NORMAL, numpoints=300);

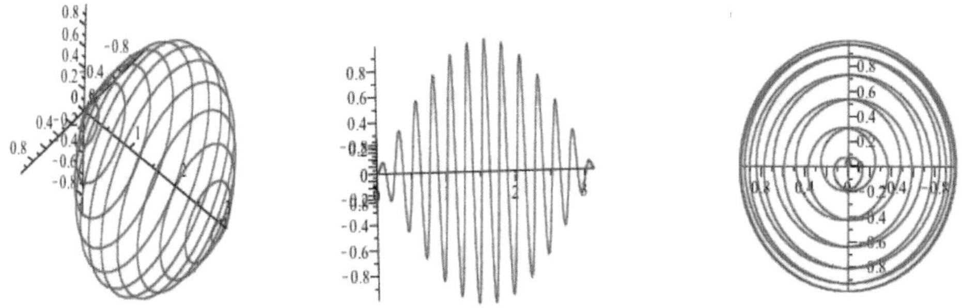

Fig. M3: three views of [sin(t)*cos(25t), t , sin(t)*sin(25t)]

12.2 DERIVATIVES AND ANTIDERIVATIVES OF VECTOR–VALUED FUNCTIONS

Derivatives of Vector–valued Functions

The derivative of a vector–valued function is another vector–valued function, and this derivative is defined much like the derivative of a scalar function. Derivatives of vector–valued functions are generally easy to compute, component–by–component, and they have a useful geometric interpretation as the vectors tangent to the graph of the vector–valued function.

Definition: The **derivative** of $\mathbf{r}(t) = \langle x(t), y(t), z(t) \rangle$, denoted $\frac{d}{dt}\mathbf{r}(t)$ or $\mathbf{r}'(t)$, is

$$\mathbf{r}'(t) = \lim_{\Delta t \to 0} \frac{r(t+\Delta t) - r(t)}{\Delta t} = \langle x'(t), y'(t), z'(t) \rangle$$

provided the limit exists and is finite. (Fig. 1)

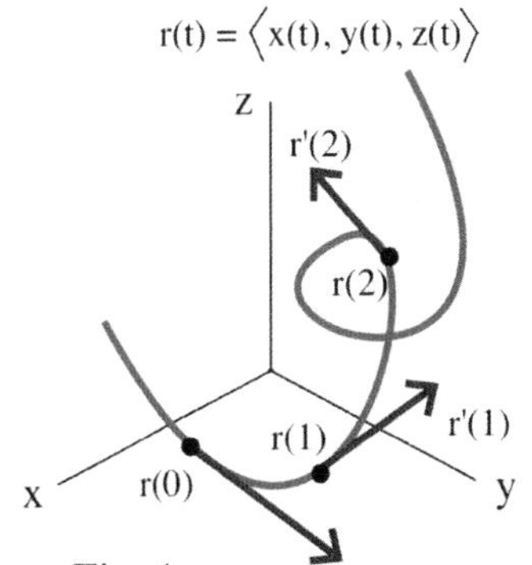

Fig. 1

A vector–valued function $\mathbf{r}(t)$ is differentiable at a point $t = t_0$ if and only if each of its component functions is differentiable at $t = t_0$, and we can calculate the derivative $\mathbf{r}'(t)$ by calculating the three derivatives $x'(t)$, $y'(t)$, and $z'(t)$.

Visualizing $\mathbf{r}'(t)$: If $\mathbf{r}(t)$ is the position of an object at time t, then the difference vector $\mathbf{r}(t + \Delta t) - \mathbf{r}(t)$ represents the **change** in position from time t to time $t + \Delta t$ (Fig. 2), and the ratio $\frac{\mathbf{r}(t+\Delta t) - \mathbf{r}(t)}{\Delta t}$ is a vector measuring the average rate of change of position during the time interval from t to $t + \Delta t$. The limit $\mathbf{r}'(t)$ of this "average rate of change" vector has two useful geometric properties (Fig. 3):

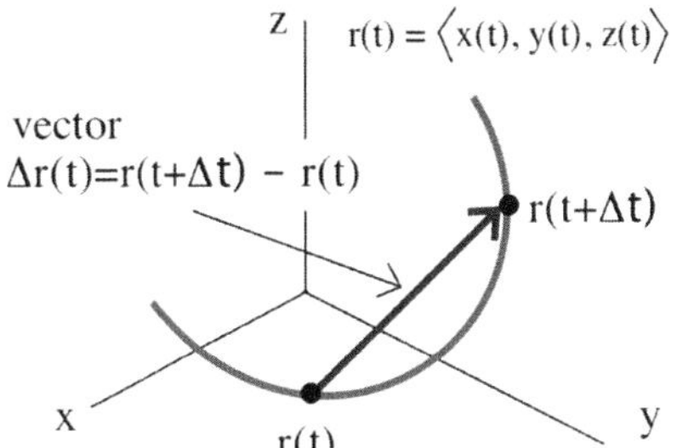

Fig. 2

- $\mathbf{r}'(t)$ is **tangent** to the graph of $\mathbf{r}(t)$, and
- the magnitude of $\mathbf{r}'(t)$ is the **speed** of the object along the path at the time t.

The vector $\mathbf{r}'(t)$ is called the velocity of $\mathbf{r}(t)$.

The vector $|\,\mathbf{r}'(t)\,|$ is called the speed of $\mathbf{r}(t)$.

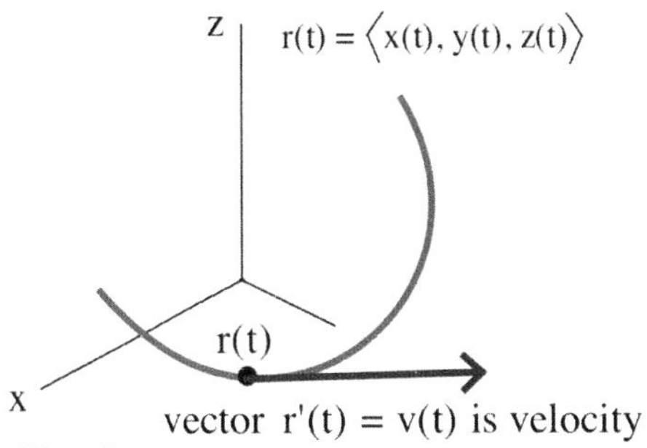

Fig. 3

Definitions: Velocity, Speed, Direction, and Acceleration

If $\mathbf{r}(t)$ is the **position** of an object at time t, then

the **velocity** of the object is $\mathbf{v}(t) = \frac{d}{dt}\mathbf{r}(t) = \mathbf{r}'(t)$ (a vector tangent to $\mathbf{r}(t)$),

the **speed** of the object is $|\mathbf{v}(t)|$ (a scalar) ,

the **direction** of travel is $\mathbf{T}(t) = \frac{\mathbf{v}(t)}{|\mathbf{v}(t)|}$ (the **unit tangent vector**) , and

the **acceleration** is $\mathbf{a}(t) = \mathbf{v}'(t) = \mathbf{r}''(t)$.

Example 1: A ladybug is crawling up a helix so its position vector is $\mathbf{r}(t) = \langle \cos(t), \sin(t), t \rangle$ as shown in Fig. 4.

(a) At the points labeled A, B, and C on the graph of $\mathbf{r}(t)$, estimate the sign (positive or negative) of each component of $\mathbf{r}'(t)$.

(b) The values of t for the point A, B, and C are $t = \pi/6, 3\pi/4$, and $7\pi/4$ respectively. Calculate $\mathbf{r}'(t)$ at the given values of t and compare the results with your estimates in part (a).

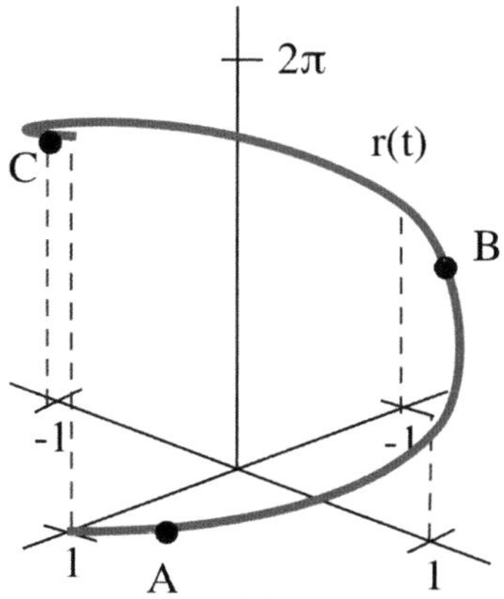

Fig. 4

Solution: (a) At A, $\mathbf{r}'(t)$ is $\langle x'(t), y'(t), z'(t) \rangle = \langle -, +, + \rangle$. At B, $\mathbf{r}'(t)$ is $\langle -, -, + \rangle$. At C, $\mathbf{r}'(t)$ is $\langle +, +, + \rangle$.

(b) $\mathbf{r}'(\pi/6)$ is $\langle x'(\pi/6), y'(\pi/6), z'(\pi/6) \rangle = \langle -\sin(\pi/6), \cos(\pi/6), 1 \rangle \approx \langle -0.5, 0.867, 1 \rangle$.
$\mathbf{r}'(3\pi/4)$ is $\langle -\sin(3\pi/4), \cos(3\pi/4), 1 \rangle \approx \langle -0.707, -0.707, 1 \rangle$.
$\mathbf{r}'(7\pi/4)$ is $\langle -\sin(7\pi/4), \cos(7\pi/4), 1 \rangle \approx \langle 0.707, 0.707, 1 \rangle$.

Practice 1: The position vector of an object at time t is $\mathbf{r}(t) = \langle t, t^2, t^3 \rangle$ as shown in Fig. 5. Calculate the position, velocity, speed, direction, and acceleration of the object when $t = 0, 1$, and 2.

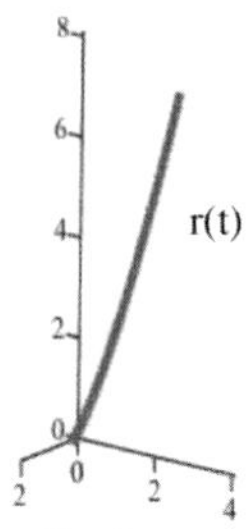

Fig. 5

Angles of Intersection Between Space Curves

The angle of intersection between two curves at a point in space is the angle between their tangent vectors (velocities) at that point of intersection, and the dot product of the tangent vectors can be used to find this angle.

Example 2: The parabolic path $\mathbf{r}(t) = \langle 1, t, t^2 \rangle$ intersects the line $\mathbf{s}(t) = \langle -2 + 3t, 6 - 4t, 2 + 2t \rangle$ (Fig. 6) at the point $(1, 2, 4)$. Find the angle of intersection of the curves at that point.

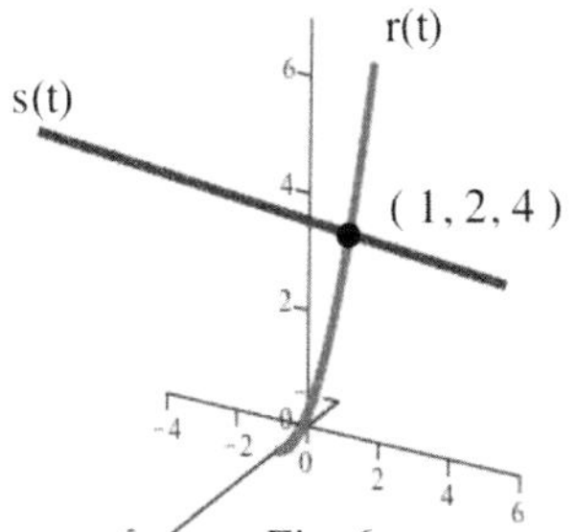

Fig. 6

Solution: The parabola goes through $(1, 2, 4)$ when $t = 2$, and the line goes through $(1, 2, 4)$ when $t = 1$. Then $\mathbf{r}'(2) = \langle 0, 1, 4 \rangle$ and $\mathbf{s}'(1) = \langle 3, -4, 2 \rangle$ so

$$\cos(\theta) = \frac{\mathbf{r}'(2) \cdot \mathbf{s}'(1)}{|\mathbf{r}'(2)|\,|\mathbf{s}'(1)|}$$

$$\approx \frac{4}{\sqrt{17}\sqrt{29}} \approx 0.180 \text{ and}$$

and $\theta \approx 1.390$ (or about 79.6^o).

Practice 2: The parabolic paths $\mathbf{r}(t) = \langle 0, t, t^2 \rangle$ and $\mathbf{s}(t) = \langle 2 - t, 1, 5 - t^2 \rangle$ (Fig. 7) intersect at the point $(0, 1, 1)$. Find the angle of intersection of the curves at that point.

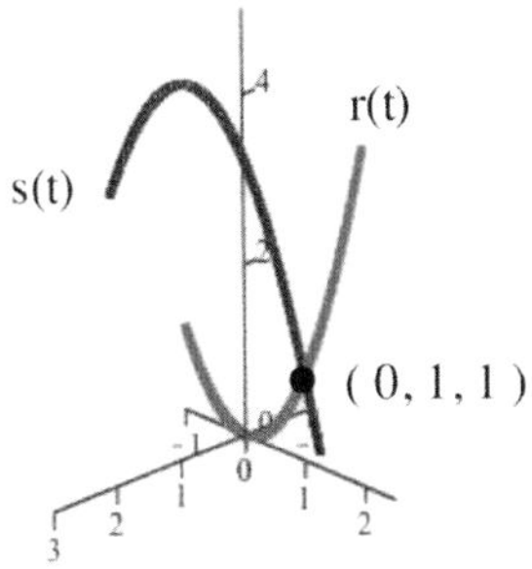

Fig. 7

Differentiation of Combinations of Vector–valued Functions

For scalar functions we have patterns for differentiating sums, differences, products, and compositions, and there are similar rules for differentiating combinations of vector–valued functions. In fact, the rules for vector–valued functions are almost identical to the corresponding rules for scalar function derivatives.

Differentiation Patterns for Vector–valued Functions

Constant: If $\mathbf{C}$ is a constant vector, then $\frac{d}{dt}\mathbf{C} = 0$ vector.

If $\mathbf{u}(t)$ and $\mathbf{v}(t)$ are differentiable vector–valued functions, k is a scalar number, and f(t) is a scalar function, then

Sum: $\frac{d}{dt}(\mathbf{u}(t) + \mathbf{v}(t)) = \frac{d}{dt}\mathbf{u}(t) + \frac{d}{dt}\mathbf{v}(t) = \mathbf{u}'(t) + \mathbf{v}'(t)$

Difference: $\frac{d}{dt}(\mathbf{u}(t) - \mathbf{v}(t)) = \frac{d}{dt}\mathbf{u}(t) - \frac{d}{dt}\mathbf{v}(t) = \mathbf{u}'(t) - \mathbf{v}'(t)$

Products: $\frac{d}{dt}(k\mathbf{u}(t)) = k\frac{d}{dt}(\mathbf{u}(t)) = k\,\mathbf{u}'(t)$

scalar $\frac{d}{dt}(\, f(t)\,\mathbf{u}(t)\,) = f(t)\frac{d}{dt}(\,\mathbf{u}(t)\,) + \frac{d\,f(t)}{dt}\,\mathbf{u}(t) = f(t)\,\mathbf{u}'(t) + f\,'(t)\,\mathbf{u}(t)$

dot $\frac{d}{dt}(\,\mathbf{u}(t)\bullet\mathbf{v}(t)\,) = \mathbf{u}(t)\bullet\frac{d\,\mathbf{v}(t)}{dt} + \frac{d\,\mathbf{u}(t)}{dt}\bullet\mathbf{v}(t) = \mathbf{u}(t)\bullet\mathbf{v}'(t) + \mathbf{u}'(t)\bullet\mathbf{v}(t)$

cross $\frac{d}{dt}(\,\mathbf{u}(t) \times \mathbf{v}(t)\,) = \mathbf{u}(t) \times \mathbf{v}'(t) + \mathbf{u}'(t) \times \mathbf{v}(t)$

Chain Rule: $\frac{d}{dt}\,\mathbf{u}(\, f(t)\,) = f\,'(t)\,\mathbf{u}'(\, f(t)\,)$

You should notice that all of the **product** differentiation patterns have the form

(first function) "times" (derivative of the second) plus (derivative of the first) "times" (second)

where the "times" is the appropriate type of multiplication, either scalar, dot, or cross.

Proofs: The proofs are very straightforward (componentwise) for the results for the derivatives of constant vectors, sums, differences and a scalar times a vector–valued function, and they are left for you.

The proofs given below for the product rules all follow the pattern of rewriting the original function as components, using our usual product rule or chain rule to differentiate the component functions, and then rewriting the results as the appropriate product of vectors.

Scalar: $\frac{d}{dt}(\, f(t)\,\mathbf{u}(t)\,) = \left\langle \frac{d}{dt}\, f(t)u_1(t)\,, \frac{d}{dt}\, f(t)u_2(t)\,, \frac{d}{dt}\, f(t)u_3(t) \right\rangle$

$$= \left\langle f(t)u_1'(t) + f\,'(t)u_1(t)\,,\ f(t)u_2'(t) + f\,'(t)u_2(t)\,,\ f(t)u_3'(t) + f\,'(t)u_3(t) \right\rangle$$

$$= \left\langle f(t)u_1'(t)\,,\ f(t)u_2'(t)\,,\ f(t)u_3'(t) \right\rangle + \left\langle f\,'(t)u_1(t)\,,\ f\,'(t)u_3(t)\,, f\,'(t)u_3(t) \right\rangle$$

$$= f(t) \left\langle u_1'(t)\,,\ u_2'(t)\,,\ u_3'(t) \right\rangle + f\,'(t) \left\langle u_1(t)\,,\ u_2(t)\,, u_3(t) \right\rangle$$

$$= f(t)\,\mathbf{u}'(t) + f\,'(t)\,\mathbf{u}(t)\,.$$

Dot: $\frac{d}{dt}(\,\mathbf{u}(t)\bullet\mathbf{v}(t)\,) = \left\langle \frac{d}{dt}\, u_1(t)v_1(t)\,,\ \frac{d}{dt}\, u_2(t)v_2(t)\,,\ \frac{d}{dt}\, u_3(t)v_3(t) \right\rangle$

$$= \left\langle u_1(t)v_1'(t) + u_1'(t)v_1(t)\,,\ u_2(t)v_2'(t) + u_2'(t)v_2(t)\,,\ u_3(t)v_3'(t) + u_3'(t)v_3(t) \right\rangle$$

$$= \left\langle u_1(t)v_1'(t)\,,\ u_2(t)v_2'(t)\,,\ u_3(t)v_3'(t) \right\rangle + \left\langle u_1'(t)v_1(t)\,,\ u_2'(t)v_2(t)\,,\ u_3'(t)v_3(t) \right\rangle$$

$$= \mathbf{u}(t)\bullet\mathbf{v}'(t) + \mathbf{u}'(t)\bullet\mathbf{v}(t)\,.$$

The pattern for the derivative of a cross product can also be proved by resorting to the definition of the cross product and showing that the components of $\frac{d}{dt}(\mathbf{u}(t) \times \mathbf{v}(t))$ match the components of $\mathbf{u}(t) \times \mathbf{v}'(t) + \mathbf{u}'(t) \times \mathbf{v}(t)$, but the process is algebraically long and is omitted.

Chain rule: $\frac{d}{dt}\mathbf{u}(f(t)) = \left\langle \frac{d}{dt} u_1(f(t)), \frac{d}{dt} u_2(f(t)), \frac{d}{dt} u_3(f(t)) \right\rangle$

$$= \left\langle f'(t)u_1'(f(t)), f'(t)u_2'(f(t)), f'(t)u_3'(f(t)) \right\rangle$$

$$= f'(t)\left\langle u_1'(f(t)), u_2'(f(t)), u_3'(f(t)) \right\rangle = f'(t)\,\mathbf{u}'(f(t)).$$

These differentiation patterns simply provide alternate, and sometimes easier, ways to compute derivatives. Occasionally they are useful for deriving results about the behavior of vector–valued functions such as the one given in the next example.

Example 3: Suppose a differentiable position vector $\mathbf{r}(t)$ of an object has constant length so $|\mathbf{r}(t)| = k$ for all t. Show that the direction of travel of the object is always perpendicular to its position.

Solution: $\mathbf{r}(t) \cdot \mathbf{r}(t) = |\mathbf{r}(t)|^2 = k^2$ for all t, so $\mathbf{r}(t) \cdot \mathbf{r}(t)$ is a constant so $\frac{d}{dt}\mathbf{r}(t) \cdot \mathbf{r}(t) = 0$.

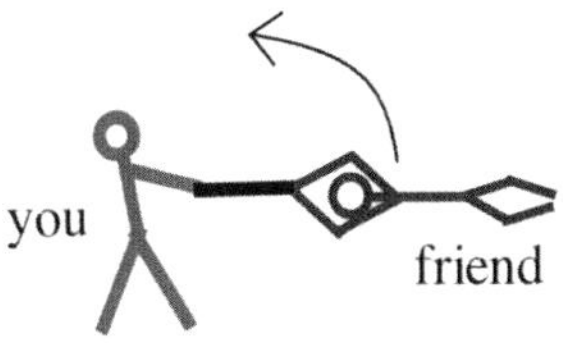

Fig. 8: Twirling a friend

But we also know that

$$\frac{d}{dt}\{\mathbf{r}(t) \cdot \mathbf{r}(t)\} = \mathbf{r}(t) \cdot \mathbf{r}'(t) + \mathbf{r}'(t) \cdot \mathbf{r}(t) = 2\mathbf{r}(t) \cdot \mathbf{r}'(t),$$

so we can conclude that $\mathbf{r}(t) \cdot \mathbf{r}'(t) = 0$ for all t.

But $\mathbf{r}(t) \cdot \mathbf{r}'(t) = 0$ means that $\mathbf{r}(t)$ is perpendicular to $\mathbf{r}'(t)$ for all t, and that means the position, $\mathbf{r}(t)$, is always perpendicular to the velocity, $\mathbf{r}'(t)$. The velocity vector $\mathbf{r}'(t)$ points in the direction of travel of the object so we have shown that the direction of travel of the object is always perpendicular to its position.

The result in Example 3 also has straightforward physical interpretations. If we are twirling someone around a central point (Fig. 8), we can take the central point to be the origin. Then the twirled person is always a constant distance from the central point and the magnitude of their position vector is a constant. The result of this example says that the twirled person's velocity vector is always perpendicular to their position vector. If we let go of the person, their motion will be a straight line that is perpendicular to the circular path they were following (Fig. 9). An equivalent situation in three

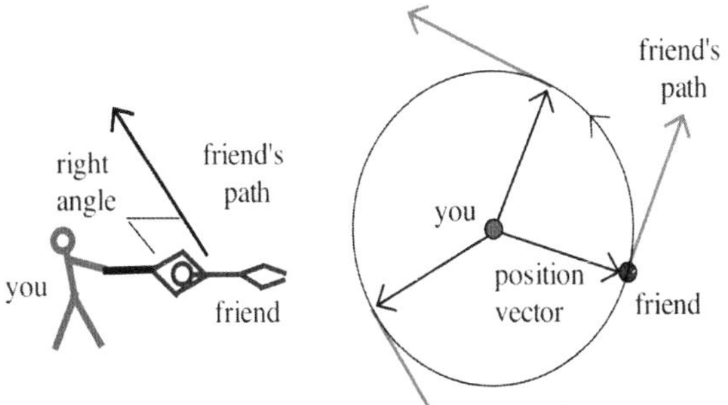

Fig. 9: Letting go

dimensions is an object moving on the surface of a sphere (Fig. 10) such as the earth (almost). If gravity is "turned off," this object will travel along a path perpendicular to the vector from the center of the earth to its position.

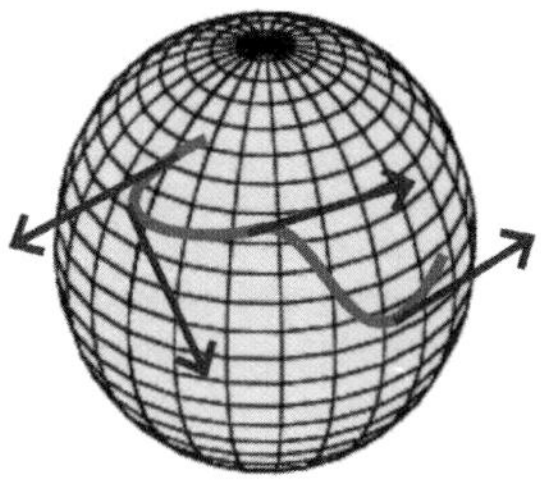

Fig. 10: Path on a sphere

Antiderivatives of Vector–valued Functions

Since the derivative of a vector–valued function is defined to be the vector formed by the derivative of each of the component functions, the antiderivative of a vector–valued function is also defined component by component.

> **Definition:** If $\mathbf{r}(t) = \langle x(t), y(t), z(t) \rangle$,
>
> then the antiderivative of $\mathbf{r}(t)$ is $\int \mathbf{r}(t)\,dt = \langle \int x(t)\,dt , \int y(t)\,dt , \int z(t)\,dt \rangle$
>
> provided that the antiderivatives of $x(t), y(t)$, and $z(t)$ exist.

Example 4: The velocity of an object is $\mathbf{v}(t) = \langle 4t , -\sin(t) , e^t \rangle$ and its position at time $t = 0$ is $\mathbf{r}(0) = \langle 2, 3, 4 \rangle$, Find a formula for $\mathbf{r}(t)$, its position at time t.

Solution: $\mathbf{v}(t) = \mathbf{r}'(t)$ so

$$\mathbf{r}(t) = \int \mathbf{r}'(t)\,dt = \langle \int 4t\,dt , \int -\sin(t)\,dt , \int e^t\,dt \rangle = \langle 2t^2 + A, \cos(t) + B, e^t + C \rangle.$$

Then we can use the initial condition that $\mathbf{r}(0) = \langle 2, 3, 4 \rangle$, to determine that $A = 2, B = 2$, and $C = 3$ so $\mathbf{r}(t) = \langle 2t^2 + 2, \cos(t) + 2, e^t + 3 \rangle$.

Practice 3: The velocity of an object is $\mathbf{v}(t) = \langle 6t^2 , \cos(t) , 12e^{3t} \rangle$ and its position at time $t = 0$ is $\mathbf{r}(0) = \langle 1, -5, 2 \rangle$, Find a formula for $\mathbf{r}(t)$, its position at time t.

The inertial guidance system on an airplane uses antiderivatives of vector–valued functions to determine the location of the airplane. The inertial guidance system starts with the initial location and velocity of the airplane and then uses lasers to measure the acceleration of the airplane in each of the x, y, and z directions several times per second. From this acceleration (change in velocity) data, the computer in the system calculates the new velocity in each direction several times per second and then uses the velocities (changes in positions) to calculate the new position of the airplane relative to the starting position.

Example 5: Fig. 11 shows the initial acceleration, velocity and position of an object along the x–axis as well as its acceleration at 1 second time intervals. Fill in the empty spaces in the table and determine the position of the object on the x–axis after 9 seconds.

time (sec)	acceleration (ft/sec)	velocity (ft/sec)	position (ft)
0	0	2	5
1	4		
2	6		
3	4		
4	2		
5	8		
6	6		
7	0		
8	0		
9	0		

Fig. 11

Solution: Acceleration is the $\frac{\text{change in velocity}}{\text{change in time}} = \frac{\text{change in velocity}}{\text{1 second}}$ so each entry in the velocity column is the previous velocity plus the change in velocity (acceleration):

at t = 1, velocity = (previous velocity) + (change in velocity) = 2 + 4 = 6

at t = 2, velocity = (previous velocity) + (change in velocity) = 6 + 6 = 12

at t = 3, velocity = (previous velocity) + (change in velocity) = 12 + 4 = 16.

The rest of the entries in the velocity column are calculated in the same way and the velocity values are shown in Fig. 12.

time (sec)	acceleration (ft/sec)	velocity (ft/sec)	position (ft)
0	0	2	5
1	4	6	11
2	6	12	23
3	4	16	39
4	2	18	57
5	8	26	83
6	6	32	115
7	0	32	147
8	0	32	179
9	0	32	211

Fig. 12

Velocity is

$$\frac{\text{change in position}}{\text{change in time}} = \frac{\text{change in position}}{\text{1 second}}$$

so each entry in the position column is the previous position value plus the change in position (velocity):

at t = 1, position = (previous position) + (change in position) = 5 + 6 = 11

at t = 2, position = (previous position) + (change in position) = 11 + 12 = 23

at t = 3, position = (previous position) + (change in position) = 23 + 16 = 39.

The rest of the entries in the position column are calculated in the same way.

Practice 4: Fig. 13 shows the initial acceleration, velocity and position of an object along the y–axis as well as its acceleration at 1 second time intervals. Fill in the empty spaces in the table and determine the position of the object on the y–axis after 9 seconds?

time (sec)	acceleration (ft/sec)	velocity (ft/sec)	position (ft)
0	0	2	5
1	4		
2	6		
3	8		
4	–3		
5	0		
6	5		
7	–2		
8	1		
9	0		

Fig. 13

PROBLEMS

In problems 1 – 4, fill in each component of $\mathbf{r}'$ with " + ", " – ", or " 0 ."

1. For $\mathbf{r}(t)$ in Fig. 14, $\mathbf{r}'(1) = \langle \ , \ , \ \rangle$, $\mathbf{r}'(2) = \langle \ , \ , \ \rangle$, and $\mathbf{r}'(3) = \langle \ , \ , \ \rangle$.

2. For $\mathbf{r}(t)$ in Fig. 14, $\mathbf{r}'(4) = \langle \ , \ , \ \rangle$, $\mathbf{r}'(5) = \langle \ , \ , \ \rangle$, and $\mathbf{r}'(6) = \langle \ , \ , \ \rangle$.

3. For $\mathbf{r}(t)$ in Fig. 15, $\mathbf{r}'(1) = \langle \ , \ , \ \rangle$, $\mathbf{r}'(2) = \langle \ , \ , \ \rangle$, and $\mathbf{r}'(3) = \langle \ , \ , \ \rangle$.

4. For $\mathbf{r}(t)$ in Fig. 15, $\mathbf{r}'(4) = \langle \ , \ , \ \rangle$, $\mathbf{r}'(5) = \langle \ , \ , \ \rangle$, and $\mathbf{r}'(6) = \langle \ , \ , \ \rangle$.

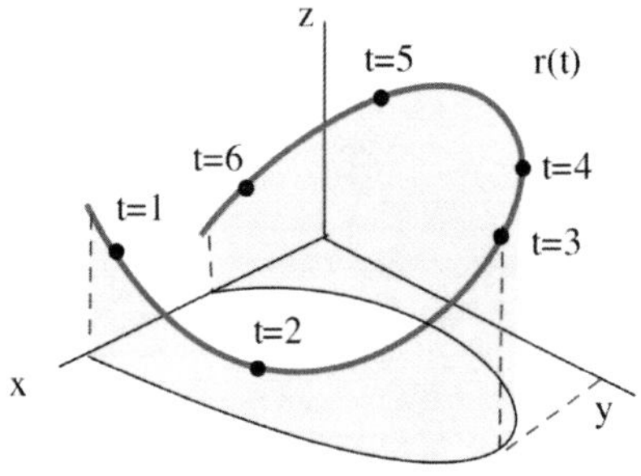

Fig. 14

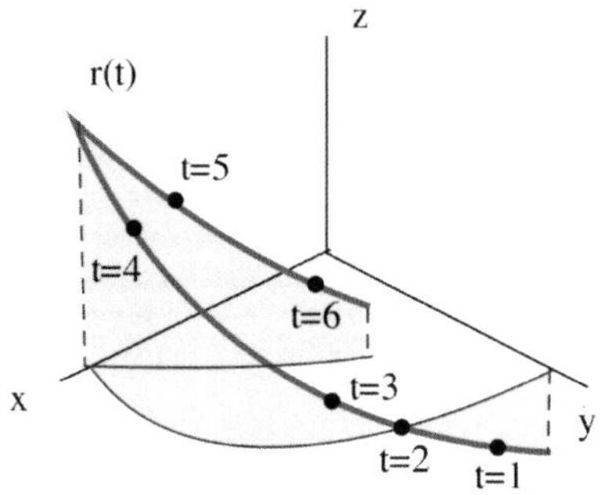

Fig. 15

In problems 5 – 8, the position vector $\mathbf{r}(t)$ is given for an object at time t. Calculate the velocity, speed, direction, and acceleration of the object at the given times.

5. $\mathbf{r}(t) = \langle t^3, 3 + 2t, t^2 \rangle$ and t = 1 and 2.

6. $\mathbf{r}(t) = \langle 5 + 3t^2, \sqrt{t}, t - t^3 \rangle$ and t = 1 and 2.

7. $\mathbf{r}(t) = (2 - t)\mathbf{i} + (4/t)\mathbf{j} + (3)\mathbf{k}$ and t = 1 and 2.

8. $\mathbf{r}(t) = (2 - t^3)\mathbf{i} + (5t)\mathbf{j} + (3 + t)\mathbf{k}$ and t = 1 and 2.

9. $\mathbf{r}(t) = \langle t^3, 7, 1 + 5t \rangle$. Calculate $\frac{d}{dt}\mathbf{r}(2t)$.

10. $\mathbf{r}(t) = \langle 1/t, 6 + 5t, t^3 \rangle$. Calculate $\frac{d}{dt}\mathbf{r}(t^2)$.

11. $\mathbf{r}(t) = \langle t, 2t^2, 3t^3 \rangle$. Calculate $\frac{d}{dt}\{\sin(t)\,\mathbf{r}(t)\}$.

12. $\mathbf{r}(t) = \langle 7 - t^2, 4, t^3 - t \rangle$. Calculate $\frac{d}{dt}\{t^3\,\mathbf{r}(t)\}$.

13. $\mathbf{r}(t) = (2 - 5t^3)\mathbf{i} + (7t)\mathbf{j} + (1 + t)\mathbf{k}$. Calculate $\frac{d}{dt}\mathbf{r}(3t)$.

14. $\mathbf{r}(t) = (1 - t^2)\mathbf{i} + (5t^3)\mathbf{j} + (3 + 2t)\mathbf{k}$. Calculate $\frac{d}{dt}\mathbf{r}(t^3)$.

In problems 15 – 18, determine $\frac{d}{dt}\{\mathbf{u} + 2\mathbf{v}\}$, $\frac{d}{dt}\{\mathbf{u}\cdot\mathbf{v}\}$, and $\frac{d}{dt}\{\mathbf{u} \times \mathbf{v}\}$ for the given vectors $\mathbf{u}(t)$ and $\mathbf{v}(t)$.

15. $\mathbf{u}(t) = \langle 0, t, t^3 \rangle$ and $\mathbf{v}(t) = \langle 1 + 5t, 4 - t, 3 \rangle$.

16. $\mathbf{u}(t) = \langle 4t, 1, 5 - t \rangle$ and $\mathbf{v}(t) = \langle t^2, 2 + 3t, t \rangle$

17. $\mathbf{u}(t) = (5t^3)\mathbf{i} + (2 - 7t)\mathbf{j} + (t + 2)\mathbf{k}$ and $\mathbf{v}(t) = (1 - 2t)\mathbf{i} + (3t)\mathbf{j} + (4)\mathbf{k}$

18. $\mathbf{u}(t) = (2t)\mathbf{i} + (4)\mathbf{k}$ and $\mathbf{v}(t) = (1)\mathbf{i} + (2t)\mathbf{j} + (3t^2)\mathbf{k}$

In problems 19 – 22, find the point and angle of intersection for the given curves.

19. $\mathbf{u}(t) = \langle 3 - t , t , t^2 \rangle$ and $\mathbf{v}(t) = \langle 0 , t , 9 \rangle$

20. $\mathbf{u}(t) = \langle 4 - t^2 , t , t^2 \rangle$ and $\mathbf{v}(t) = \langle 3 , t^2 , \sqrt{t} \rangle$

21. $\mathbf{u}(t) = (5t^2)\mathbf{i} + (9)\mathbf{j} + (2 - t)\mathbf{k}$ and $\mathbf{v}(s) = (2 + s)\mathbf{i} + (3s)\mathbf{j} + (6 - s)\mathbf{k}$

22. $\mathbf{u}(t) = (2 + t)\mathbf{i} + (7 - t)\mathbf{j} + (t + 4)\mathbf{k}$ and $\mathbf{v}(s) = (3s)\mathbf{i} + (s + 1)\mathbf{j} + (s^3)\mathbf{k}$

23. The vectors $\mathbf{u}(t) = \langle 0 , t , t^2 \rangle$ and $\mathbf{v}(t) = \langle t , 2t , 0 \rangle$ form two sides of a parallelogram. How fast is the area of the parallelogram changing when $t = 1$. When $t = 2$?

24. The vectors $\mathbf{r}(t) = \langle 2t , 1 , 0 \rangle$ and $\mathbf{s}(t) = \langle 1 , 0 , 3 \rangle$ form two sides of a parallelogram. How fast is the area of the parallelogram changing when $t = 1$. When $t = 2$?

25. The vectors $\mathbf{u}(t) = \langle 1 , t , 3 \rangle$ and $\mathbf{v}(t) = \langle 2t , 0 , 0 \rangle$ form two sides of a triangle. How fast is the area of the triangle changing when $t = 1$. When $t = 2$?

26. The vectors $\mathbf{r}(t) = \langle t^2 , t , 1 \rangle$ and $\mathbf{s}(t) = \langle t , t^2 , 0 \rangle$ form two sides of a triangle. How fast is the area of the triangle changing when $t = 1$. When $t = 2$?

27. The vectors $\mathbf{u}(t) = \langle 1 , 0 , 0 \rangle$, $\mathbf{v}(t) = \langle 1 , t , 0 \rangle$ and $\mathbf{s}(t) = \langle 0 , t , 3t \rangle$ form three sides of a tetrahedron. (a) How fast is the volume of the tetrahedron changing when $t = 1$. When $t = 2$?
(b) How fast is the surface area of the tetrahedron changing when $t = 1$. When $t = 2$?

28. The vectors $\mathbf{u}(t) = \langle 2t , 0 , 0 \rangle$, $\mathbf{v}(t) = \langle 0 , 3t , 0 \rangle$ and $\mathbf{s}(t) = \langle 0 , 0 , 4t \rangle$ form three sides of a tetrahedron. (a) How fast is the volume of the tetrahedron changing when $t = 1$. When $t = 2$?
(b) How fast is the surface area of the tetrahedron changing when $t = 1$. When $t = 2$?

In problems 29 – 32, use the given information to find a formula for $\mathbf{r}(t)$.

29. $\mathbf{r}'(t) = \langle 12t , 12t^2 , 6e^t \rangle$ and $\mathbf{r}(0) = \langle 1 , 2 , 3 \rangle$.

30. $\mathbf{r}'(t) = \langle 3 + 4t , \cos(t) , 1 - 6t \rangle$ and $\mathbf{r}(0) = \langle 7 , 2 , 5 \rangle$.

31. $\mathbf{r}'(t) = (6t^2)\mathbf{i} + (4)\mathbf{j} + (8t - 5)\mathbf{k}$ and $\mathbf{r}(1) = 6\mathbf{i} + 2\mathbf{j} - 3\mathbf{k}$.

32. $\mathbf{r}'(t) = (5t^2)\mathbf{i} + (8t)\mathbf{j} + (2 - t)\mathbf{k}$ and $\mathbf{r}(2) = (3)\mathbf{i} + (7)\mathbf{j} + (0)\mathbf{k}$.

33. Fill in the rest of the **i** coordinate entries for **r** and **r** ' in the table in Fig. 16.

34. Fill in the rest of the **j** coordinate entries for **r** and **r** ' in the table in Fig. 16.

35. Fill in the rest of the **i** coordinate entries for **r** and **r** ' in the table in Fig. 17.

36. Fill in the rest of the **j** coordinate entries for **r** and **r** ' in the table in Fig. 17.

t	**r** '' (t)	**r** ' (t)	**r**(t)
0	⟨ 0, 2, 5 ⟩	⟨ 1 , 2 , 3 ⟩	⟨ 0 , 3 , 1 ⟩
1	⟨ 4, 1, 3 ⟩	⟨ , , 6 ⟩	⟨ , , 7 ⟩
2	⟨ 6, 0, 1 ⟩	⟨ , , 7 ⟩	⟨ , , 14 ⟩
3	⟨ 4, –2, 0 ⟩	⟨ , , 7 ⟩	⟨ , , 21 ⟩
4	⟨ 2, 0, 2 ⟩	⟨ , , 9 ⟩	⟨ , , 30 ⟩
5	⟨ 8, 3, 4 ⟩	⟨ , , 13 ⟩	⟨ , , 43 ⟩

Fig. 16

t	**r** '' (t)	**r** ' (t)	**r**(t)
0	⟨ 1, 2, 3 ⟩	⟨ , , 4 ⟩	⟨ , , 2 ⟩
1	⟨ 4, 2, 2 ⟩	⟨ , , 6 ⟩	⟨ , , 8 ⟩
2	⟨ 3, 1, 0 ⟩	⟨ 8 , 9 , 6 ⟩	⟨ 30, 20 , 14 ⟩
3	⟨ 2, 3, 1 ⟩	⟨ , , 7 ⟩	⟨ , , 21 ⟩
4	⟨ 1, 4, 0 ⟩	⟨ , , 7 ⟩	⟨ , , 28 ⟩
5	⟨ 0, 1, 3 ⟩	⟨ , , 10 ⟩	⟨ , , 38 ⟩

Fig. 17

37. State and prove a differentiation rule for $\frac{d}{dt}\left(\frac{\mathbf{u}(t)}{f(t)}\right)$.

38. Prove that $\frac{d}{dt}(\,\mathbf{u}(t) \times \mathbf{v}(t)\,) = -\frac{d}{dt}(\,\mathbf{v}(t) \times \mathbf{u}(t)\,)$.

Practice Answers

Practice 1: $\mathbf{r}(t) = \langle t, t^2, t^3 \rangle$. $\mathbf{v}(t) = \text{velocity} = \mathbf{r}\,'(t) = \langle 1, 2t, 3t^2 \rangle$ so $\mathbf{v}(0) = \langle 1, 0, 0 \rangle$, $\mathbf{v}(1) = \langle 1, 2, 3 \rangle$, $\mathbf{v}(2) = \langle 1, 4, 12 \rangle$.

$\mathbf{sp}(t) = \text{speed} = |\,\mathbf{v}(t)\,|$ so $\mathbf{sp}(0) = 1$, $\mathbf{sp}(1) = \sqrt{14}$, $\mathbf{sp}(2) = \sqrt{161}$.

$\mathbf{dir}(t) = \text{direction} = \frac{\mathbf{v}(t)}{|\,\mathbf{v}(t)\,|}$. $\mathbf{dir}(0) = \langle 1, 0, 0 \rangle$, $\mathbf{dir}(1) = \frac{1}{\sqrt{14}} \langle 1, 2, 3 \rangle$, $\mathbf{dir}(2) = \frac{1}{\sqrt{161}} \langle 1, 4, 12 \rangle$.

$\mathbf{a}(t) = \text{acceleration} = \mathbf{r}\,''(t) = \langle 0, 2, 6t \rangle$ so $\mathbf{a}(0) = \langle 0, 2, 0 \rangle$, $\mathbf{a}(1) = \langle 0, 2, 6 \rangle$, $\mathbf{a}(2) = \langle 0, 2, 12 \rangle$.

Practice 2: The paths $\mathbf{r}(t) = \langle 0, t, t^2 \rangle$ and $\mathbf{s}(t) = \langle 2 - t, 1, 5 - t^2 \rangle$ intersect at $\mathbf{r}(1) = \mathbf{s}(2) = (0,1,1)$.

$\mathbf{r}\,'(t) = \langle 0, 1, 2t \rangle$ and $\mathbf{s}\,'(t) = \langle -1, 0, -2t \rangle$ so $\mathbf{r}\,'(1) = \langle 0, 1, 2 \rangle$ and $\mathbf{s}\,'(2) = \langle -1, 0, -4 \rangle$.

$\cos(\theta) = \frac{\mathbf{r}\,'(1) \cdot \mathbf{s}\,'(2)}{|\mathbf{r}\,'(1)|\,|\mathbf{s}\,'(2)|} = \frac{-8}{\sqrt{5}\sqrt{17}} = -0.868$ so the angle between $\mathbf{r}\,'(1)$ and $\mathbf{s}\,'(1)$ is

$\theta \approx 2.922$ (or 150.2°)

Practice 3: $\mathbf{r}\,'(t) = \mathbf{v}(t) = \langle 6t^2, \cos(t), 12e^{3t} \rangle$ so $\mathbf{r}(t) = \langle 2t^3 + A, \sin(t) + B, 4e^{3t} + C \rangle$.

Since $\mathbf{r}(0) = \langle 1, -5, 2 \rangle = \langle 2(0)^3 + A, \sin(0) + B, 4e^{3(0)} + C \rangle = \langle A, B, 4 + C \rangle$, we have $A = 1, B = -5$, and $C = -2$. Then $\mathbf{r}(t) = \langle 2t^3 + 1, \sin(t) - 5, 4e^{3t} - 2 \rangle$.

Practice 4: Velocity entries: 2, 6, 12, 20, 17, 17, 22, 20, 21, 21

Acceleration entries: 5, 11, 23, 43, 60, 77, 99, 119, 140, 161

12.3 ARC LENGTH AND CURVATURE OF SPACE CURVES

In earlier sections we have emphasized the dynamic nature of vector–valued functions by considering them as the path of a moving object. This is a very fruitful approach, but sometimes it is useful to consider a space curve as a static object and to investigate some of its geometric properties. This section considers two geometric aspects of space curves: arc length (how long is it along the curve from one point to another point?) and curvature (how quickly does the curve bend?).

Arc Length

In Section 9.4 we went through a careful derivation of an integral formula for finding the length of a parametric curve $(x(t), y(t))$ from $t = a$ to $t = b$ (Fig. 1): we

(1) partitioned the interval $[a, b]$ for the variable t and found the points $(x(t_i), y(t_i))$,

(2) found the lengths of the line segments between consecutive points along the curve

(3) added the lengths of the line segments (an approximation of the length of the curve) and got a Riemann sum

(4) took the limit of the Riemann sum to an integral formula for the length of the curve

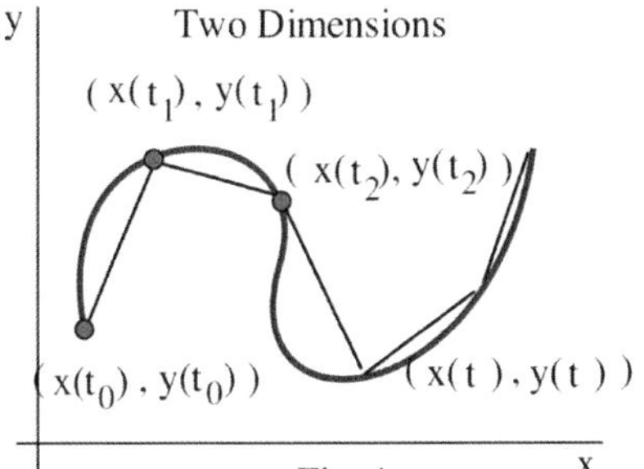

Fig. 1

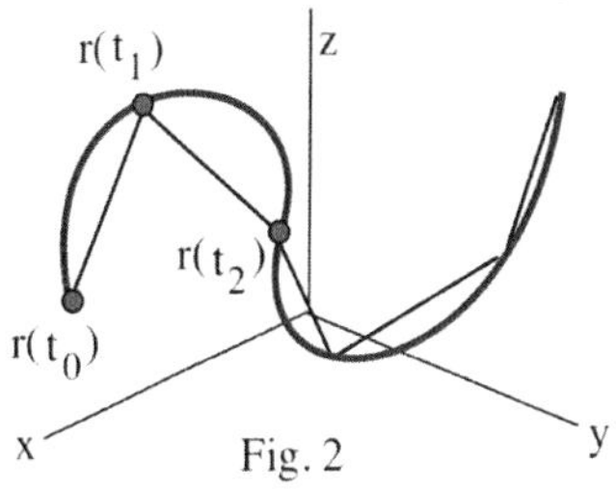

Fig. 2

A very similar process also works for finding the length of a curve given by a vector–valued function in three dimensions, a space curve (Fig. 2), and we define the result of that process to be the length of a space curve.

Definition: **Distance Along the Graph of a Vector–Valued Function (Arc Length of a Space Curve)**

If $\mathbf{r}(t) = \langle x(t), y(t), z(t) \rangle$ and $x'(t), y'(t), z'(t)$ are continuous

then the **distance traveled, L, along the graph** of $\mathbf{r}(t)$ from $t = a$ to $t = b$ is

$$\textbf{distance traveled L} = \int_{t=a}^{t=b} \sqrt{ \left(\frac{dx}{dt}\right)^2 + \left(\frac{dy}{dt}\right)^2 + \left(\frac{dz}{dt}\right)^2 } \, dt$$

If we travel along each part of the curve $\mathbf{r}(t)$ exactly once, then the arc length of the curve is the distance traveled:

$$\textbf{arc length = distance traveled L} = \int_{t=a}^{t=b} |\mathbf{r}'(t)| \, dt = \int_{t=a}^{t=b} |\mathbf{v}(t)| \, dt .$$

Example 1: Represent the length of the helix $\mathbf{r}(t) = \langle \cos(t), \sin(t), t \rangle$ from $t = 0$ to $t = 2\pi$. (Fig. 3)

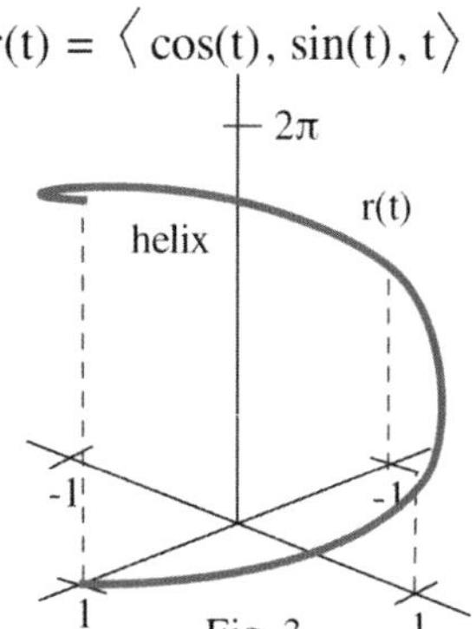

Fig. 3

Solution: $$L = \int_{t=0}^{t=2\pi} \sqrt{(-\sin(t))^2 + (\cos(t))^2 + (1)^2}\ dt$$

$$= \int_{t=0}^{t=2\pi} \sqrt{\sin^2(t) + \cos^2(t) + 1}\ dt = \int_{t=0}^{t=2\pi} \sqrt{2}\ dt = 2\pi\sqrt{2} \approx 8.89 .$$

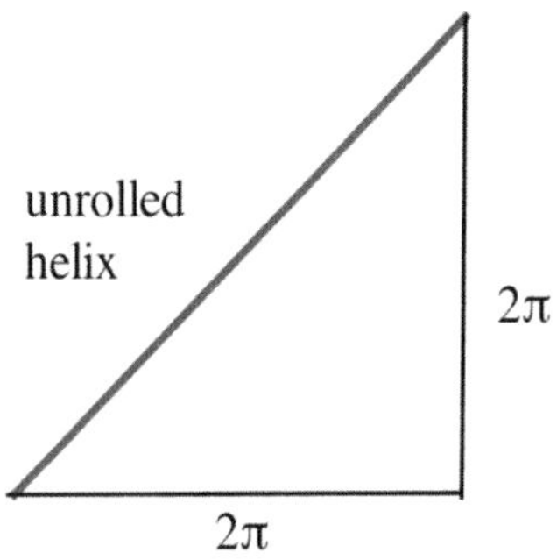

Fig. 4

(Actually, we can do this particular problem without calculus. If we "unroll" the helix (Fig. 4) we get a right triangle with base 2π (the circumference of the circle with radius 1) and height 2π (the value of $z(t)$ when $t = 2\pi$). The length of the helix is the length of the hypotenuse of this triangle: hypotenuse = $\sqrt{(2\pi)^2 + (2\pi)^2} = 2\pi\sqrt{2}$.)

Practice 1: Represent the length of the graph of $\mathbf{r}(t) = \langle t, t^2, t^3 \rangle$ from $t = 0$ to $t = 2$ as an integral and use Simpson's Rule with $n = 20$ to approximate the value of the integral.

Example 2: Represent the length of the graph of $\mathbf{r}(t) = \langle \cos(t), \cos^2(t), 0 \rangle$ from $t = 0$ to $t = 2\pi$ as an integral and use numerical integration on your calculator to approximate the value of the integral.

Solution: The graph of $\mathbf{r}(t)$ is part of a parabola (Fig. 5), and the distance traveled along the parabola is

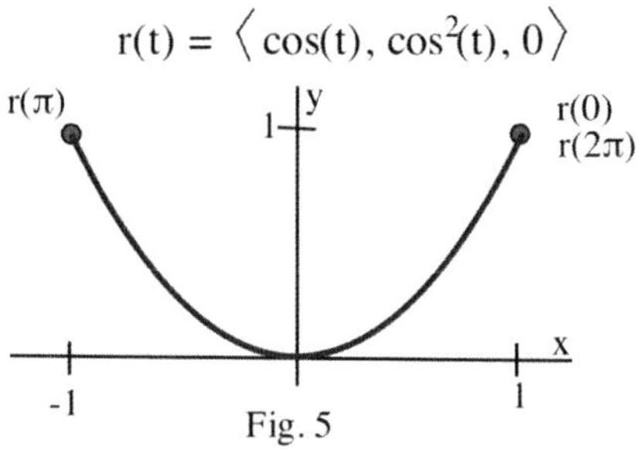

Fig. 5

$$\text{distance} = \int_{t=0}^{t=2\pi} \sqrt{(-\sin(t))^2+(-2\cos(t)\sin(t))^2+(0)^2}\ dt$$

$$= \int_{t=0}^{t=2\pi} \sqrt{\sin^2(t) + 4\cos^2(t)\sin^2(t)}\ dt \approx 5.916 \text{ (using numerical integration on a calculator).}$$

But in this example, the distance is NOT the length of the graph. As t goes from 0 to π we travel along the parabola from the point $(1, 1, 0)$ to the origin and on to $(-1, 1, 0)$. As t goes from π to 2π we travel back along the parabola to the starting point $(1, 1, 0)$. As t goes from 0 to 2π we cover the parabola twice so the length of the parabola is half of the distance traveled:

$$\text{length} = \frac{1}{2}(\text{ distance travelled }) \approx \frac{1}{2}(5.916) = 2.958 .$$

We could have calculated the length of the curve as the value of integral from $t = 0$ to $t = \pi$.

Parameterizing a Curve with respect to Arc Length

So far in our dealings with parametric space curves and vector–valued functions we have treated the curves of functions of the variable t and often thought about t as representing time. We have referred to the position vector $\mathbf{r}(t) = \langle x(t), y(t), z(t) \rangle$ as representing the position (x(t), y(t), z(t)) of an object at time t. With a space curve, however, it is sometimes more useful to represent a location on the curve as a function of "how far along the curve" we are. For example, if we are giving someone directions to a good picnic spot in the mountains, we might describe the location (Fig. 6) as "drive 5.3 miles along the road from the turnoff" indicating that the driver should travel 5.3 miles from the beginning of the mountain road. This "how far along the road or curve" method avoids the obvious drawbacks of directions such as "drive 9 minutes at 35 miles per hour." Similarly, interstate highways are often marked with signs indicating how far we are from the the beginning of the road or the point where the road entered our state. It is usually more useful to describe the location of a knot in a wire as "17 inches from the end of the wire." That description does not depend on how fast we move along the wire or the orientation of the wire in space or even on the shape of the curve.

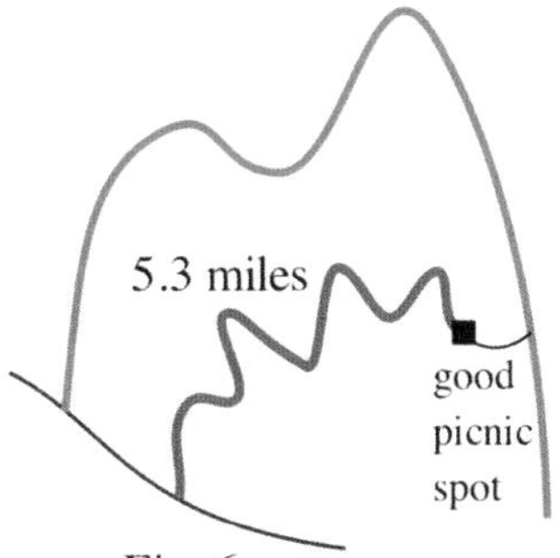

Fig. 6

The benefits of giving directions in terms of "how far along a road or wire" are the same for describing a location on a curve as "how far along the curve." The description of locations along a curve in terms of distance along the curve is called a **parameterization of the curve in terms of arc length**.

Definition: **Arc Length Function s(t)**

For a curve that begins at $\mathbf{r}(a) = \langle x(a), y(a), z(a) \rangle$ with continuous x ', y ' and z ', the distance along the curve $\mathbf{r}(t) = \langle x(t), y(t), z(t) \rangle$ at time t is the arc length function s(t) with

$$s(t) = \int_{u=a}^{u=t} \sqrt{\left(\frac{dx(u)}{du}\right)^2 + \left(\frac{dy(u)}{du}\right)^2 + \left(\frac{dz(u)}{du}\right)^2}\, du = \int_{u=a}^{u=t} |\mathbf{r}'(u)|\, du .$$

Example 3: Fig. 7 illustrates the path $\mathbf{r}$ of a salmon swimming up a river marked with dots at 1 mile intervals.

(a) Label the location of $\mathbf{r}(4)$ with an "X" for $\mathbf{r}$ parameterized in terms of arc length.

(b) Label the location of $\mathbf{r}(4)$ with an "O" for $\mathbf{r}$ parameterized in terms of time.

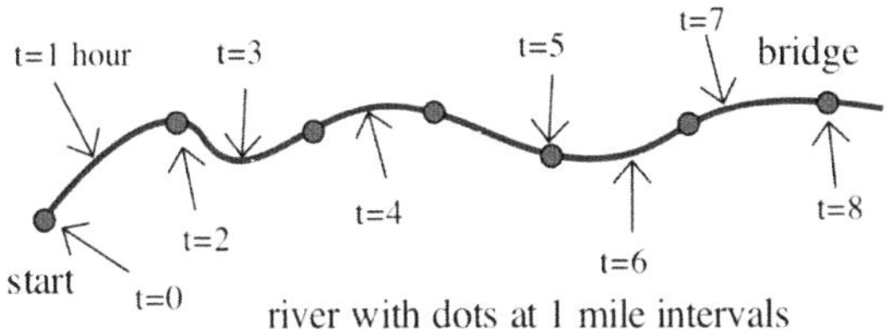

river with dots at 1 mile intervals
Fig. 7

(c) For an arc length parameterization, find A so $\mathbf{r}(A)$ = bridge.

(d) During which time interval was the fish swimming the fastest?

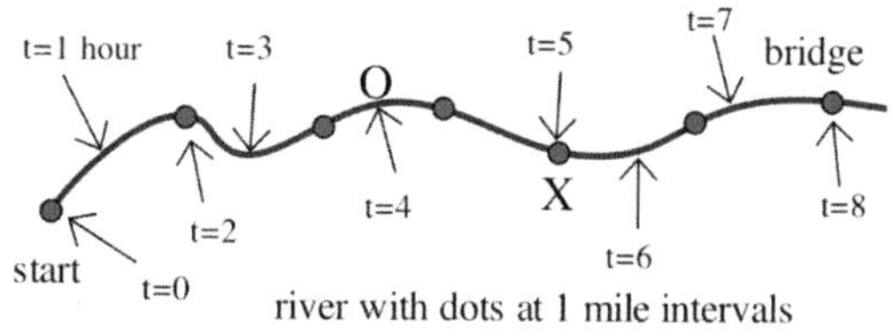

Fig. 8

Solution: (a) and (b) Fig. 8 shows the correct locations of the "X" and the "O."

(c) The bridge is 6 miles from the beginning of the river so $A = 6$.

(d) The fish swims the greatest distance between $t = 4$ and $t = 5$, so it was swimming fastest during that 1 hour time interval.

Practice 2: For the salmon in Fig. 7

(a) label the location of $\mathbf{r}(3)$ with an "S" for $\mathbf{r}$ parameterized in terms of arc length.

(b) Label the location of $\mathbf{r}(3)$ with an "T" for $\mathbf{r}$ parameterized in terms of time.

(c) For a time parameterization, find B so $\mathbf{r}(B)$ = bridge.

(d) During which time interval was the fish swimming the slowest?

For most curves, it is difficult to find a simple formula for the arc length function $s(t)$. But sometimes we do get such a nice result for $s(t)$ that we can solve for $t(s)$, t in terms of s, and then we can rewrite the original parameterization $\mathbf{r}(t) = \langle x(t), y(t), z(t) \rangle$ as $\mathbf{r}(\,t(s)\,) = \langle x(\,t(s)\,), y(\,t(s)\,), z(\,t(s)\,) \rangle$.

Example 4: Write an arc length parameterization of the helix $\mathbf{r}(t) = \langle 3t, 4\cos(t), 4\sin(t) \rangle$ using $\mathbf{r}(0) = \langle 0, 0, 0 \rangle$ as the starting point.

Solution: $\mathbf{r}'(t) = \langle 3, -4\sin(t), 4\cos(t) \rangle$ for all t so

$$|\,\mathbf{r}'(t)\,| = \sqrt{(3)^2 + (-4\sin(t))^2 + (4\cos(t))^2} = \sqrt{9 + 16\sin^2(t) + 16\cos^2(t)} = 5 \text{ for all t.}$$

Then $s = s(t) = \int_{u=0}^{u=t} |\,\mathbf{r}'(u)\,|\,du = \int_{u=0}^{u=t} 5\,du = 5t$ so $t = \frac{s}{5}$. By substituting $t(s) = \frac{s}{5}$ for t,

the original parameterization $\mathbf{r}(t) = \langle 3t, 4\cos(t), 4\sin(t) \rangle$ becomes

$$\mathbf{r}(\,t(s))\,) = \langle 3t(s), 4\cos(\,t(s)\,), 4\sin(\,t(s)\,) \rangle = \langle 3\tfrac{s}{5}, 4\cos(\tfrac{s}{5}), 4\sin(\tfrac{s}{5}) \rangle, \text{ a function of s alone.}$$

Practice 3: Write an arc length parameterization of the line $\mathbf{r}(t) = \langle 8t, t, 4t \rangle$ using $\mathbf{r}(0) = \langle 0, 0, 0 \rangle$ as the starting point.

The conversions from "time parameterization" to "arc length parameterization" in the example and practice problems were relatively easy because the object was moving along each curve at a constant speed ($|\mathbf{r}'(t)|$ was constant). Usually this conversion is not that easy, but most of the time the arc length parameterization for a curve will be given so we will not need to translate to get from a "time parameterization" to an "arc length parameterization."

Curvature

Fig. 9 shows a space curve $\mathbf{r}(t)$ and unit tangent vectors $\mathbf{T}(t) = \frac{\mathbf{r}'(t)}{|\mathbf{r}'(t)|}$ at several equally spaced (in terms of arc length) points along $\mathbf{r}(t)$. When the curve twists and bends sharply in the left part of the graph, the unit tangent vectors change direction rapidly from point to point. When the curve is almost straight and bends slowly, the unit tangent vectors also change direction slowly. This geometric pattern between the "bendedness" of a curve and the rate of change (with respect to arc length) of the direction of the unit tangent vectors leads to our definition of the **curvature** of a space curve at a point.

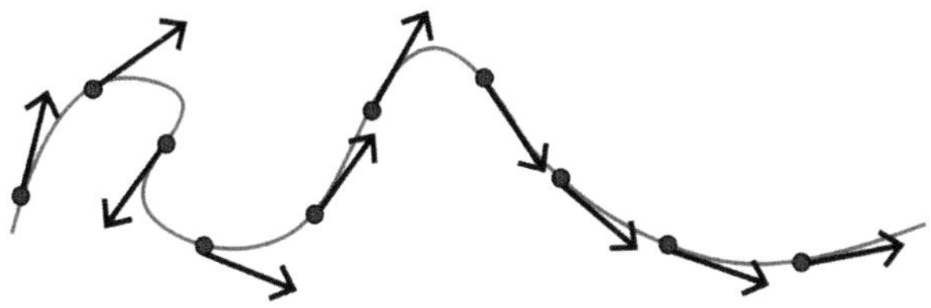

curve and unit tangent vectors

Fig. 9

Definition: Curvature

If $\mathbf{r}$ is a space curve with unit tangent vector $\mathbf{T}$ and arc length parameterization $\mathbf{s}$,

then the **curvature** of $\mathbf{r}$ is

$$\kappa = \left| \frac{d\mathbf{T}}{ds} \right| . \quad (\kappa \text{ is the Greek letter "kappa"})$$

The curvature of a space curve is defined to be the magnitude of the rate of change of direction of the unit tangent vectors with respect to arc length. This definition of curvature is nicely motivated geometrically, but it is difficult to use for computations if we do not have an arc length parameterization of $\mathbf{r}$. However, the Chain Rule and the Fundamental Theorem of Calculus provide us with an easier way to actually calculate the curvature of a space curve $\mathbf{r}(t)$.

By the Chain Rule, $\frac{d\mathbf{T}}{dt} = \frac{d\mathbf{T}}{ds} \cdot \frac{ds}{dt}$ so $\left| \frac{d\mathbf{T}}{ds} \right| = \left| \frac{d\mathbf{T}/dt}{ds/dt} \right| = \left| \frac{\mathbf{T}'(t)}{ds/dt} \right|$. From the Fundamental Theorem of Calculus and the definition of $s(t) = \int_{u=a}^{u=t} |\mathbf{r}'(u)|\, du$, we know that $\frac{d\,s(t)}{dt} = |\mathbf{r}'(t)|$.

By putting these two results together, we get a much easier to use formula for the curvature of a space curve.

A Formula for Curvature: $\kappa = \dfrac{|\mathbf{T}'(t)|}{|\mathbf{r}'(t)|}$

Example 5: For a positive number A, the graph of $\mathbf{r}(t) = \langle A\cdot\cos(t), A\cdot\sin(t), 0 \rangle$ is a circle with radius A in the xy–plane. Find the curvature of this circle.

Solution: $\mathbf{r}'(t) = \langle -A\cdot\sin(t), A\cdot\cos(t), 0 \rangle$ so $|\mathbf{r}'(t)| = \sqrt{A^2\sin^2(t) + A^2\cos^2(t) + 0} = A$.

$$\mathbf{T}(t) = \frac{\mathbf{r}'(t)}{|\mathbf{r}'(t)|} = \frac{\mathbf{r}'(t)}{A} = \langle -\sin(t), \cos(t), 0 \rangle \text{ so } \mathbf{T}'(t) = \langle -\cos(t), -\sin(t), 0 \rangle \text{ and } |\mathbf{T}'(t)| = 1.$$

Then, for all t, $\kappa = \left| \dfrac{\mathbf{T}'(t)}{\mathbf{r}'(t)} \right| = \dfrac{1}{A}$. The curvature of a circle of radius A is $\kappa = \dfrac{1}{A}$.

This agrees with our geometric idea of curvature:

- when the radius of the circle is large (Fig. 10a), the circle bends slowly and the curvature κ is small
- when the radius of the circle is small (Fig. 10b), the circle bends quickly and the curvature κ is large.

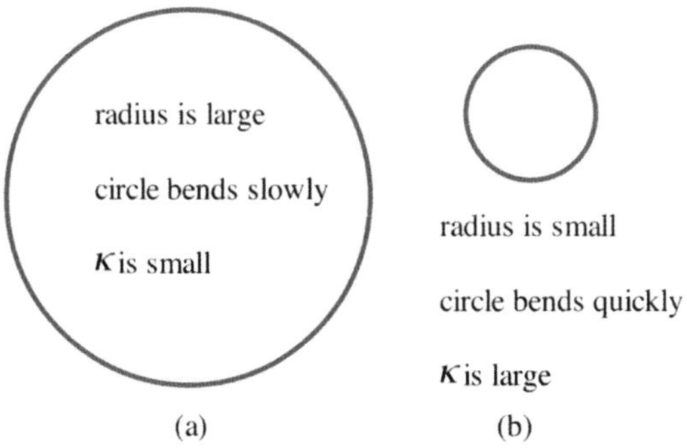

Fig. 10

This pattern for curvature of circles leads to the definition of the radius of curvature of a curve.

Definition: The **radius of curvature** of $\mathbf{r}(t)$ is $\dfrac{1}{\text{curvature of } \mathbf{r}(t)} = \dfrac{1}{\kappa}$.

Practice 4: For A, B, and C not equal to 0, show that the **line** $\mathbf{r}(t) = \langle At, Bt, Ct \rangle$ has curvature $\kappa = 0$.

It was relatively straightforward to calculate the curvature in the example and practice problem because $|\mathbf{r}'(t)|$ was a constant. When $|\mathbf{r}'(t)|$ is not constant, it can be difficult to calculate $\mathbf{T}'(t)$, and some other formulas for curvature are often easier. The following formula for curvature looks complicated, but in practice it is often the easiest one to use.

"Easiest" Formula for Curvature in 3D: $\kappa = \dfrac{|\mathbf{r}' \times \mathbf{r}''|}{|\mathbf{r}'|^3}$

A proof that this formula follows from the formula $\kappa = \dfrac{|\mathbf{T}'(t)|}{|\mathbf{r}'(t)|}$ is given in an appendix after the problem set.

Example 6: Use the formula $\kappa = \dfrac{|\mathbf{r}' \times \mathbf{r}''|}{|\mathbf{r}'|^3}$ to determine the curvature and the radius of curvature of $\mathbf{r}(t) = \langle t, t^2, t^3 \rangle$ when $t = 0, 1,$ and 2.

Solution: $\mathbf{r}'(t) = \langle 1, 2t, 3t^2 \rangle$ so $|\mathbf{r}'(t)| = \sqrt{1 + 4t^2 + 9t^4}$ and $\mathbf{r}''(t) = \langle 0, 2, 6t \rangle$.

Then $\mathbf{r}' \times \mathbf{r}'' = \begin{vmatrix} \mathbf{i} & \mathbf{j} & \mathbf{k} \\ 1 & 2t & 3t^2 \\ 0 & 2 & 6t \end{vmatrix} = \mathbf{i}\begin{vmatrix} 2t & 3t^2 \\ 2 & 6t \end{vmatrix} - \mathbf{j}\begin{vmatrix} 1 & 3t^2 \\ 0 & 6t \end{vmatrix} + \mathbf{k}\begin{vmatrix} 1 & 2t \\ 0 & 2 \end{vmatrix}$

$= (6t^2)\mathbf{i} - (6t)\mathbf{j} + (2)\mathbf{k}$ and

$|\mathbf{r}' \times \mathbf{r}''| = \sqrt{(6t^2)^2 + (-6t)^2 + (2)^2} = \sqrt{36t^4 + 36t^2 + 4}$.

Putting this all together, we have $\kappa = \dfrac{\sqrt{36t^4 + 36t^2 + 4}}{(1 + 4t^2 + 9t^4)^{3/2}}$.

When $t = 0$, $\kappa = \dfrac{\sqrt{4}}{1} = 2$ so the radius of curvature is $\dfrac{1}{2}$.

When $t = 1$, $\kappa = \dfrac{\sqrt{76}}{(14)^{3/2}} \approx 0.166$ so the radius of curvature is approximately $\dfrac{1}{0.166} \approx 6.02$.

When $t = 2$, $\kappa = \dfrac{\sqrt{724}}{(161)^{3/2}} \approx 0.013$ so the radius of curvature is approximately 76.9.

For $\mathbf{r}(t) = \langle t, t^2, t^3 \rangle$, as t grows larger (and is positive), the curvature κ becomes smaller.

Practice 5: Use the formula $\kappa = \dfrac{|\mathbf{r}' \times \mathbf{r}''|}{|\mathbf{r}'|^3}$ to determine the curvature of $\mathbf{r}(t) = \langle t, \sin(t), 0 \rangle$ when $t = 0, \pi/4,$ and $\pi/2$.

Curvature in Two Dimensions: $\mathbf{r}(t) = \langle x(t), y(t), 0 \rangle$ and $y = f(x)$

Every curve confined to the xy–plane can be thought of as a curve in space whose z–coordinate is always 0, and that approach leads to alternate formulas for the curvature of the graph.

If the curve we are dealing with is given parametrically in two dimensions as $(x(t), y(t))$ then we can consider it as the vector–valued function $\mathbf{r}(t) = \langle x(t), y(t), 0 \rangle$ in three dimensions, and the curvature formula $\kappa = \dfrac{|\mathbf{r}' \times \mathbf{r}''|}{|\mathbf{r}'|^3}$ "simplifies" to the following.

> The graph of $(x(t), y(t))$ has curvature $\kappa = \dfrac{| x'y'' - x''y' |}{((x')^2 + (y')^2)^{3/2}}$
>
> where the derivatives of x and y are with respect to t.

If $y = f(x)$, then the curve can be parameterized in two dimensions using $x(t) = t$ and $y(t) = f(x) = f(t)$. With this parameterization $\mathbf{r}(t) = \langle x(t), y(t), 0 \rangle = \langle t, y(t), 0 \rangle$ and we have $x'(t) = 1$ and $x''(t) = 0$ so the previous pattern reduces to

> If $y = f(x)$, then $\kappa = \dfrac{| y'' |}{(1 + (y')^2)^{3/2}}$ where the derivatives are with respect to x.

These last two formulas for curvature are typically easier to use than the previous ones, but **they are only valid for two–dimensional graphs**.

Example 7: Use the appropriate formula to determine the curvature of $y = x^2$ when $x = 0, 1$ and 2.

Solution: We can use the formula $\kappa = \dfrac{| y'' |}{(1 + (y')^2)^{3/2}}$ with $y' = 2x$ and $y'' = 2$. Then

$$\kappa = \frac{| 2 |}{(1 + (2x)^2)^{3/2}} = \frac{2}{(1 + 4x^2)^{3/2}} .$$

When $x = 0$, $\kappa = \dfrac{2}{(1 + 4x^2)^{3/2}} = \dfrac{2}{(1)^{3/2}} = 2$. When $x = 1$, $\kappa = \dfrac{2}{(1 + 4x^2)^{3/2}} = \dfrac{2}{(5)^{3/2}} \approx 0.179$.

When $x = 2$, $\kappa = \dfrac{2}{(1 + 4x^2)^{3/2}} = \dfrac{2}{(17)^{3/2}} \approx 0.029$.

Practice 6: Use the appropriate 2–dimensional formula to determine the curvature of $y = \sin(x)$ when $x = 0, 1$ and 2. (Your answers should agree with your answers to Practice 5.)

PROBLEMS

In problems 1 – 4, determine the length of the helices for $0 \le t \le 2\pi$.

1. $\mathbf{r}(t) = \langle 2\cdot\cos(t), 2\cdot\sin(t), t \rangle$
2. $\mathbf{r}(t) = \langle 3\cdot\cos(t), 3\cdot\sin(t), t \rangle$
3. $\mathbf{r}(t) = \langle 4\cdot\cos(t), 4\cdot\sin(t), t \rangle$
4. $\mathbf{r}(t) = \langle R\cdot\cos(t), R\cdot\sin(t), t \rangle$

In problems 5 – 12, determine the length of the "modified helices" for $0 \le t \le 2\pi$. If necessary, use a calculator to approximate the arc length integrals.

5. $\mathbf{r}(t) = \langle 2\cdot\cos(t), 3\cdot\sin(t), t \rangle$
6. $\mathbf{r}(t) = \langle 2\cdot\cos(t), 5\cdot\sin(t), t \rangle$
7. $\mathbf{r}(t) = \langle A\cdot\cos(t), B\cdot\sin(t), t \rangle$
8. $\mathbf{r}(t) = \langle \cos(2t), \sin(2t), t \rangle$
9. $\mathbf{r}(t) = \langle t\cdot\cos(t), t\cdot\sin(t), t \rangle$
10. $\mathbf{r}(t) = \langle 2t\cdot\cos(t), 2t\cdot\sin(t), t \rangle$
11. $\mathbf{r}(t) = \langle 2t\cdot\cos(t), t\cdot\sin(t), t \rangle$
12. $\mathbf{r}(t) = \langle t^2\cdot\cos(t), t^2\cdot\sin(t), t \rangle$

In problems 13 – 16, determine the length of the Bezier curves for $0 \le t \le 1$. If necessary, use a calculator to approximate the arc length integrals.

13. $\mathbf{r}(t) = \langle 3(1-t)^2t + 3(1-t)t^2, 9(1-t)t^2 + 2t^3, (1-t)^3 + 6(1-t)^2t \rangle$
14. $\mathbf{r}(t) = \langle 2(1-t)^3 + 6(1-t)t^2, 3(1-t)^2t + 3(1-t)t^2 + 3t^3, 3(1-t)^2t + 3(1-t)t^2 \rangle$
15. The Bezier curve determined by the control points $P_0 = (2, 0, 0)$, $P_1 = (0, 1, 1)$, $P_2 = (2, 1, 1)$, and $P_3 = (0, 3, 0)$.
16. The Bezier curve determined by the control points $P_0 = (2, 0, 0)$, $P_1 = (0, 3, 2)$, $P_2 = (2, 0, 3)$, and $P_3 = (0, 3, 0)$.

In problems 17 – 22, determine the curvature of the given curves at the specified points.

17. $\mathbf{r}(t) = \langle \cos(t), \sin(t), t \rangle$ when $t = 0, \pi/4$, and $\pi/2$.
18. $\mathbf{r}(t) = \langle 3\cdot\cos(t), 3\cdot\sin(t), t \rangle$ when $t = 0, \pi/4$, and $\pi/2$.
19. $\mathbf{r}(t) = \langle R\cdot\cos(t), R\cdot\sin(t), t \rangle$ when $t = 0, \pi/4$, and $\pi/2$.
20. $\mathbf{r}(t) = \langle 5 + 3t, 2 - t, 3 - 2t \rangle$ when $t = 0, 2$, and 7.
21. $\mathbf{r}(t) = \langle 3(1-t)^2t + 3(1-t)t^2, 9(1-t)t^2 + 2t^3, (1-t)^3 + 6(1-t)^2t \rangle$ when $t = 0.2$ and 0.5.
22. $\mathbf{r}(t) = \langle 2(1-t)^3 + 6(1-t)t^2, 3(1-t)^2t + 3(1-t)t^2 + 3t^3, 3(1-t)^2t + 3(1-t)t^2 \rangle$ when $t = 0.2$ and 0.5.

In problems 23 – 26, determine the curvature and the radius of curvature of the given curves at the specified points.

23. $\mathbf{r}(t) = \langle 3\cdot\cos(t), 5\cdot\sin(t) \rangle$ when $t = 0, \pi/4$, and $\pi/2$.

24. $\mathbf{r}(t) = \langle 2\cdot\cos(t), 7\cdot\sin(t) \rangle$ when $t = 0, \pi/4$, and $\pi/2$.

25. $\mathbf{r}(t) = \langle A\cdot\cos(t), B\cdot\sin(t) \rangle$ when $t = 0, \pi/4$, and $\pi/2$.

26. $\mathbf{r}(t) = \langle t\cdot\cos(t), t\cdot\sin(t) \rangle$ when $t = 1, 2$, and 3 .

27. Determine the curvature of $y = 3x + 5$ when $x = 1, 2$, and 3. For what value of x is the curvature of $y = 3x + 5$ maximum?

28. Determine the curvature of $y = Ax + B$ when $x = 1, 2$, and 3. For what value of x is the curvature of $y = Ax + B$ maximum?

29. Determine the curvature of $y = x^2$ when $x = 1, 2$, and 3. For what value of x is the curvature of $y = x^2$ maximum? For what value of x is the radius of curvature of $y = x^2$ minimum?

30. Determine the curvature of $y = x^3 - x$ when $x = 0, 1$, and 2. For what value of x is the curvature of $y = x^3 - x$ maximum? For what value of x is the radius of curvature of $y = x^3 - x$ minimum?

Practice Answers

Practice 1: $|\mathbf{r}'(t)| = \sqrt{1^2 + (2t)^2 + (3t^2)^2} = \sqrt{1 + 4t^2 + 9t^4}$. Then

$$\mathbf{L} = \int_{t=0}^{t=2} |\mathbf{r}'(t)|\, dt = \int_{t=0}^{t=2} \sqrt{1 + 4t^2 + 9t^4}\, dt \approx 9.57 \text{ (using Simpson's Rule with } n = 20).$$

Practice 2: (a) and (b) Fig. 11 shows the correct locations of the "S" and the "T."

(c) The bridge is 8 hours from the beginning of the river so $B = 8$.

(d) The fish swims the smallest distance between $t = 2$ and $t = 3$, so it was swimming slowest during that 1 hour time interval.

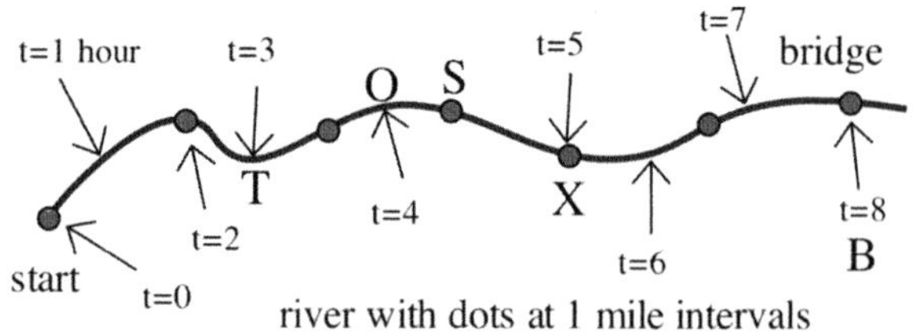

river with dots at 1 mile intervals
Fig. 11

Practice 3: $|\mathbf{r}'(t)| = \sqrt{(8)^2 + (1)^2 + (4)^2} = 9$ for all t.

Then $s = s(t) = \int_{u=0}^{u=t} |\mathbf{r}'(u)|\,du = \int_{u=0}^{u=t} 9\,du = 9t$ so $t = \frac{s}{9}$. By substituting $t(s) = \frac{s}{9}$ for t,

the original parameterization $\mathbf{r}(t) = \langle 8t, t, 4t \rangle$ becomes

$\mathbf{r}(\,t(s)\,) = \langle 8t(s), t(s), 4t(s) \rangle = \langle 8\frac{s}{9}, \frac{s}{9}, 4\frac{s}{9} \rangle$, a function of s alone.

Practice 4: $\mathbf{r}'(t) = \langle A, B, C \rangle$ so $|\mathbf{r}'(t)| = \sqrt{A^2 + B^2 + C^2}$.

$\mathbf{T}(t) = \frac{\mathbf{r}'(t)}{|\mathbf{r}'(t)|} = \frac{1}{\sqrt{A^2 + B^2 + C^2}} \langle A, B, C \rangle$ which is a constant vector so $\mathbf{T}'(t) = \mathbf{0}$.

Then, for all t, $\kappa = \left| \frac{\mathbf{T}'(t)}{\mathbf{r}'(t)} \right| = \frac{0}{\sqrt{A^2 + B^2 + C^2}} = 0$.

The curvature of a the line $\mathbf{r}(t) = \langle At, Bt, Ct \rangle$ is $\kappa = 0$.

Practice 5: $\mathbf{r}'(t) = \langle 1, \cos(t), 0 \rangle$ so $|\mathbf{r}'(t)| = \sqrt{1 + \cos^2(t)}$ and $\mathbf{r}''(t) = \langle 0, -\sin(t), 0 \rangle$.

Then $\mathbf{r}' \times \mathbf{r}'' = \begin{vmatrix} \mathbf{i} & \mathbf{j} & \mathbf{k} \\ 1 & \cos(t) & 0 \\ 0 & -\sin(t) & 0 \end{vmatrix} = 0\mathbf{i} - 0\mathbf{j} + (-\sin(t))\mathbf{k}$ and $|\mathbf{r}' \times \mathbf{r}''| = |-\sin(t)|$.

Putting this together, we have $\kappa = \frac{|-\sin(t)|}{(1 + \cos^2(t))^{3/2}}$.

When $t = 0$, $\kappa = \frac{0}{(1 + 1)^{3/2}} = 0$. When $t = \pi/4$, $\kappa = \frac{\sqrt{2}/2}{(1 + 1/2)^{3/2}} \approx 0.385$.

When $t = \pi/2$, $\kappa = \frac{1}{(1 + 0)^{3/2}} = 1$

Practice 6: $y = \sin(x)$, $y'(x) = \cos(x)$, and $y''(x) = -\sin(x)$. Then

$$\kappa = \frac{|y''|}{(1 + (y')^2)^{3/2}} = \frac{|-\sin(x)|}{(1 + \cos^2(x))^{3/2}}$$ which is the same result we got in Practice 5.

Appendix: A proof that $\kappa = \dfrac{|\mathbf{T}'(t)|}{|\mathbf{r}'(t)|} = \dfrac{|\mathbf{r}' \times \mathbf{r}''|}{|\mathbf{r}'|^3}$

Since $\mathbf{T}(t) = \dfrac{\mathbf{r}'(t)}{|\mathbf{r}'(t)|}$ and $|\mathbf{r}'(t)| = \dfrac{ds}{dt}$ we know that $\mathbf{r}'(t) = |\mathbf{r}'(t)|\,\mathbf{T}(t) = \dfrac{ds}{dt}\,\mathbf{T}(t)$.

Then, by the Product rule for Derivatives, $\mathbf{r}''(t) = \dfrac{d}{dt}\left\{ \dfrac{ds}{dt}\,\mathbf{T}(t) \right\} = \dfrac{d^2 s}{dt^2}\,\mathbf{T}(t) + \dfrac{ds}{dt}\,\mathbf{T}'(t)$.

Replacing $\mathbf{r}'$ and $\mathbf{r}''$ with these results and using the distributive pattern $\mathbf{A}\times(\mathbf{B} + \mathbf{C}) = \mathbf{A}\times\mathbf{B} + \mathbf{A}\times\mathbf{C}$ for the cross product, we have

$$\mathbf{r}' \times \mathbf{r}'' = \frac{ds}{dt}\,\mathbf{T}(t) \times \left\{ \frac{d^2 s}{dt^2}\,\mathbf{T} + \frac{ds}{dt}\,\mathbf{T}' \right\} = \frac{ds}{dt}\frac{d^2 s}{dt^2}\left\{ \mathbf{T} \times \mathbf{T} \right\} + \left\{ \frac{ds}{dt} \right\}^2 \left\{ \mathbf{T} \times \mathbf{T}' \right\} .$$

We know for every vector $\mathbf{V}$ that $\mathbf{V} \times \mathbf{V} = \mathbf{0}$ (the zero vector) so $\mathbf{T} \times \mathbf{T} = \mathbf{0}$.
We also know that $|\mathbf{T}(t)| = 1$ for all t so from Example 3 of Section 12.2 we can conclude that $\mathbf{T}$ is perpendicular to $\mathbf{T}'$. Then $|\mathbf{T} \times \mathbf{T}'| = |\mathbf{T}|\,|\mathbf{T}'|\,|\sin(\theta)| = |\mathbf{T}|\,|\mathbf{T}'| = |\mathbf{T}'|$. Using these results together with the previous result for $\mathbf{r}' \times \mathbf{r}''$ we have

$$|\mathbf{r}' \times \mathbf{r}''| = \left\{ \frac{ds}{dt} \right\}^2 |\mathbf{T}'| = |\mathbf{r}'|^2\,|\mathbf{T}'| \quad \text{so} \quad |\mathbf{T}'| = \frac{|\mathbf{r}' \times \mathbf{r}''|}{|\mathbf{r}'|^2} .$$

Finally, $\dfrac{|\mathbf{T}'|}{|\mathbf{r}'|} = \dfrac{|\mathbf{r}' \times \mathbf{r}''|}{|\mathbf{r}'|^3}$, the result we wanted.

12.4 CYLINDRICAL & SPHERICAL COORDINATE SYSTEMS IN 3D

Most of our work in two dimensions used the rectangular coordinate system, but we also examined the polar coordinate system (Fig. 1), and for some uses the polar coordinate system was more effective and efficient. A similar situation occurs in three dimensions. Mostly we use the 3–dimensional xyz–coordinate system, but there are two alternate systems, called cylindrical coordinates and spherical coordinates, that are sometimes better. In two dimensions, the rectangular and the polar coordinate systems each located a point by means of two numbers, but each system used those two numbers in different ways. In three dimensions, each of the coordinate systems locates a point using three numbers, and each system uses those three numbers in different ways. Fig. 2 illustrates how three numbers are used to locate the point P in each of the different systems, and the rest of this section examines the cylindrical and spherical coordinate systems.

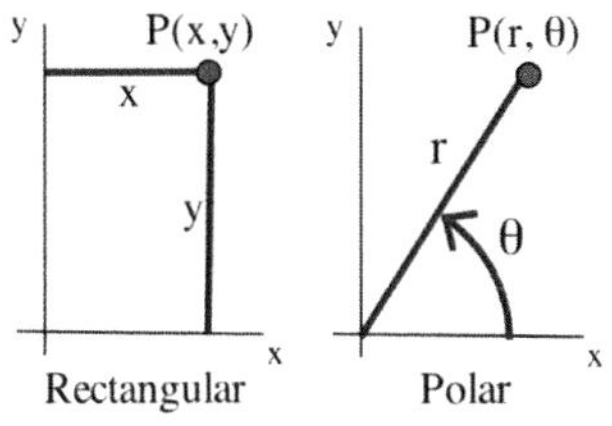

Fig. 1

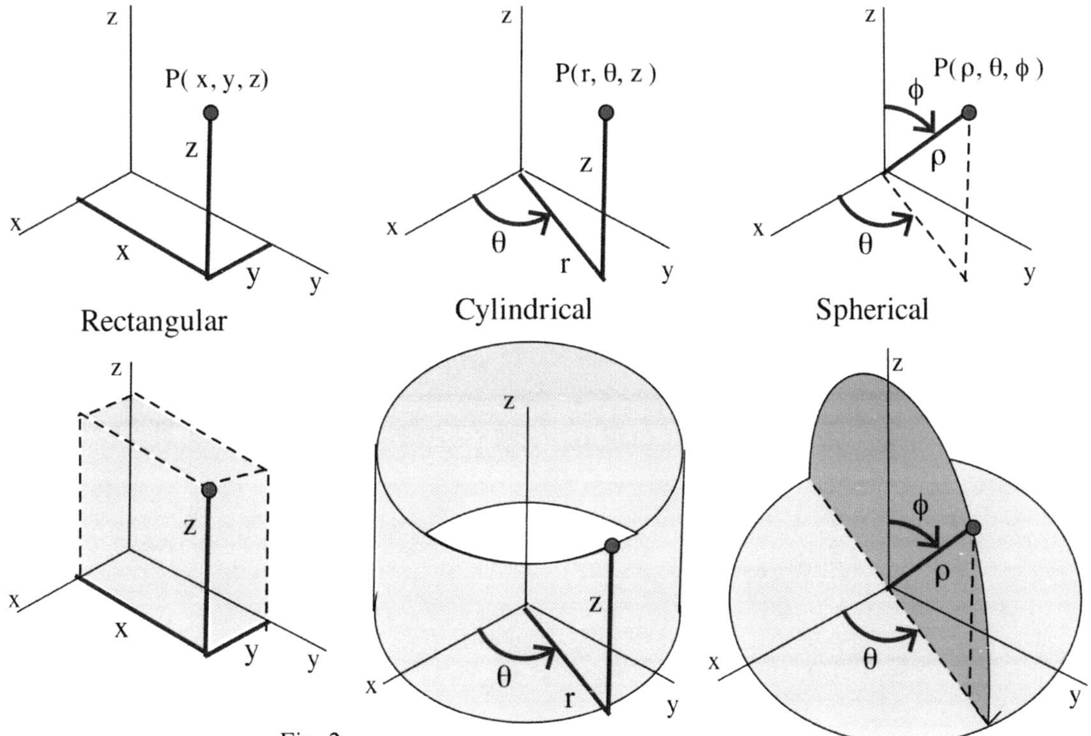

Fig. 2

CYLINDRICAL COORDINATES

Cylindrical coordinates are basically "polar coordinates with altitude." The cylindrical coordinates (r, θ, z) specify the point P that is z units above the point on the xy–plane whose polar coordinates are r and θ (Fig. 3).

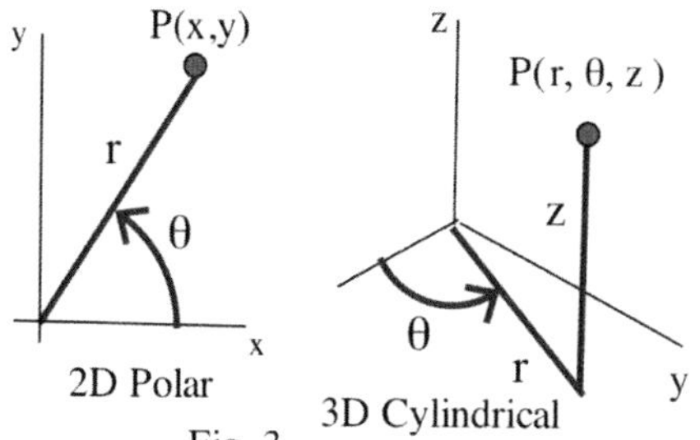

Fig. 3

Example 1: Plot the points given by the cylindrical coordinates $A(3, \pi/3, 1)$, $B(1, 0, -2)$, and $C(2, 180^{\circ}, -1)$.

Solution: The points are plotted in Fig. 4.

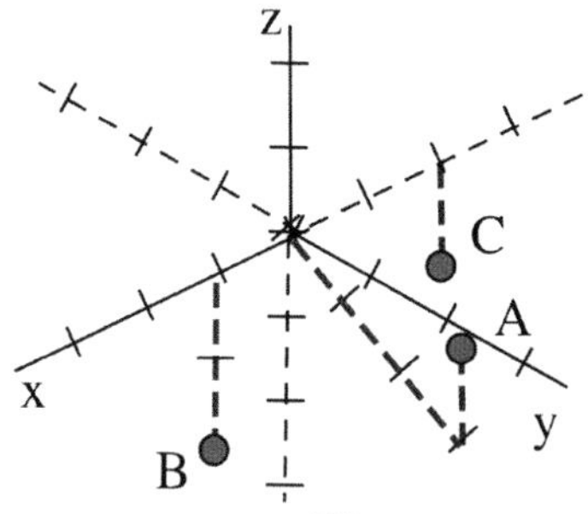

Fig. 4

Practice 1: On Fig. 4, plot the points given by the cylindrical coordinates $P(3, \pi/6, -1)$, $Q(3, \pi/2, 2))$, and $R(0, \pi, 3)$.

We can start to develop an understanding of the effect of each variable in the ordered triple by holding two of the variables fixed and letting the other one vary. Fig. 5 shows the results in the rectangular coordinate system of fixing x and y ($x = 1, y = 2$) and letting z vary: we get a vertical line parallel to the z–axis. Similarly, in the rectangular coordinate system, when we fix x and z ($x = 1, z = 3$) and let y vary we get a line parallel to the y–axis. We can try the same process in the cylindrical coordinate system.

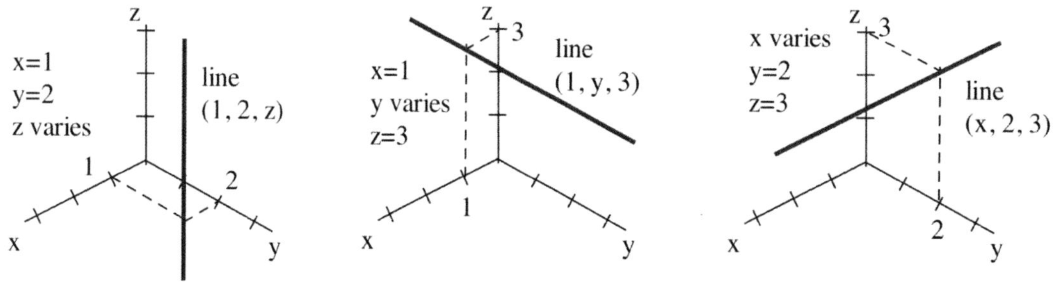

Fig. 5: Rectangular coordinates, two variables fixed

Fixing two of the variables and letting the other one vary:

In the cylindrical coordinate system, if we fix r and θ ($r = 2, \theta = \pi/3 = 60^{\circ}$) and let z vary (Fig. 6a) we get a line parallel to the z–axis.

If we fix r and z ($r = 2, z = 3$) and let θ vary (Fig. 6b) we get a circle of radius 1 centered around the z–axis at a height of 3 units above the xy plane.

It we fix θ and z ($= \pi/3 = 60^{\circ}, z = 3$) and let r vary (Fig. 6c) we get a line that is always 3 units above the xy plane.

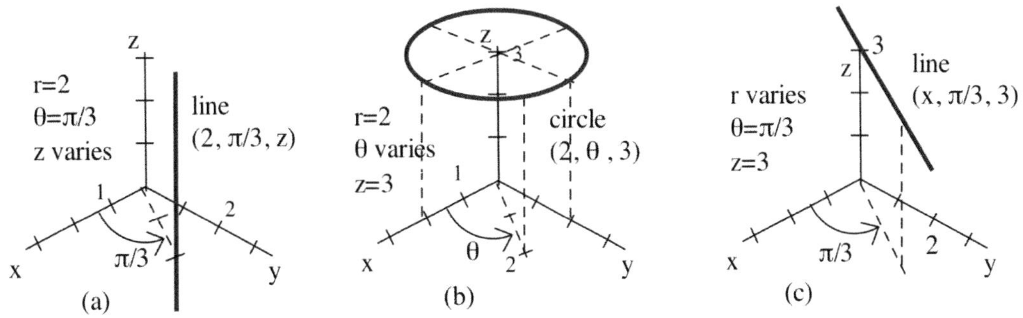

Fig. 6: Cylindrical coordinates, two variables fixed

Fixing the value of only one of the variables and letting the other two vary can also be informative.

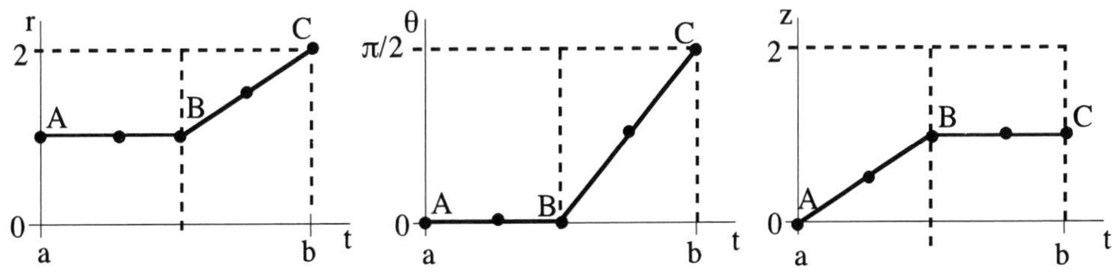

Fig. 8

If we fix r $(r = 2)$ and let θ and z vary, the result (Fig. 7a) is a cylinder, the reason this is called the cylindrical coordinate system.

If we fix θ $(\theta = \pi/3)$ and let r and z vary, the result (Fig. 7b) is a plane.

If we fix z $(z = 3)$ and let r and θ vary, the result (Fig. 7c) is a plane parallel to the xy plane.

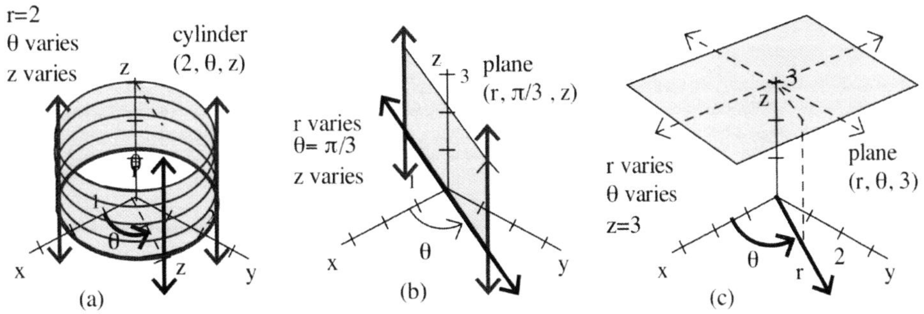

Fig. 7: Cylindrical coordinates, one variable fixed

Practice 2: **An Alternate View: Cylindrical coordinates are "rectangular coordinates on a door."** We began the discussion of cylindrical coordinates by describing them as "polar coordinates with altitude." A student said she found it easier to think of cylindrical coordinates as "rectangular coordinates on a door." By means of a labeled sketch show how this new description makes good sense.

Parametric Curves Using Cylindrical Coordinates

If we consider each of the rectangular coordinate variables $x, y,$ and z to be a function of a parameter t, then $(x, y, z) = (x(t), y(t), z(t))$ describes the location of a point at time t. When the values of t are an interval of numbers, the graph of $(x(t), y(t), z(t))$ is a path or curve in three dimensional space. Similarly, if we consider each of the cylindrical coordinate variables $r, \theta,$ and z to be a function of a parameter t, then $(r, \theta, z) = (r(t), \theta(t), z(t))$ describes the location of a point, in cylindrical coordinates, at time t. When the values of t are an interval of numbers, the graph of $(r(t), \theta(t), z(t))$ is again a path or curve in three dimensional space.

Example 2: The graphs in Fig. 8 show values of $r, \theta,$ and z as functions of t. Use those values to graph the path $(r(t), \theta(t), z(t))$ for $a \le t \le b$.

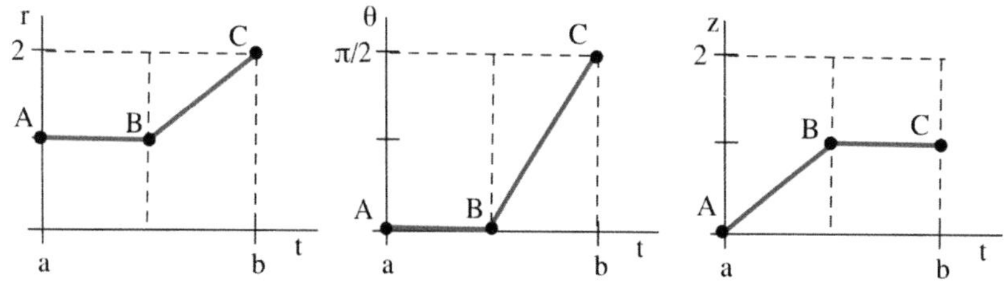

Fig. 8

Solution: Fig. 9 shows the result of plotting several points $(r(t), \theta(t), z(t))$ and connecting them.

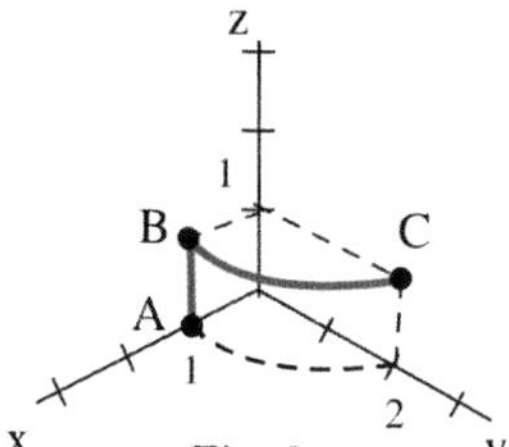

Fig. 9

Practice 3: The graphs in Fig. 10 show values of $r, \theta,$ and z as functions of t. Use those values to graph the path $(r(t), \theta(t), z(t))$ for $a \le t \le b$.

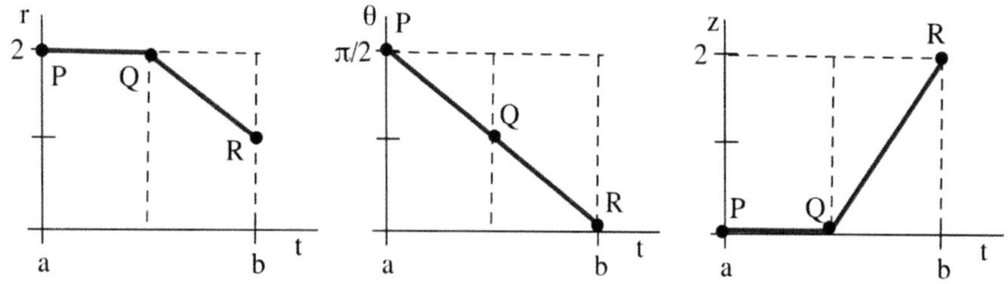

Fig. 10

Converting Cylindrical To And From Rectangular Coordinates

Typically we stay in one coordinate system for any particular use, but occasionally it may be necessary to convert information from one system to another, and it is rather straightforward to convert between rectangular and cylindrical coordinates. The important conversion information is contained in Fig. 11 and is summarized below.

Converting Between Cylindrical and Rectangular Systems

If P(x, y, z) and P(r, θ, z) represent the same point in rectangular and cylindrical coordinates (Fig. 11),

then (cylindrical to rectangular) $x = r \cdot \cos(\theta)$ $y = r \cdot \sin(\theta)$

(rectangular to cylindrical) $r^2 = x^2 + y^2$ $\tan(\theta) = y/x$

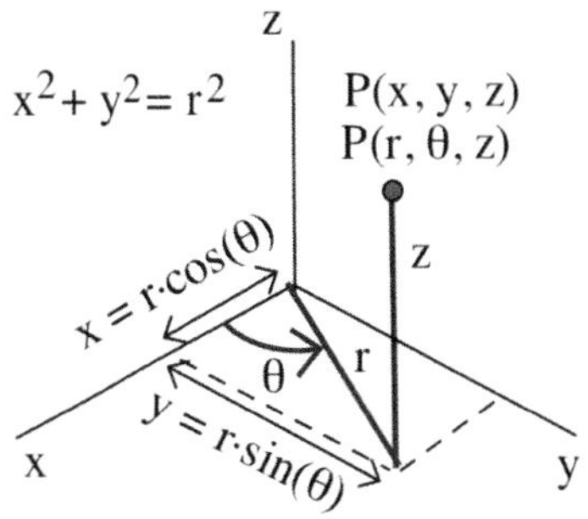

Fig. 11

Example 3: (a) Write the rectangular coordinate location A(3, 4, 2) in the cylindrical coordinate system.

(b) Write the cylindrical coordinate location B(2, π/6, 3) in the rectangular coordinate system.

Solution: (a) $r^2 = x^2 + y^2 = 3^2 + 4^2 = 5^2$. $\tan(\theta) = y/x = 4/3$ so $\theta = \arctan(4/3) \approx 0.927$ (53.1°) .

The cylindrical coordinates of A are approximately (5, 0.927, 2).

(b) $x = r \cdot \cos(\pi/6) = 2(\sqrt{3/2}) \approx 1.73$, $y = r \cdot \sin(\pi/6) = 2(0.5) = 1$. The rectangular coordinates of B are approximately (1.73, 1, 3).

Practice 4: (a) Write the rectangular coordinate location P(–5, 12, 1) in the cylindrical coordinate system.

(b) Write the cylindrical coordinate location Q(4, 3π/4, 3) in the rectangular coordinate system.

Example 4: (a) Rewrite the rectangular coordinate equation $x^2 + y^2 = 6z$ as an equation in cylindrical coordinates.

(b) Rewrite the cylindrical coordinate equation $r = 6 \cdot \cos(\theta)$ as an equation in rectangular coordinates.

Solution: (a) $x^2 + y^2 = r^2$ so $x^2 + y^2 = 6z$ becomes $r^2 = 6z$.

(b) Multiplying each side of $r = 6 \cdot \cos(\theta)$ by r we have $r^2 = 6r \cdot \cos(\theta)$ so $x^2 + y^2 = 6x$ or, moving the 6x to the left side and completing the square, $(x-3)^2 + y^2 = 9$. The graph of $(x-3)^2 + y^2 = 9$ is a circular cylinder generated by moving the circle with radius 3 and center (3, 0, 0) parallel to the z–axis.

SPHERICAL COORDINATES

A point P with spherical coordinates (ρ, θ, φ) (the Greek letters ρ = rho and φ = phi are pronounced as "row" and "fee" or "fie") is located in three dimensions as shown in Fig. 12: ρ is the distance of P from the origin, and φ is the angle the segment from the origin to P makes with the positive z–axis.

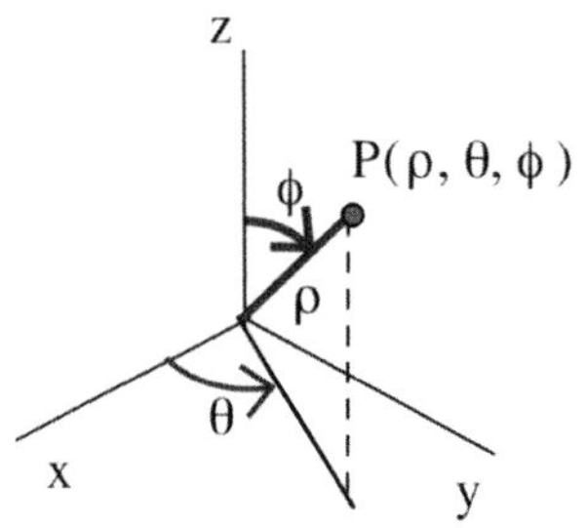

Fig. 12

Example 5: Plot the points given by the spherical coordinates $A(3, \pi/3, \pi/2)$, $B(2, 0, \pi/3)$, and $C(2, 180^{\circ}, 0)$.

Solution: The points are plotted in Fig. 13.

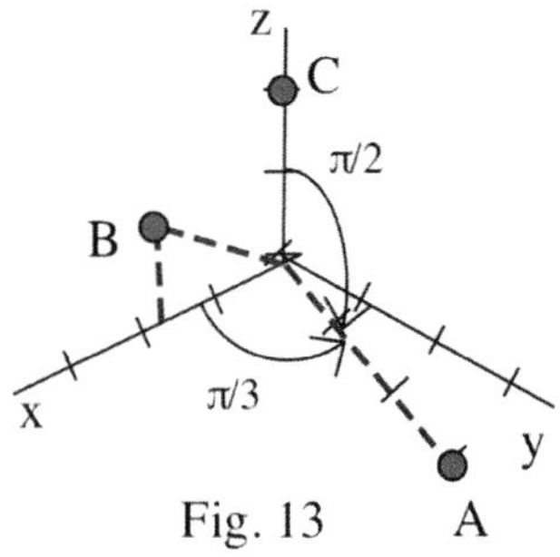

Fig. 13

Practice 5: On Fig. 13, plot the points given by the spherical coordinates $P(3, 3\pi/4, \pi/6)$, $Q(3, 0, \pi/4)$, and $R(2, \pi/6, 0)$.

Fixing two of the variables and letting the other one vary:

In the spherical coordinate system, if we fix ρ and θ $(\rho = 2, \theta = \pi/3)$ and let φ vary (Fig. 14a) we get a circle with radius 2 and center at the origin.

If we fix ρ and φ $(\rho = 2, \varphi = \pi/4)$ and let θ vary (Fig. 14b) we get a circle with radius $\sqrt{2}$ and center at $(0, 0, \sqrt{2})$.

If we fix θ and φ $(\theta = \pi/3, \varphi = \pi/4)$ and let ρ vary (Fig. 14c) we get a straight line through the origin.

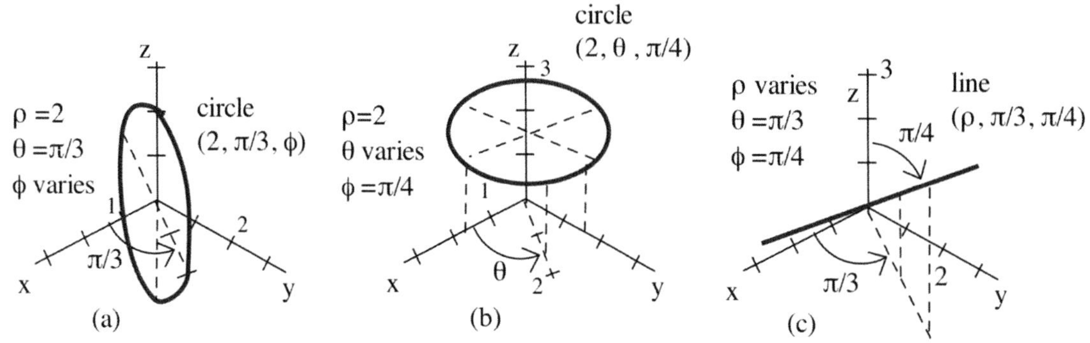

Fig. 14: Spherical coordinates, two variables fixed

Fixing one of the variables and letting the other two vary:

If we fix ρ ($\rho = 2$) and let θ and φ vary (Fig. 15a), we get a sphere, the reason this is called the spherical coordinate system, with radius 2 and center at the origin.

If we fix θ ($\theta = \pi/3$) and let ρ and φ vary (Fig. 15b), we get get a plane.

If we fix φ ($\varphi = \pi/4$) and let ρ and θ vary (Fig. 15c), we get a cone around the z–axis.

It is possible to sketch the path of a point when the variables ρ, θ and φ are functions of a parameter t, but it is difficult to do "by hand" and we omit it. The Appendix at the end of this section illustrates some commands in the language Maple to create these graphs.

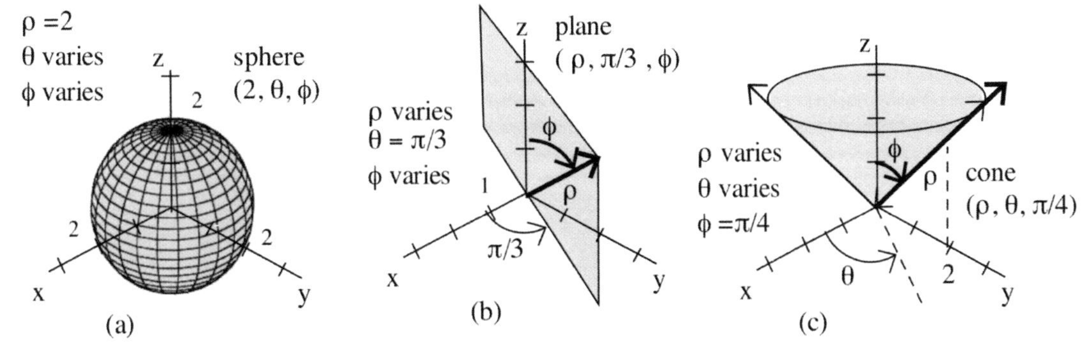

Fig. 15: Spherical coordinates, one variable fixed

Converting Spherical To And From Rectangular Coordinates

The conversions between rectangular and spherical coordinates are mostly a matter of applied trigonometry, and the essential conversion information is contained in Fig. 16.

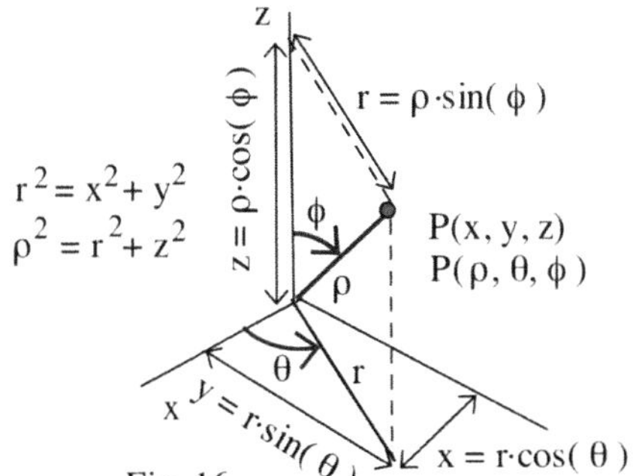

Fig. 16

The conversion from rectangular coordinates (x, y, z) to spherical coordinates (ρ, θ, φ) is straightforward:

r is the distance of the point (x, y, z) from the origin so

$$\rho^2 = x^2 + y^2 + z^2 ,$$

$$\tan(\theta) = \frac{y}{x} , \text{ and}$$

$$\cos(\varphi) = \frac{z}{\sqrt{x^2+y^2+z^2}} = \frac{z}{\rho} .$$

For the spherical to rectangular conversion it helps to calculate the value of $r = \rho \cdot \sin(\varphi)$ in Fig. 16. Then

$$x = r \cdot \cos(\theta) = \rho \cdot \sin(\varphi) \cdot \cos(\theta), \quad y = r \cdot \sin(\theta) = \rho \cdot \sin(\varphi) \cdot \sin(\theta), \text{ and } z = \rho \cdot \cos(\varphi) .$$

Converting Between Spherical and Rectangular Systems

If $P(x, y, z)$ and $P(\rho, \theta, \varphi)$ represent the same point in rectangular and spherical coordinates (Fig. 16),

then (spherical to rectangular) $x = \rho\cdot\sin(\varphi)\cdot\cos(\theta)$ $y = \rho\cdot\sin(\varphi)\cdot\sin(\theta)$ $z = \rho\cdot\cos(\varphi)$

(rectangular to spherical) $\rho^2 = x^2 + y^2 + z^2$ $\tan(\theta) = \frac{y}{x}$ $\cos(\varphi) = \frac{z}{\sqrt{x^2+y^2+z^2}} = \frac{z}{\rho}$

Example 6: (a) Write the rectangular coordinate location $A(3, 6, 2)$ in the spherical coordinate system.
(b) Write the spherical coordinate location $B(2, \pi/6, \pi/4)$ in the rectangular coordinate system.

Solution: (a) $\rho^2 = x^2 + y^2 + z^2 = 3^2 + 6^2 + 2^2 = 7^2$. $\tan(\theta) = 6/3 = 2$ so $\theta = \arctan(2) \approx 1.107$ (63.4^o). $\cos(\varphi) = z/\rho = 2/7$ so $\varphi = \arccos(2/7) \approx 1.281$ (73.4^o). The spherical coordinates of A are approximately $(7, 1.107, 1.281)$.

(b) $x = 2\cdot\sin(\pi/4)\cdot\cos(\pi/6) = 2(\frac{\sqrt{2}}{2})(\sqrt{\frac{3}{2}}) = \sqrt{3}$.

$y = 2\cdot\sin(\pi/4)\cdot\sin(\pi/6) = 2(\frac{\sqrt{2}}{2})(\frac{1}{2}) = \frac{\sqrt{2}}{2}$. $z = 2\cdot\cos(\pi/4) = 2(\frac{\sqrt{2}}{2}) = \sqrt{2}$.

The rectangular coordinates of B are approximately $(\sqrt{3}, \sqrt{2}/2, \sqrt{2}) \approx (1.73, 0.707, 1.414)$.

Practice 6: (a) Write the rectangular coordinate location $P(2, 9, 6)$ in the spherical coordinate system.
(b) Write the spherical coordinate location $Q(4, 3\pi/4, \pi/2)$ in the rectangular coordinate system.

Example 7: (a) Rewrite the rectangular coordinate equation $x^2 + y^2 = 6z$ as an equation in spherical coordinates.
(b) Rewrite the spherical coordinate equation $\rho = 6\cdot\cos(\varphi)$ as an equation in rectangular coordinates.

Solution: (a) $x^2 + y^2 = 6z$ becomes $\rho^2\cdot\sin^2(\varphi)\cdot\cos^2(\theta) + \rho^2\cdot\sin^2(\varphi)\cdot\sin^2(\theta) = 6\rho\cdot\cos(\varphi)$ or $\rho^2\cdot\sin^2(\varphi) = 6\rho\cdot\cos(\varphi)$.

(b) Multiplying each side of $\rho = 6\cdot\cos(\varphi)$ by ρ we have $\rho^2 = 6\rho\cdot\cos(\theta)$ so $x^2 + y^2 + z^2 = 6z$ or, moving the 6z to the left side and completing the square, $x^2 + y^2 + (z-3)^2 = 9$. The graph of $x^2 + y^2 + (z-3)^2 = 9$ is a sphere with radius 3 and center at $(0, 0, 3)$.

PROBLEMS

Cylindrical coordinates

In problems 1 – 4, plot the points whose cylindrical coordinates are given.

1. $A(3, \pi/2, -2), B(2, \pi/6, 3), C(0, 30^o, 3)$
2. $D(3, 3\pi/2, 2), E(4, 0, -2), F(2, \pi, 1)$
3. $P(1, 45^o, -2), Q(3, 3\pi/2, 2), R(3, \pi/3, 0)$
4. $S(1, 2\pi/3, 2), T(0, 30^o, -2), U(2, \pi/4, 3)$

In problems 5 – 10, values are given for two of the r, θ, z variables. Plot the graph as the third variable takes all possible values.

5. $r = 3, \theta = 0$
6. $r = 1, \theta = \pi/6$
7. $r = 1, z = 3$
8. $r = 3, z = -2$
9. $\theta = \pi/4, z = 2$
10. $\theta = \pi, z = 3$

In problems 11 – 16, a value is given for one of the r, θ, z variables. Plot the graph as the other two variables take all possible values.

11. $r = 1$
12. $r = 3$
13. $\theta = \pi/2$
14. $\theta = \pi/6$
15. $z = 2$
16. $z = -1$

In Problems 17 – 20, separate parametric graphs of the r, θ, z variables are given. Using the information in these graphs, sketch the graph of (r, θ, z) in the cylindrical coordinate system.

17. See Fig. 17.
18. See Fig. 18.
19. See Fig. 19.
20. See Fig. 20.

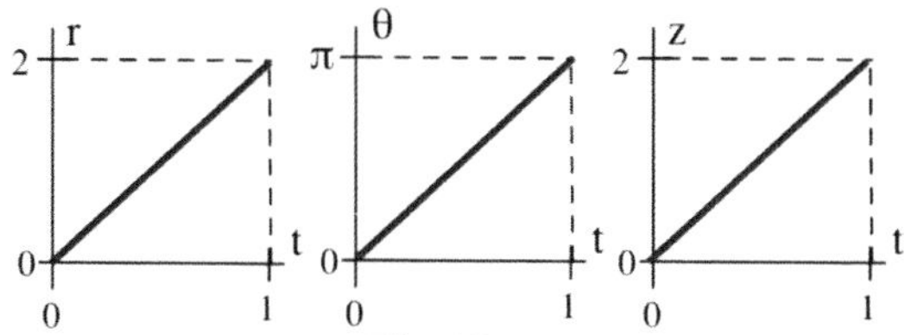

Fig. 17

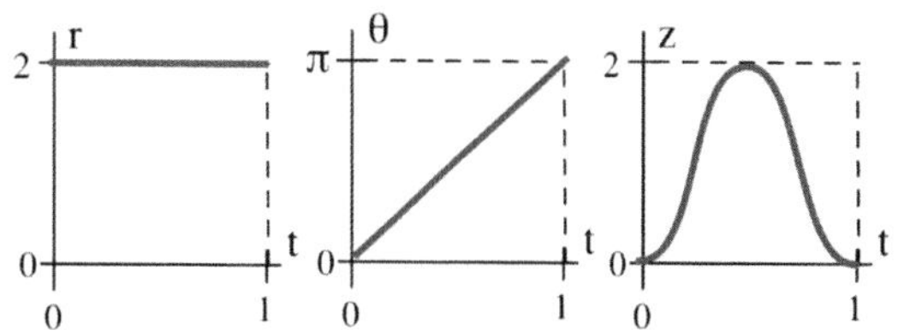

Fig. 18

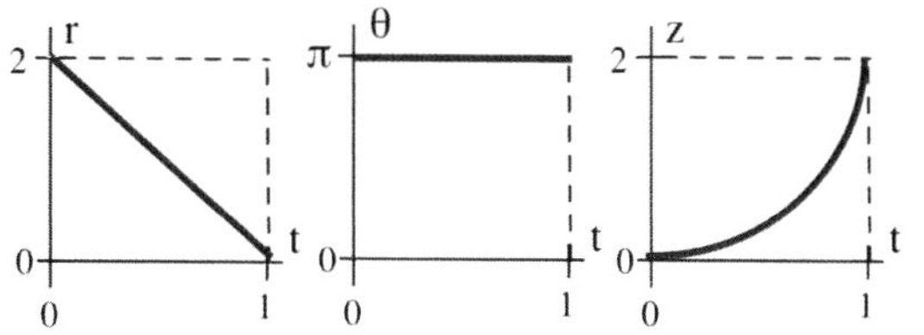

Fig. 19

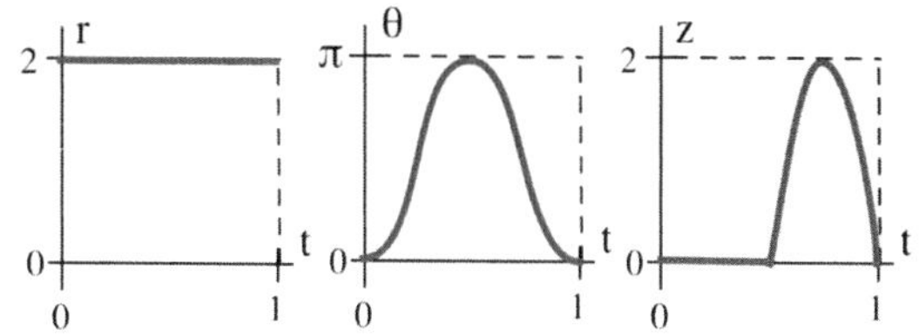

Fig. 20

In problems 21 – 24, convert the cylindrical coordinates of the given point to rectangular coordinates.

21. $(5, \pi/6, 3)$
22. $(4, \pi/3, 5)$
23. $(3, 35^{\circ}, -2)$
24. $(6, 75^{\circ}, -4)$

In problems 25 – 28, convert the rectangular coordinates of the given point to cylindrical coordinates.

25. $(1, 2, 3)$
26. $(5, 2, -3)$
27. $(-4, 3, -1)$
28. $(7, -5, 3)$

In problems 29 – 32, convert the cylindrical coordinate equations to equations in rectangular coordinates.

29. (a) $r^2 = 4r \cdot \sin(\theta) - 1$ (b) $r = 7$
30. (a) $r^2 = 8r \cdot \cos(\theta) + 3$ (b) $r = 1$
31. (a) $r = 5 \cdot \cos(\theta)$ (b) $z = r^2$
32. (a) $r = 5 \cdot \sec(\theta)$ (b) $z = r^2 \sin(\theta) \cos(\theta)$

In problems 33 – 36, convert the rectangular coordinate equations to equations in cylindrical coordinates.

33. (a) $z = x^2 + y^2 - 3x + 2y$ (b) $x = 3$

34. (a) $z = 3x^2 + 3y^2$ (b) $y = 2$

35. (a) $z = x^2 + 5y^2$ (b) $x + y + z = 5$

36. (a) $y = x^2$ (b) $x + 5y = z$

Spherical coordinates

In problems 37 – 40, plot the points whose spherical coordinates are given.

37. $A(3, \pi/2, \pi/4), B(2, 0, \pi/6), C(1, \pi, 90^o)$

38. $D(3, \pi/6, 0), E(3, \pi/2, \pi/6), F(2, 60^o, 20^o)$

39. $P(2, 2\pi/3, 2\pi/3), Q(3, 1, 1), R(0, 71^o, 7\pi/13)$

40. $S(3, \pi/3, \pi), T(0, \pi/3, \pi/7), U(2, 2, 1)$

In problems 41 – 46, values are given for two of the ρ, θ, φ variables. Plot the graph as the third variable takes all possible values.

41. $\rho = 3, \theta = 0$

42. $\rho = 1, \theta = \pi/6$

43. $\rho = 1, \varphi = \pi/4$

44. $\rho = 3, \varphi = \pi/2$

45. $\theta = \pi/4, \varphi = \pi/2$

46. $\theta = \pi, \varphi = \pi/6$

In problems 47 – 52, a value is given for one of the ρ, θ, φ variables. Plot the graph as the other two variables take all possible values.

47. $\rho = 1$

48. $\rho = 3$

49. $\theta = \pi/2$

50. $\theta = \pi/6$

51. $\varphi = 0$

52. $\varphi = \pi/2$

In problems 53 – 56, convert the spherical coordinates of the given point to rectangular coordinates.

53. $(5, \pi/2, \pi/3)$

54. $(3, \pi/3, \pi/6)$

55. $(4, 45^o, 30^o)$

56. $(7, 90^o, 45^o)$

In problems 57 – 60, convert the rectangular coordinates of the given point to spherical coordinates.

57. $(1, 2, 3)$

58. $(5, 2, 7)$

59. $(-5, 3, 2)$

60. $(3, -4, -2)$

In problems 61 – 64, convert the spherical coordinate equations to equations in rectangular coordinates.

61. (a) $\rho = 5$ (b) $\theta = \pi/2$

62. (a) $\rho = 3$ (b) $\varphi = \pi/2$

63. (a) $\rho = 5 \cdot \sin(\varphi) \cdot \cos(\theta)$ (b) $\rho = 3 \cdot \sec(\varphi)$

64. $\rho \cdot \sin(\varphi) = 5\cos(\theta)$

In problems 65 – 68, convert the rectangular coordinate equation to an equation in spherical coordinates.

65. (a) $x^2 + y^2 + z^2 = 9$ (b) $x + z = 5$

66. (a) $x^2 + y^2 = 9$ (b) $y + z = 2$

67. (a) $z = 2x^2 + 2y^2$ (b) $z^2 = 25 - x^2$

68. (a) $z = 3$ (b) $z^2 = 25 - x^2 - y^2$

PRACTICE ANSWERS

Practice 1: The points $P(3, \pi/6, -1)$, $Q(3, \pi/2, 2))$, and $R(0, \pi, 3)$ are shown on Fig. 21.

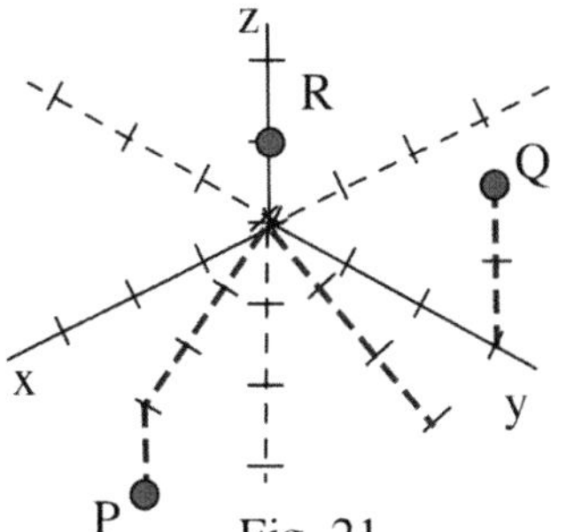

Fig. 21

Practice 2: Fig. 22 illustrates cylindrical coordinates as "rectangular coordinates on a door" by treating r and z as the rectangular coordinates (r along the bottom edge of the door and z as the amount above the bottom edge) and θ as the angular opening of the door.

Sometimes this is an excellent way to think of cylindrical coordinates.

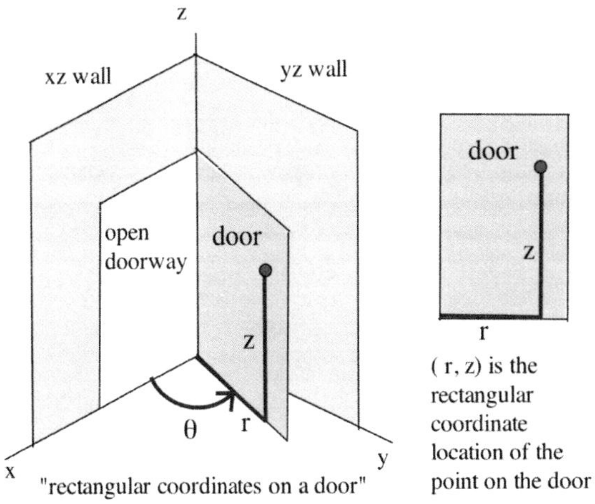

Fig. 22

Practice 3: The (approximate) graph of the path is shown in Fig. 23.

Practice 4: (a) $(x,y,z) = (-5,12,1)$:

$r^2 = x^2 + y^2 = (-5)^2+12^2 = 13^2$. $\tan(\theta) = 12/(-5)$

so $\theta = \arctan(-12/5) \approx -1.176$ (-67.4^o)

$\theta = \arctan(-12/5) \approx -1.176$ (-67.4^o) but the point $(-5,12)$ is in the second quadrant so the angle with the positive x–axis is $-1.176 + \pi \approx 1.966$ (112.6^o). $z = 1$.

The cylindrical coordinates of P are approximately $(13, 1.966, 1)$.

(b) $x = r\cdot\cos(3\pi/4) = 4(-\frac{\sqrt{2}}{2}) = -2\sqrt{2}$. $y = r\cdot\sin(3\pi/4) = 4(\frac{\sqrt{2}}{2}) = 2\sqrt{2}$.

The rectangular coordinates of Q are approximately $(-2\sqrt{2}\ , 2\sqrt{2}\ , 3)$.

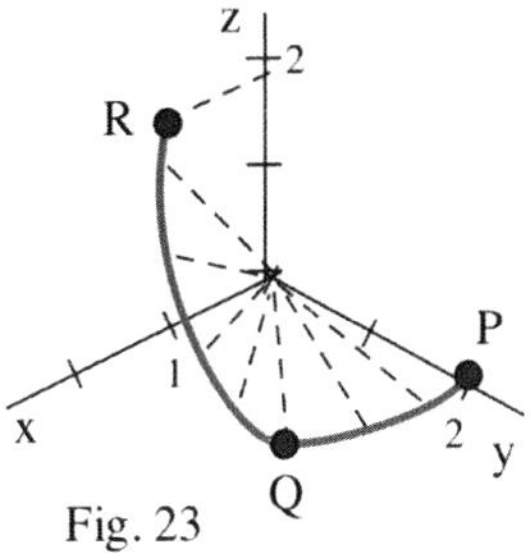

Fig. 23

Practice 5: The points $P(3, 3\pi/4, \pi/6)$, $Q(3, 0, \pi/4)$, and $R(2, \pi/6, 0)$ are shown on Fig. 24.

Practice 6: (a) $(x,y,z) = (2,9,6)$. $\rho^2 = x^2 + y^2 + z^2 = 2^2 + 9^2 + 6^2 = 11^2$.

$\tan(\theta) = 9/2 = 4.5$ so $\theta = \arctan(4.5) \approx 1.352$ (77.5^o).

$\cos(\varphi) = z/\rho = 6/11$ so $\varphi = \arccos(6/11) \approx 0.994$ (56.9^o).

The spherical coordinates of A are approximately $(11, 1.352, 0.994)$.

(b) $(\rho,\theta,\varphi) = (4, 3\pi/4, \pi/2)$. $x = 4\cdot\sin(\pi/2)\cdot\cos(3\pi/4) = 4(1)(-\frac{\sqrt{2}}{2}) = -2\sqrt{2}$.

$y = 4\cdot\sin(\pi/2)\cdot\sin(3\pi/4) = 4(1)(\frac{\sqrt{2}}{2}) = 2\sqrt{2}$. $z = 4\cdot\cos(3\pi/4) = 4(0) = 0$.

The rectangular coordinates of B are approximately

$(-2\sqrt{2}\ , 2\sqrt{2}\ , 0) \approx (-1.414, 1.414, 0)$.

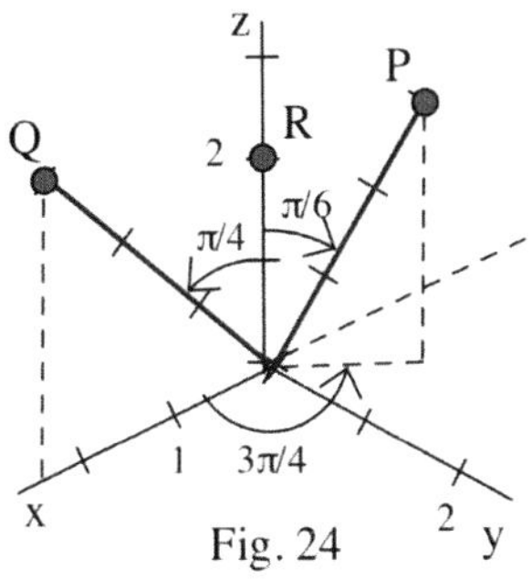

Fig. 24

Appendix: MAPLE Commands for Graphs in Cylindrical & Spherical Coordinates

The computer language Maple as the ability to plot points whose locations are specified in cylindrical or spherical coordinates as well as paths and surfaces in cylindrical or spherical coordinates.

The following commands illustrate how to begin using Maple to graph paths and surfaces in cylindrical and spherical coordinates.

Begin your session with the following command to load the special commands we need:

with(plots); then press the *enter* key.

Now try the following, and some of your own too.

Cylindrical coordinate paths given parametrically:

cylinderplot([3, theta, cos(theta)], theta=0..2*Pi, z = –1..1, axes = NORMAL);

cylinderplot([3, theta, cos(theta)], theta=0..2*Pi, z = –1..1, grid = [30,30], axes = NORMAL);

Cylindrical coordinate surfaces:

cylinderplot(1, theta = 0..2*Pi, z = –1..1, axes = NORMAL); a cylinder

cylinderplot((z+3*cos(2*theta), theta=0..Pi, z=0..3, axes = NORMAL); strange, but nice

Spherical coordinate surfaces:

sphereplot(1, theta=0..2*Pi, phi=0..Pi, grid = [30,30], axes = NORMAL); a sphere

sphereplot([t, 0.3, r], t=4..5, r=0..Pi/2, axes = NORMAL);

sphereplot([r=0..5, t=0..2*Pi, Pi/4], axes = NORMAL);

sphereplot([p*t, exp(t/10, p^2], t=0..Pi, p=–2..2, axes = NORMAL); strange

sphereplot((3*sin(x)^2–1)/2, x = –Pi..Pi, y = 0..Pi);

In Chapter 13 we will be working with surfaces z = f(x,y). The Maple option for how we view these surfaces is "orientation = [theta, phi]" which specifies our viewing angle along the ray whose spherical coordinates have angles theta and phi (given in degrees). Try viewing the paraboloid of revolution $z = x^2 + y^2$ with several different orientations:

plo3d(x^2 + y^2, x= –3..3, y= –3..3, orientation = [10,30]); (try this command with some other orientations)

Now position the cursor on the graph, press (and hold) down the mouse button, and slowly move the mouse. This will rotate a "box" containing the graph (and display the orientation coordinates). When the button is released, the graph will be redrawn with the new orientation.

12.1 Selected Answers

The graphs for the odd problems 1 to 15 are given.

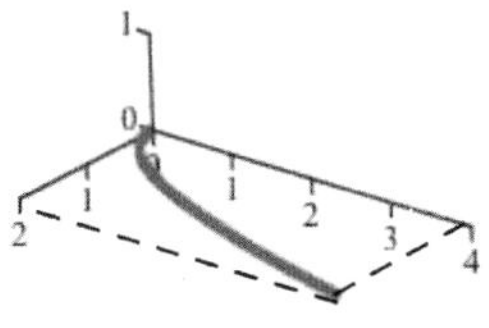

Fig. Problem 1

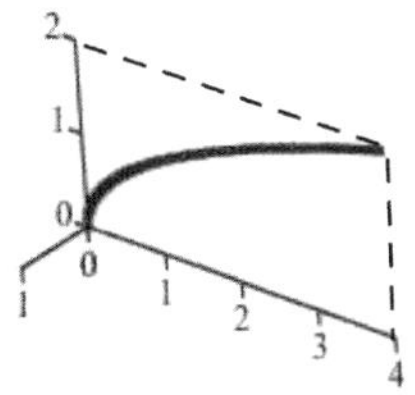

Fig. Problem 3

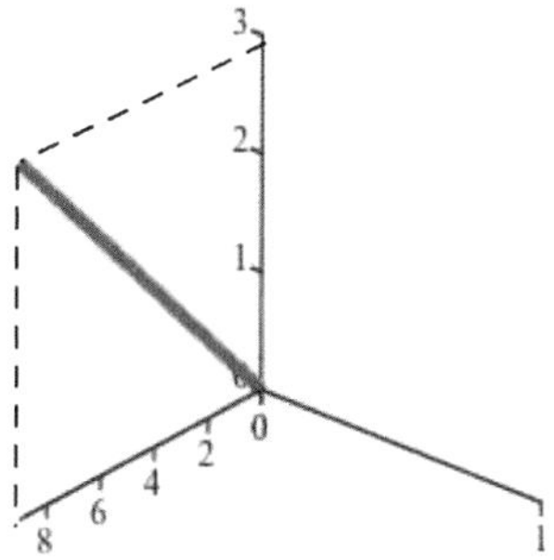

Fig. Problem 5

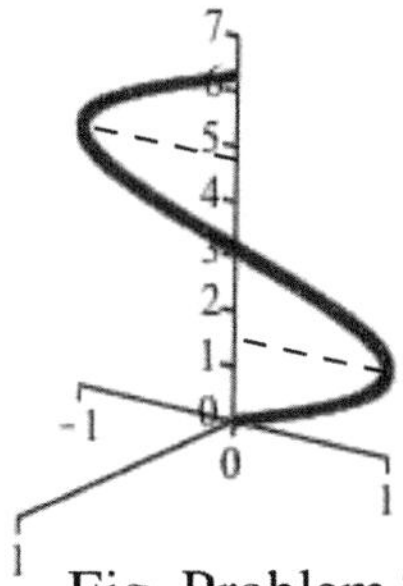

Fig. Problem 7

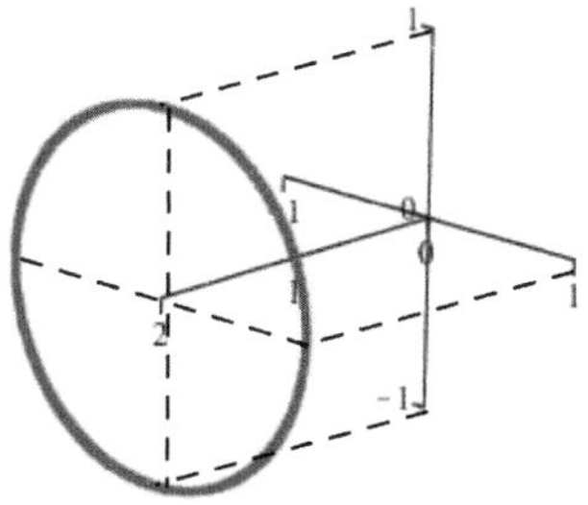

Fig. Problem 9

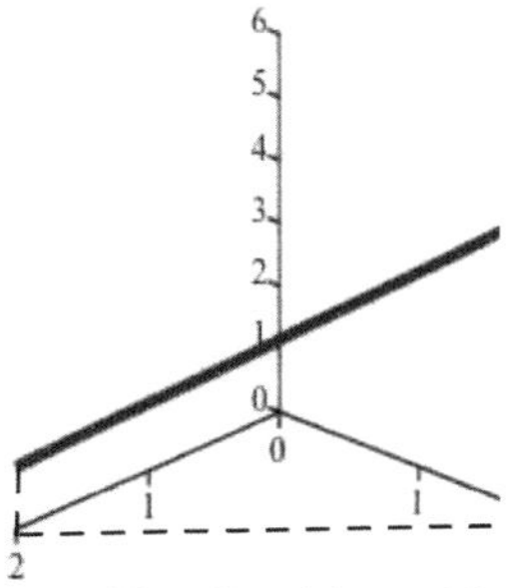

Fig. Problem 11

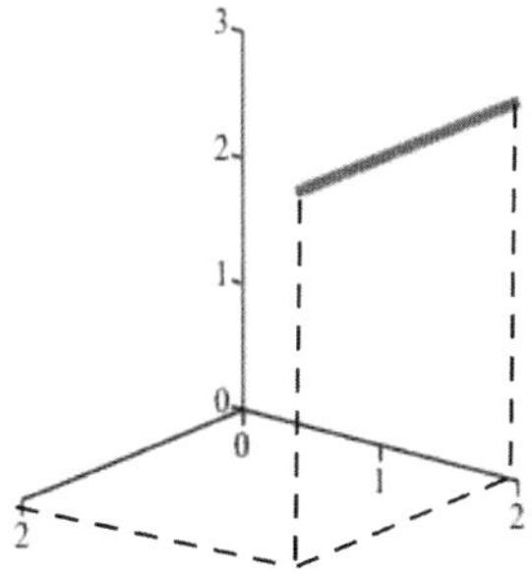

Fig. Problem 13

The graphs for problems 17 to 20 are given.

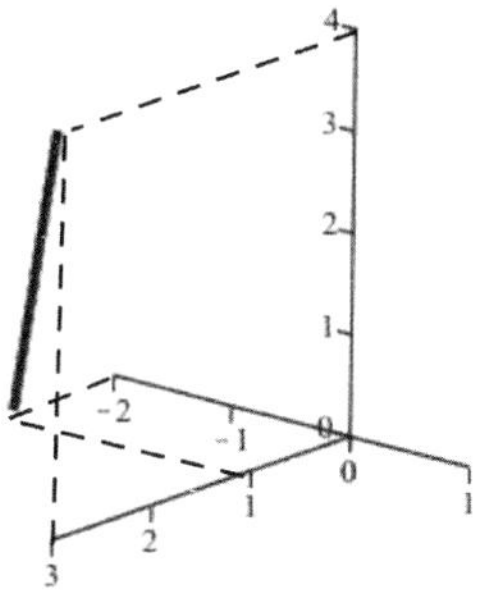

Fig. Problem 15

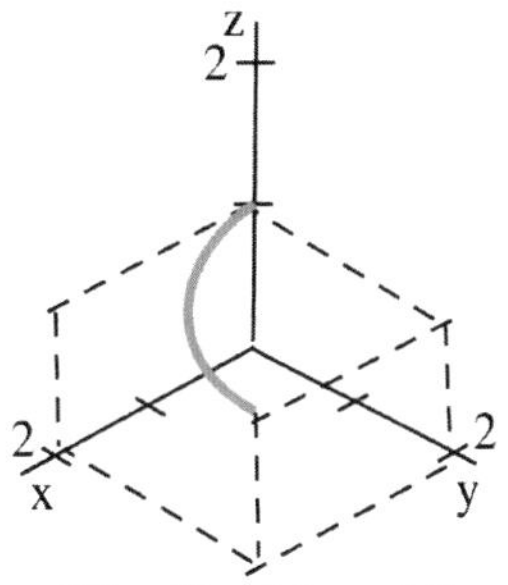

Fig. Problem 17

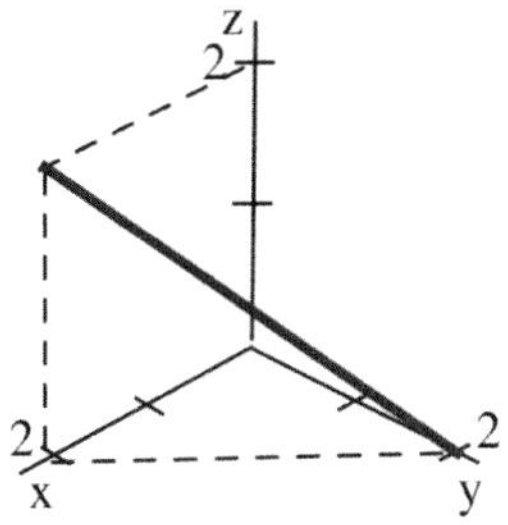

Fig. Problem 18

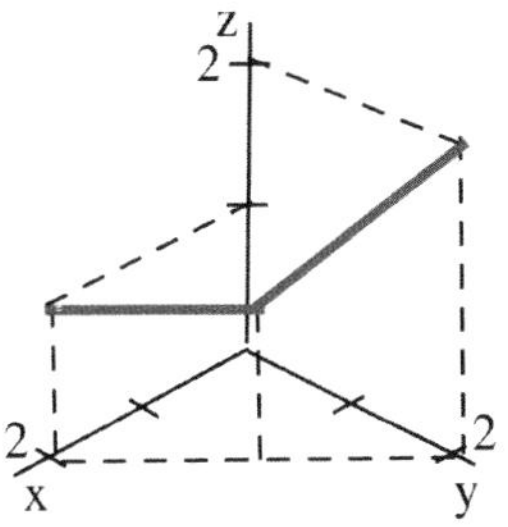

Fig. Problem 19

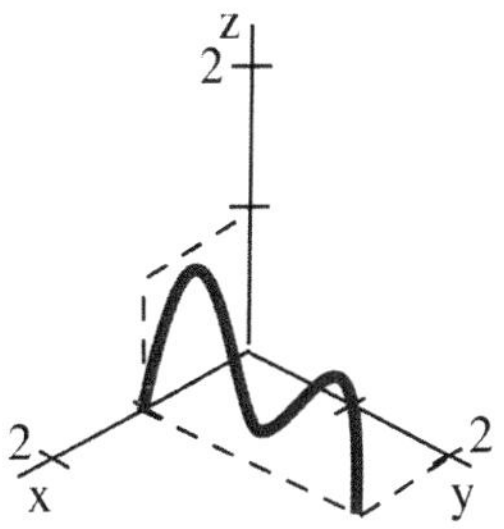

Fig. Problem 20

21. $\langle 3, 9/2, 1/2 \rangle$

22. $\langle 1, 0, 0 \rangle$

23. does not exist

24. $\langle 1, 8, 1 \rangle$

25. $\langle 0, 3, 0 \rangle$

26. does not exist

27. $\langle \pi/2, 0, 0 \rangle$

28. $\langle 1, 1, 1 \rangle$

29. continuous for all $t \neq 1$

30. continuous for all $t \neq 3$

31. continuous for all $t \geq -7$ and $t \neq 2$

32. continuous for all $t \geq 1$

33. continuous for all $t \neq -1, 0, 2$

34. continuous for all $t \neq -3, 2$

35. continuous for all $t > 0$

36. continuous for all $t \neq 0$

37. $x(t) = (1-t)^3(0) + 3(1-t)^2 \cdot t(1) + 3(1-t) \cdot t^2(1) + t^3(0) = -3t^2 + 3t$

$y(t) = (1-t)^3(0) + 3(1-t)^2 \cdot t(0) + 3(1-t) \cdot t^2(2) + t^3(2) = -4t^3 + 6t^2$

$z(t) = (1-t)^3(1) + 3(1-t)^2 \cdot t(1) + 3(1-t) \cdot t^2(0) + t^3(0) = 2t^3 - 3t^2 + 1$

38. $x(t) = (1-t)^3(0) + 3(1-t)^2 \cdot t(1) + 3(1-t) \cdot t^2(0) + t^3(1) = 4t^3 - 6t^2 + 3t$

$y(t) = (1-t)^3(0) + 3(1-t)^2 \cdot t(0) + 3(1-t) \cdot t^2(2) + t^3(3) = -3t^3 + 6t^2$

$z(t) = (1-t)^3(2) + 3(1-t)^2 \cdot t(1) + 3(1-t) \cdot t^2(0) + t^3(0) = t^3 - 3t + 2$

39. $x(t) = (1-t)^3(0) + 3(1-t)^2 \cdot t(0) + 3(1-t) \cdot t^2(2) + t^3(2) = -4t^3 + 6t^2$

$y(t) = (1-t)^3(1) + 3(1-t)^2 \cdot t(0) + 3(1-t) \cdot t^2(1) + t^3(0) = -4t^3 + 6t^2 - 3t + 1$

$z(t) = (1-t)^3(2) + 3(1-t)^2 \cdot t(2) + 3(1-t) \cdot t^2(0) + t^3(0) = 4t^3 - 6t^2 + 2$

40. $x(t) = (1-t)^3(0) + 3(1-t)^2 \cdot t(0) + 3(1-t) \cdot t^2(2) + t^3(1) = -5t^3 + 6t^2$

$y(t) = (1-t)^3(1) + 3(1-t)^2 \cdot t(0) + 3(1-t) \cdot t^2(0) + t^3(1) = 3t^2 - 3t + 1$

$z(t) = (1-t)^3(2) + 3(1-t)^2 \cdot t(2) + 3(1-t) \cdot t^2(0) + t^3(0) = 4t^3 - 6t^2 + 2$

The graphs for problems 41 to 44 are given.

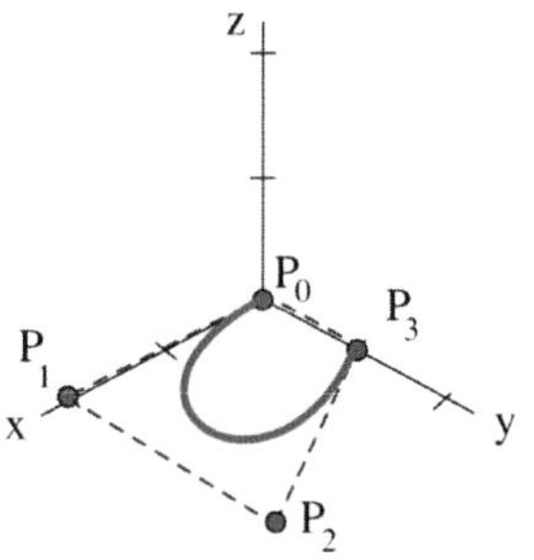

Fig. Problem 41

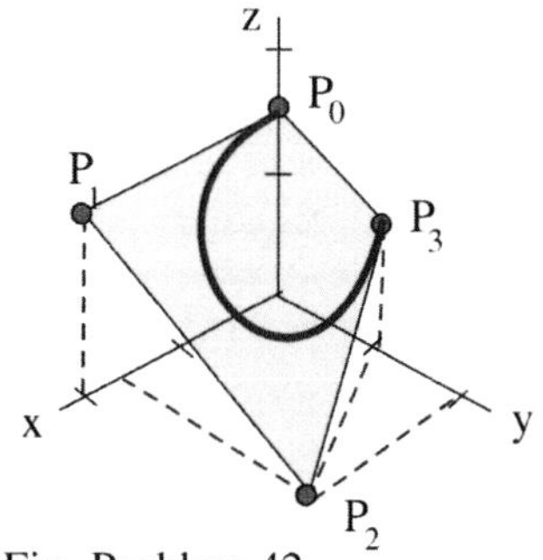

Fig. Problem 42

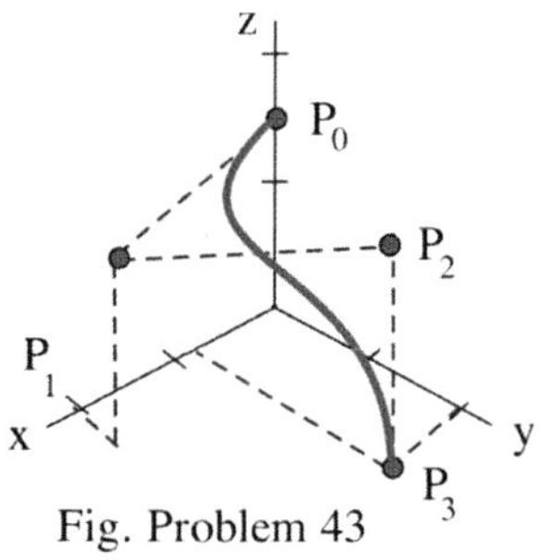

Fig. Problem 43

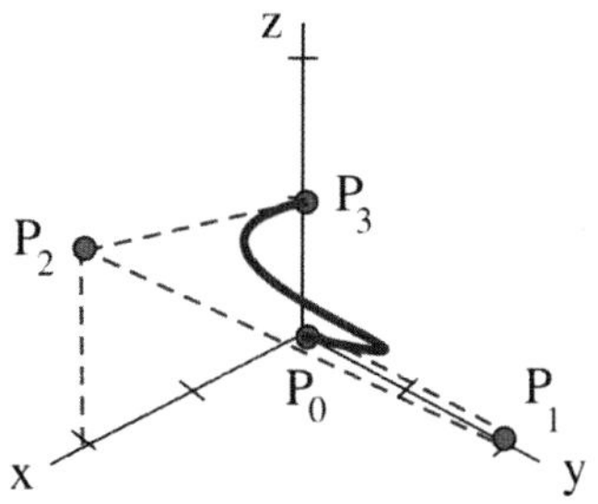

Fig. Problem 44

45. See Fig. 54

46. See Fig. 55

47. See Fig. 56

	A	B	C
x	D	D	D
y	I	I	I
z	D	D	D

Fig. 54

	A	B	C
x	D	D	D
y	I	I	I
z	D	I	D

Fig. 55

	A	B	C
x	D	I	D
y	I	I	I
z	D	D	D

Fig. 56

12.2 Selected Answers

1. $\mathbf{r}'(1) = \langle -, +, - \rangle$, $\mathbf{r}'(2) = \langle -, +, 0 \rangle$, and $\mathbf{r}'(3) = \langle -, 0, + \rangle$.

3. $\mathbf{r}'(1) = \langle +, -, + \rangle$, $\mathbf{r}'(2) = \langle +, -, + \rangle$, and $\mathbf{r}'(3) = \langle +, -, + \rangle$.

5. $\mathbf{r}'(t) = \langle 3t^2, 2, 2t \rangle$, $\mathbf{r}''(t) = \langle 6t, 0, 2 \rangle$

$\mathbf{v}(1) = \mathbf{r}'(1) = \langle 3, 2, 2 \rangle$, speed(1) $= |\mathbf{v}(1)| = \sqrt{17}$, direction(1) $= \dfrac{\mathbf{r}'(1)}{|\mathbf{r}'(1)|} = \dfrac{1}{\sqrt{17}} \langle 3, 2, 2 \rangle$

$\mathbf{v}(2) = \mathbf{r}'(2) = \langle 12, 2, 4 \rangle$, speed(2) $= |\mathbf{v}(2)| = \sqrt{164}$, direction(2) $= \dfrac{\mathbf{r}'(2)}{|\mathbf{r}'(2)|} = \dfrac{1}{\sqrt{164}} \langle 12, 2, 4 \rangle$.

$\mathbf{a}(1) = \mathbf{r}''(1) = \langle 6, 0, 2 \rangle$, $\mathbf{a}(2) = \mathbf{r}''(2) = \langle 12, 0, 2 \rangle$.

7. $\mathbf{r}'(t) = \langle -1, -4/t^2, 0 \rangle$, $\mathbf{r}''(t) = \langle 0, 8/t^3, 0 \rangle$

$\mathbf{v}(1) = \mathbf{r}'(1) = \langle -1, -4, 0 \rangle$, speed(1) $= |\mathbf{v}(1)| = \sqrt{17}$, direction(1) $= \dfrac{\mathbf{r}'(1)}{|\mathbf{r}'(1)|} = \dfrac{1}{\sqrt{17}} \langle -1, -4, 0 \rangle$

$\mathbf{v}(2) = \mathbf{r}'(2) = \langle -1, -1, 0 \rangle$, speed(2) $= |\mathbf{v}(2)| = \sqrt{2}$, direction(2) $= \dfrac{\mathbf{r}'(2)}{|\mathbf{r}'(2)|} = \dfrac{1}{\sqrt{2}} \langle -1, -1, 0 \rangle$.

$\mathbf{a}(1) = \mathbf{r}''(1) = \langle 0, 8, 0 \rangle$, $\mathbf{a}(2) = \mathbf{r}''(2) = \langle 0, 1, 0 \rangle$.

9. $\mathbf{r}(t) = \langle t^3, 7, 1 + 5t \rangle$, $\mathbf{r}'(t) = \langle 3t^2, 0, 5 \rangle$.

$\frac{d}{dt}\mathbf{r}(2t) = \mathbf{r}'(2t)\,\frac{d}{dt}(2t) = \langle 3(2t)^2, 0, 5 \rangle (2) = \langle 24t^2, 0, 10 \rangle$.

Or, $\mathbf{r}(2t) = \langle (2t)^3, 7, 1 + 5(2t) \rangle = \langle 8t^3, 7, 1 + 10t \rangle$ so $\frac{d}{dt}\mathbf{r}(2t) = \langle 24t^2, 0, 10 \rangle$.

11. $\mathbf{r}(t) = \langle t, 2t^2, 3t^3 \rangle$, $\mathbf{r}'(t) = \langle 1, 4t, 9t^2 \rangle$.

$\frac{d}{dt}\{\sin(t)\,\mathbf{r}(t)\} = \sin(t)\,\mathbf{r}'(t) + \cos(t)\,\mathbf{r}(t) = \sin(t)\langle 1, 4t, 9t^2 \rangle + \cos(t)\langle t, 2t^2, 3t^3 \rangle$.

$= \langle \sin(t) + t\cos(t), 4t\sin(t) + 2t^2\cos(t), 9t^2\sin(t) + 3t^3\cos(t) \rangle$.

13. $\mathbf{r}(t) = (2 - 5t^3)\mathbf{i} + (7t)\mathbf{j} + (1 + t)\mathbf{k} = \langle 2 - 5t^3, 7t, 1 + t \rangle$, $\mathbf{r}'(t) = \langle -15t^2, 7, 1 \rangle$.

$\frac{d}{dt}\mathbf{r}(3t) = \mathbf{r}'(3t)\,\frac{d}{dt}(3t) = \langle -15(3t)^2, 7, 1 \rangle (3) = \langle -405t^2, 21, 3 \rangle$.

Or, $\frac{d}{dt}\mathbf{r}(3t) = \frac{d}{dt}\langle 2 - 5(3t)^3, 7(3t), 1 + (3t) \rangle$

$= \frac{d}{dt}\langle 2 - 135t^3, 21t, 1 + 3t \rangle = \langle -405t^2, 21, 3 \rangle$.

15. $\frac{d}{dt}\{\mathbf{u} + 2\mathbf{v}\} = \frac{d}{dt}\langle 2 + 10t, 8 - t, 6 + t^3 \rangle = \langle 10, -1, 3t^2 \rangle$

$\frac{d}{dt}\{\mathbf{u}\cdot\mathbf{v}\} = \frac{d}{dt}\{4t - t^2 + 3t^3\} = 4 - 2t + 9t^2$

$\frac{d}{dt}\{\mathbf{u} \times \mathbf{v}\} = \frac{d}{dt}\langle 3t - 4t^3 + t^4, t^3 + 5t^4, -t - 5t^2 \rangle = \langle 3 - 12t^2 + 4t^3, 3t^2 + 20t^3, -1 - 10t \rangle$.

17. $\frac{d}{dt}\{\mathbf{u}+2\mathbf{v}\} = \frac{d}{dt}\langle 5t^3-4t+2, 2-t, t+10\rangle = \langle 15t^2-4, -1, 1\rangle$

$\frac{d}{dt}\{\mathbf{u}\cdot\mathbf{v}\} = \frac{d}{dt}\{-10t^4+5t^3-21t^2+10t+8\} = -40t^3+15t^2-42t+10$

$\frac{d}{dt}\{\mathbf{u}\times\mathbf{v}\} = \frac{d}{dt}\langle -3t^2-34t+8, -20t^3-2t^2-3t+2, 15t^4-14t^2+11t-2\rangle$

$= \langle -6t-34, -60t^2-4t-3, 60t^3-28t+11\rangle$.

19. The curves intersect at the point (0,3,9) when t = 3: $\mathbf{u}(3) = \langle 0, 3, 9\rangle = \mathbf{v}(3)$.

$\mathbf{u}'(3) = \langle -1, 1, 6\rangle$ and $\mathbf{v}'(3) = \langle 0, 1, 0\rangle$.

$\cos(\theta) = \frac{\mathbf{u}'(3)\cdot\mathbf{v}'(3)}{|\mathbf{u}'(3)||\mathbf{v}'(3)|} = \frac{1}{\sqrt{38}} \approx 0.162$ so $\theta \approx 1.408$ ($\approx 80.7^o$)

21. The curves intersect at the point (5,9,3) when t=–1 and s=3: $\mathbf{u}(-1) = \langle 5, 9, 3\rangle = \mathbf{v}(3)$.

$\mathbf{u}'(-1) = \langle -10, 0, -1\rangle$ and $\mathbf{v}'(3) = \langle 1, 3, -1\rangle$.

$\cos(\theta) = \frac{\mathbf{u}'(-1)\cdot\mathbf{v}'(3)}{|\mathbf{u}'(-1)||\mathbf{v}'(3)|} = \frac{-9}{\sqrt{101}\sqrt{11}} \approx -0.270$ so $\theta \approx 1.844$ ($\approx 105.7^o$)

23. Area of the parallelogram $= |\mathbf{u}\times\mathbf{v}| = |\langle -2t^3, t^3, -t^2\rangle| = \sqrt{5t^6+t^4}$

Rate of change of area $= \frac{d}{dt}\sqrt{5t^6+t^4} = \frac{15t^5+2t^3}{\sqrt{5t^6+t^4}} = \frac{17}{\sqrt{6}}$ when t = 1 and $\frac{496}{\sqrt{336}}$ when t = 2.

25. Area of the triangle $= \frac{1}{2}|\mathbf{u}\times\mathbf{v}| = \frac{1}{2}|\langle 0, 6t, -2t^2\rangle| = \frac{1}{2}\sqrt{36t^2+4t^4} = \sqrt{9t^2+t^4}$

Rate of change of area $= \frac{d}{dt}\sqrt{9t^2+t^4} = \frac{9t+2t^3}{\sqrt{9t^2+t^4}} = \frac{11}{\sqrt{10}}$ when t = 1 and $\frac{34}{\sqrt{52}}$ when t = 2.

27. Volume of the tetrahedron $= \frac{1}{6}|\mathbf{u}\cdot(\mathbf{v}\times\mathbf{s})| = \frac{1}{6}|3t^2| = \frac{1}{2}t^2$.

Rate of change of volume $= \frac{d}{dt}\frac{1}{2}t^2 = t = 1$ when t = 1 and 2 when t = 2.

29. $\mathbf{r}(t) = \langle 6t^2+1, 4t^3+2, 6e^t-3\rangle$

30. $\mathbf{r}(t) = \langle 3t+2t^2+7, \sin(t)+2, t-3t^2+5\rangle$

31. $\mathbf{r}(t) = \langle 2t^3+4, 4t-2, 4t^2-5t-2\rangle$

33 and 34. See Fig. 18

35 and 36. See Fig. 19

t	$\mathbf{r}''(t)$	$\mathbf{r}'(t)$	$\mathbf{r}(t)$
0	⟨0, 2, 5⟩	⟨1, 2, 3⟩	⟨0, 3, 1⟩
1	⟨4, 1, 3⟩	⟨5, 3, 6⟩	⟨5, 6, 7⟩
2	⟨6, 0, 1⟩	⟨11, 3, 7⟩	⟨16, 9, 14⟩
3	⟨4, –2, 0⟩	⟨15, 1, 7⟩	⟨31, 10, 21⟩
4	⟨2, 0, 2⟩	⟨17, 1, 9⟩	⟨48, 11, 30⟩
5	⟨8, 3, 4⟩	⟨25, 4, 13⟩	⟨73, 15, 43⟩

Fig. 18

t	$\mathbf{r}''(t)$	$\mathbf{r}'(t)$	$\mathbf{r}(t)$
0	⟨1, 2, 3⟩	⟨1, 6, 4⟩	⟨17, 3, 2⟩
1	⟨4, 2, 2⟩	⟨5, 8, 6⟩	⟨22, 11, 8⟩
2	⟨3, 1, 0⟩	⟨8, 9, 6⟩	⟨30, 20, 14⟩
3	⟨2, 3, 1⟩	⟨10, 12, 7⟩	⟨40, 32, 21⟩
4	⟨1, 4, 0⟩	⟨11, 16, 7⟩	⟨51, 48, 28⟩
5	⟨0, 1, 3⟩	⟨11, 17, 10⟩	⟨62, 65, 38⟩

Fig. 19

12.3 Selected Answers

1. $\mathbf{L} = \int_{t=a}^{t=b} |\mathbf{r}'(t)|\,dt = \int_{t=0}^{t=2\pi} \sqrt{4\sin^2(t) + 4\cos^2(t) + 1}\;dt = \int_{t=0}^{t=2\pi} \sqrt{5}\;dt = 2\pi\cdot\sqrt{5} \approx 14.05$

2. $\mathbf{L} = \int_{t=a}^{t=b} |\mathbf{r}'(t)|\,dt = \int_{t=0}^{t=2\pi} \sqrt{9\sin^2(t) + 9\cos^2(t) + 1}\;dt = \int_{t=0}^{t=2\pi} \sqrt{10}\;dt = 2\pi\cdot\sqrt{10} \approx 19.87$

3. $\mathbf{L} = 2\pi\sqrt{17} \approx 25.91$

4. $\mathbf{L} = 2\pi\sqrt{R^2 + 1}$

5. $\mathbf{L} = \int_{t=a}^{t=b} |\mathbf{r}'(t)|\,dt = \int_{t=0}^{t=2\pi} \sqrt{4\sin^2(t) + 9\cos^2(t) + 1}\;dt \approx 17.08$

6. $\mathbf{L} = \int_{t=a}^{t=b} |\mathbf{r}'(t)|\,dt = \int_{t=0}^{t=2\pi} \sqrt{4\sin^2(t) + 25\cos^2(t) + 1}\;dt \approx 23.93$

7. $\mathbf{L} = \int_{t=a}^{t=b} |\mathbf{r}'(t)|\,dt = \int_{t=0}^{t=2\pi} \sqrt{A^2\sin^2(t) + B^2\cos^2(t) + 1}\;dt$

8. $\mathbf{L} = \int_{t=a}^{t=b} |\mathbf{r}'(t)|\,dt = \int_{t=0}^{t=2\pi} \sqrt{4\sin^2(t) + 4\cos^2(t) + 1}\;dt = \int_{t=0}^{t=2\pi} \sqrt{5}\;dt = 2\pi\cdot\sqrt{5} \approx 14.05$

9. $\mathbf{L} = \int_{t=a}^{t=b} |\mathbf{r}'(t)|\,dt = \int_{t=0}^{t=2\pi} \sqrt{t^2 + 2}\;dt \approx 22.43$

11. $\mathbf{L} = \int_{t=a}^{t=b} |\mathbf{r}'(t)|\,dt = \int_{t=0}^{t=2\pi} \sqrt{(-2t\sin(t) + 2\cos(t))^2 + (t\cos(t) + \sin(t))^2 + 1}\;dt \approx 34.02$

13. $x'(t) = 3 - 6t,\ y'(t) = 18t - 21t^2,\ z'(t) = 3 - 18t + 5t^2$.

$$\mathbf{L} = \int_{t=a}^{t=b} |\mathbf{r}'(t)|\,dt = \int_{t=0}^{t=1} \sqrt{(x'(t))^2 + (y'(t))^2 + (z'(t))^2}\;dt$$

15. $x'(t) = -6 + 24t - 24t^2,\ y'(t) = 3 - 6t + 9t^2,\ z'(t) = 3 - 6t$.

$$\mathbf{L} = \int_{t=a}^{t=b} |\mathbf{r}'(t)|\,dt = \int_{t=0}^{t=1} \sqrt{(x'(t))^2 + (y'(t))^2 + (z'(t))^2}\;dt$$

17. $\mathbf{r}'(t) = \langle -\sin(t), \cos(t), 1 \rangle$, $\mathbf{r}''(t) = \langle -\cos(t), -\sin(t), 0 \rangle$, $\mathbf{r}'(t) \times \mathbf{r}''(t) = \langle \sin(t), -\cos(t), 1 \rangle$.

Then $|\mathbf{r}'(t)| = \sqrt{2}$ and $|\mathbf{r}'(t) \times \mathbf{r}''(t)| = \sqrt{2}$, so

$$\kappa = \frac{|\mathbf{r}' \times \mathbf{r}''|}{|\mathbf{r}'|^3} = \frac{\sqrt{2}}{(\sqrt{2})^3} = \frac{1}{2} \text{ for all values of } t.$$

19. $\mathbf{r}'(t) = \langle -R\sin(t), R\cos(t), 1 \rangle$, $\mathbf{r}''(t) = \langle -R\cos(t), -R\sin(t), 0 \rangle$, $\mathbf{r}'(t) \times \mathbf{r}''(t) = \langle R\sin(t), -R\cos(t), R^2 \rangle$.

Then $|\mathbf{r}'(t)| = \sqrt{R^2+1}$ and $|\mathbf{r}'(t) \times \mathbf{r}''(t)| = |R|\sqrt{R^2+1}$, so

$$\kappa = \frac{|\mathbf{r}' \times \mathbf{r}''|}{|\mathbf{r}'|^3} = \frac{|R|\sqrt{R^2+1}}{(\sqrt{R^2+1})^3} = \frac{|R|}{R^2+1} \text{ for all values of } t.$$

21. $x'(t) = -6 - 18t + 9t^2$, $y'(t) = 18t - 21t^2$, $z'(t) = 3 - 18t + 15t^2$.

$x''(t) = -18 + 18t$, $y''(t) = 18 - 42t$, $z''(t) = -18 + 30t$.

When $t = 0.2$, $x' = 2.76$, $y' = 2.76$, $z' = 0$, $x'' = -17.28$, $y'' = -26.4$, $z'' = -12$ so

$\mathbf{r}' \times \mathbf{r}'' = -33.12\mathbf{i} + 33.12\mathbf{j} - 25.1712\mathbf{k}$, $|\mathbf{r}' \times \mathbf{r}''| \approx 53.17$, $|\mathbf{r}'| \approx 3.90$

and $\kappa = \dfrac{|\mathbf{r}' \times \mathbf{r}''|}{|\mathbf{r}'|^3} = \dfrac{53.17}{(3.90)^3} = 0.896$. The radius of curvature is $\dfrac{1}{\kappa} \approx 1.12$.

When $t = 0.5$, $x' = -0.75$, $y' = 3.75$, $z' = -2.25$, $x'' = -9$, $y'' = -3$, $z'' = -3$

$\mathbf{r}' \times \mathbf{r}'' = -18\mathbf{i} + 18\mathbf{j} + 36\mathbf{k}$, $|\mathbf{r}' \times \mathbf{r}''| \approx 44.09$, $|\mathbf{r}'| \approx 4.44$

and $\kappa = \dfrac{|\mathbf{r}' \times \mathbf{r}''|}{|\mathbf{r}'|^3} = \dfrac{44.09}{(4.44)^3} = 0.504$. The radius of curvature is $\dfrac{1}{\kappa} \approx 1.98$.

23. $x'(t) = -3\sin(t)$, $y'(t) = 5\cos(t)$, $x''(t) = -3\cos(t)$, $y''(t) = -5\sin(t)$. Then

$$\kappa = \frac{|x'y'' - x''y'|}{((x')^2 + (y')^2)^{3/2}} = \frac{|15\sin^2(t) - -15\cos^2(t)|}{(9\sin^2(t) + 25\cos^2(t))^{3/2}} .$$

When $t = 0$, $\kappa = \dfrac{15}{(9\sin^2(0) + 25\cos^2(0))^{3/2}} = \dfrac{15}{125} = 0.12$. Radius of curvature = $\dfrac{25}{3} \approx 8.33$.

When $t = \dfrac{\pi}{4}$, $\kappa = \dfrac{15}{(9\sin^2(\pi/4) + 25\cos^2(\pi/4))^{3/2}} = \dfrac{15}{17^{3/2}} = 0.214$. Radius of curvature ≈ 4.67 .

When $t = \dfrac{\pi}{2}$, $\kappa = \dfrac{15}{(9\sin^2(\pi/2) + 25\cos^2(\pi/2))^{3/2}} = \dfrac{15}{27} = 0.555$. Radius of curvature $= \dfrac{9}{5} = 1.8$.

25. $x'(t) = -A\sin(t)$, $y'(t) = B\cos(t)$, $x''(t) = -A\cos(t)$, $y''(t) = -B\sin(t)$. Then

$$\kappa = \frac{|x'y'' - x''y'|}{((x')^2 + (y')^2)^{3/2}} = \frac{|AB\sin^2(t) + AB\cos^2(t)|}{(A^2\sin^2(t) + B^2\cos^2(t))^{3/2}} .$$

When $t = 0$, $\kappa = \dfrac{|AB|}{(A^2\sin^2(0) + B^2\cos^2(0))^{3/2}} = \dfrac{|AB|}{|B|^3} = \dfrac{|A|}{|B|^2}$. Radius of curvature $= \dfrac{1}{\kappa} = \dfrac{|B|^2}{|A|}$.

When $t = \dfrac{\pi}{4}$, $\kappa = \dfrac{|AB|}{(A^2\sin^2(\pi/4) + B^2\cos^2(\pi/4))^{3/2}} = \dfrac{|AB|}{((A^2 + B^2)/2)^{3/2}} =$. Radius of curvature $= \dfrac{1}{\kappa}$.

When $t = \dfrac{\pi}{2}$, $\kappa = \dfrac{|AB|}{(A^2\sin^2(\pi/2) + B^2\cos^2(\pi/2))^{3/2}} = \dfrac{|AB|}{|A|^3} = \dfrac{|B|}{|A|^2}$. Radius of curvature $= \dfrac{1}{\kappa} = \dfrac{|A|^2}{|B|}$.

27. $y' = 3$ and $y'' = 0$ so $\kappa = \dfrac{|y''|}{(1 + (y')^2)^{3/2}} = \dfrac{0}{10^{3/2}} = 0$. (As we might expect, the curvature of the straight line $y = 3x + 5$ is 0.)

29. $y' = 2x$ and $y'' = 2$ so $\kappa = \dfrac{|y''|}{(1 + (y')^2)^{3/2}} = \dfrac{2}{(1 + 4x^2)^{3/2}}$.

When $x = 1$, $\kappa = \dfrac{2}{5^{3/2}} \approx 0.1789$. When $x = 2$, $\kappa = \dfrac{2}{17^{3/2}} \approx 0.0285$. When $x = 3$, $\kappa = \dfrac{2}{37^{3/2}} \approx 0.0089$.

12.4 Selected Answers

1 – 23 Odd: The answers are shown in the figures.

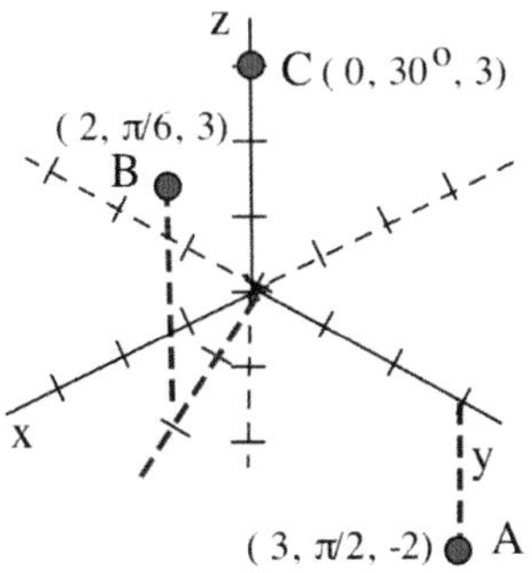

Fig. Problem 1

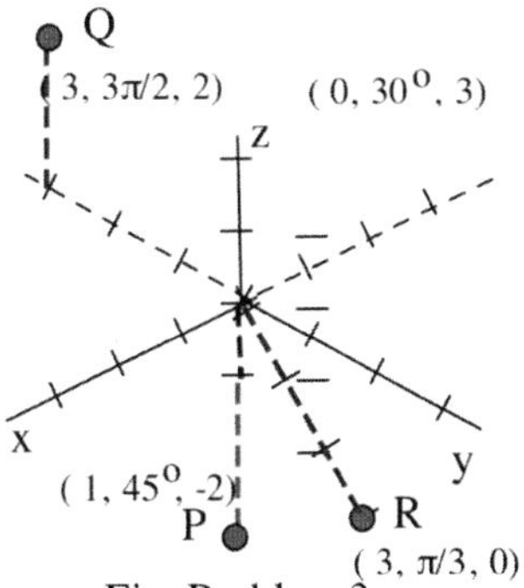

Fig. Problem 3

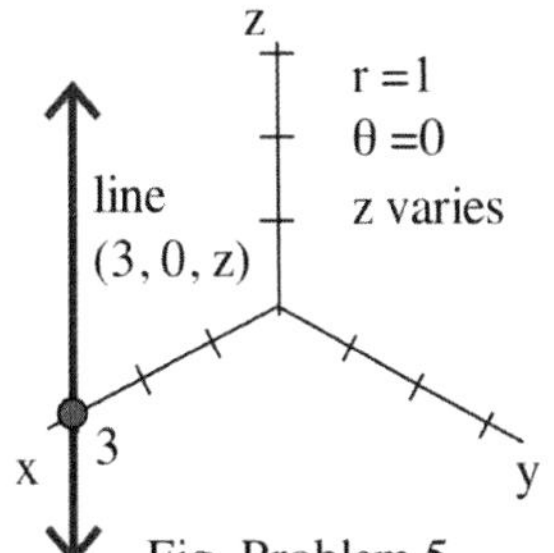

Fig. Problem 5

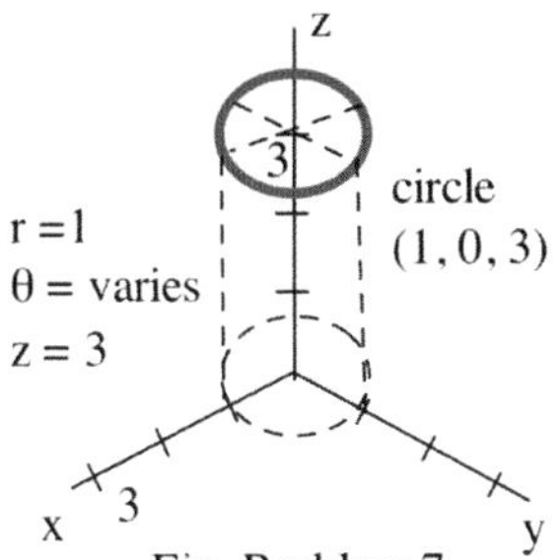

Fig. Problem 7

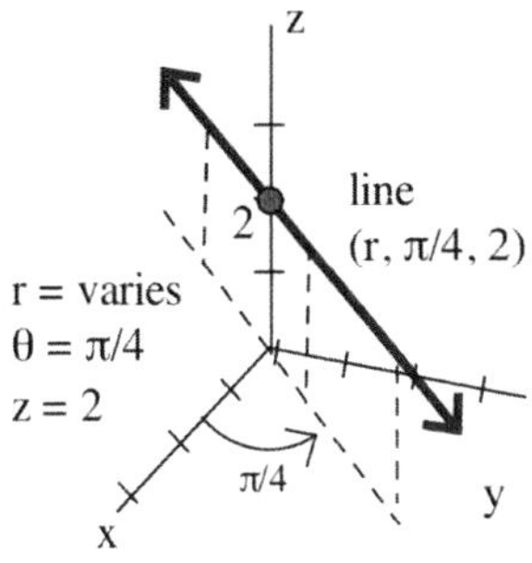

Fig. Problem 9

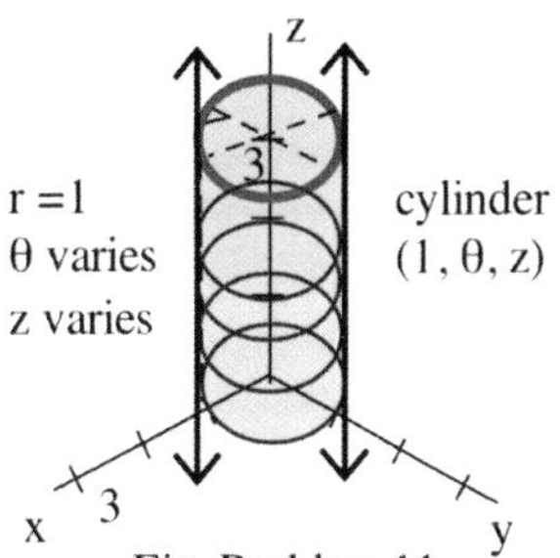

Fig. Problem 11

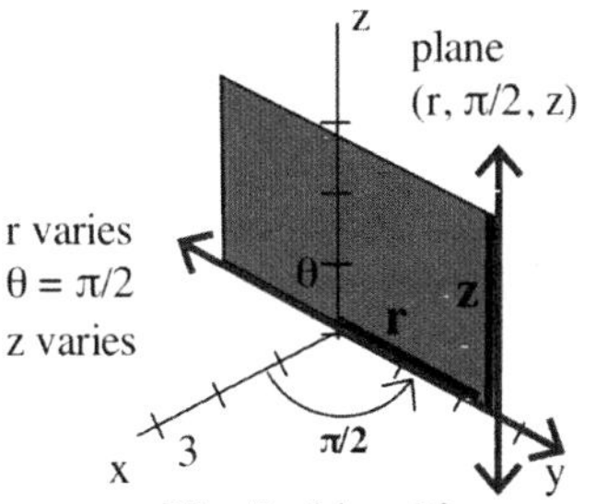

Fig. Problem 13

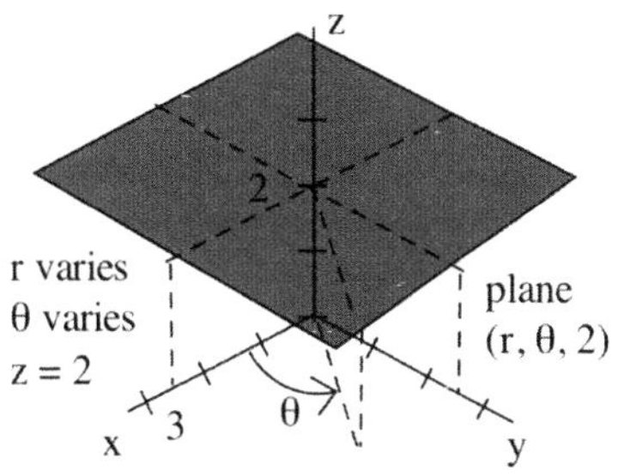

Fig. Problem 15

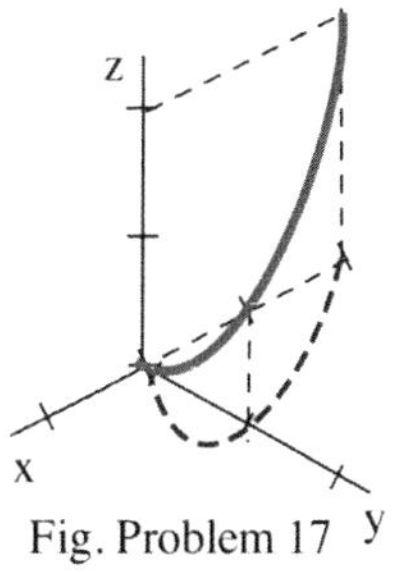

Fig. Problem 17

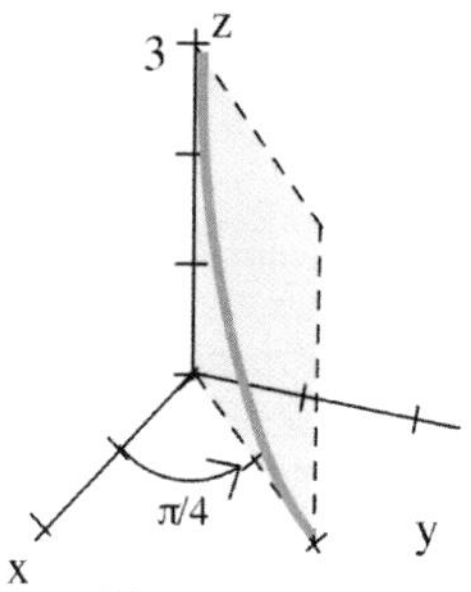

Fig. Problem 19

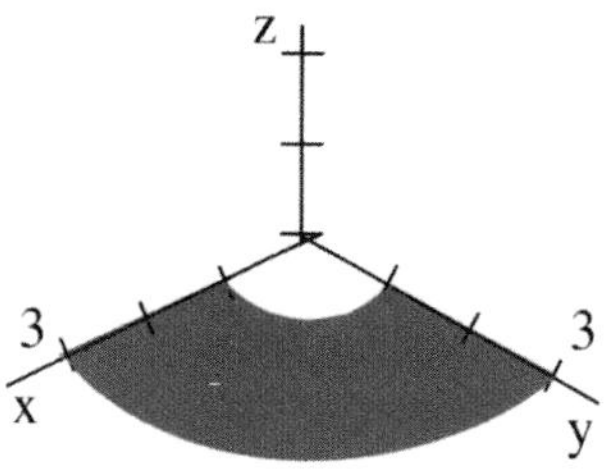

Fig. Problem 21

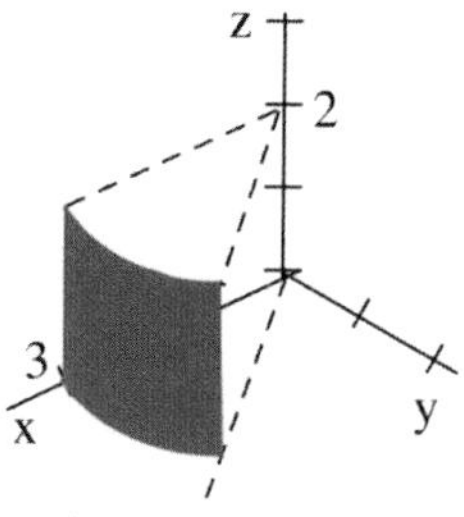

Fig. Problem 23

25. $x = 5\cos(\pi/6) \approx 4.33$, $y = 5\sin(\pi/6) = 2.5$, $z = 3$.

27. $x = 3\cos(35^o) \approx 2.46$, $y = 3\sin(35^o) \approx 1.72$, $z = -2$.

29. $r = \sqrt{x^2 + y^2} = \sqrt{1^2 + 2^2} = \sqrt{5}$,

$\theta = \arctan(y/x) = \arctan(2/1) \approx 1.107$, $z = 3$.

31. $r^2 = x^2 + y^2 = 5^2 = 25$,

$\theta = \arctan(y/x) = \arctan(-3/4) \approx -0.644$,

$z = -1$. To get the correct location, we need to use $r = -5$.

33. (a) $x^2 + y^2 = 4y - 1$ or $x^2 + (y-2)^2 = 3$. (b) $\sqrt{x^2 + y^2} = 7$ or $x^2 + y^2 = 49$

35. (a) $x^2 + y^2 = 5x$ or $(x - 5/2)^2 + y^2 = 25/4$. (b) $z = x^2 + y^2$

37. (a) $z = r^2 - 3r\cdot\cos(\theta) + 2r\cdot\sin(\theta)$ (b) $r\cdot\cos(\theta) = 3$

39. $z = r^2\cos^2(\theta) + 5r^2\sin^2(\theta)$ (b) $r\cdot\cos(\theta) + r\cdot\sin(\theta) + z = 5$

41 – 59 Odd: The answers are shown in the figures.

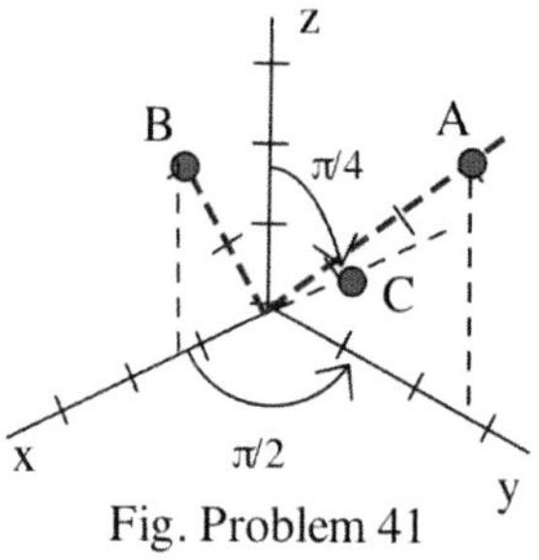

Fig. Problem 41

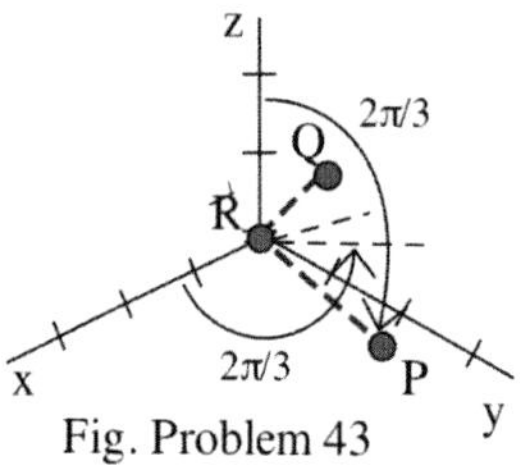

Fig. Problem 43

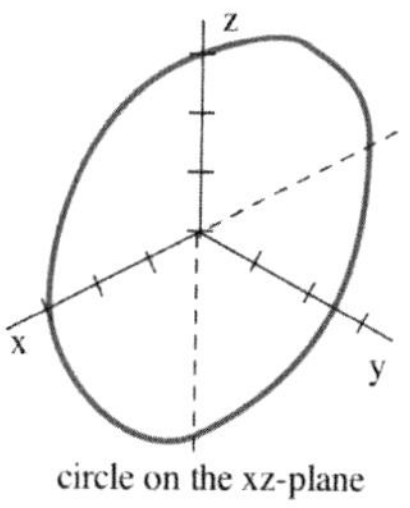

circle on the xz-plane
Fig. Problem 45

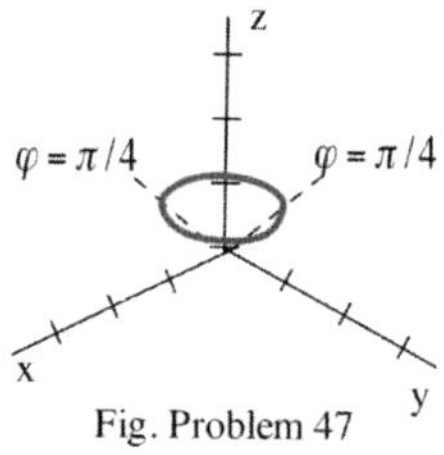

Fig. Problem 47

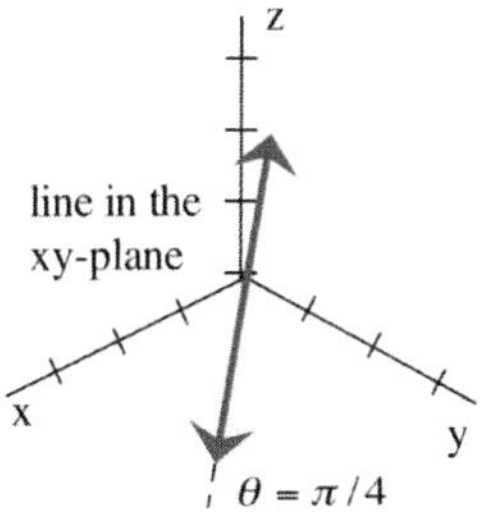

Fig. Problem 49

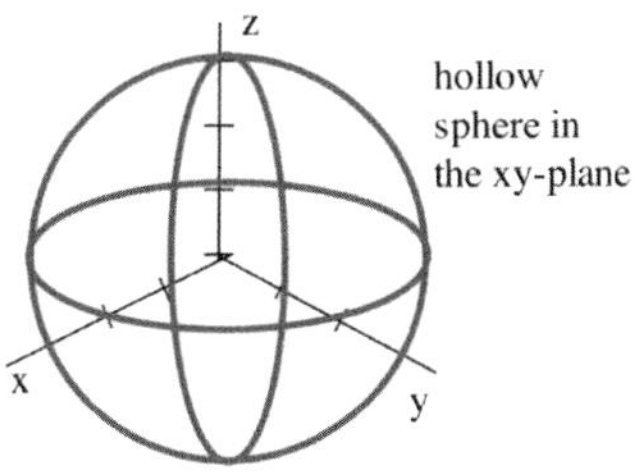

Fig. Problem 51

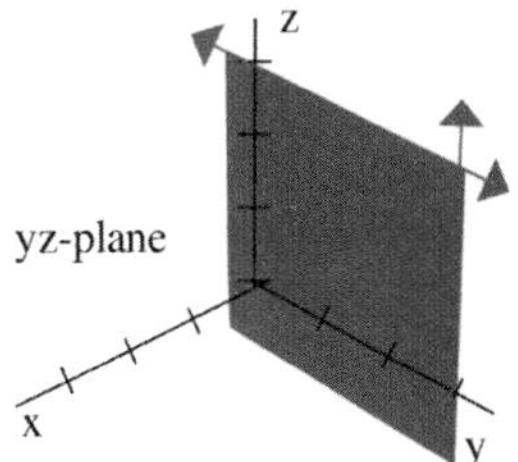

Fig. Problem 53

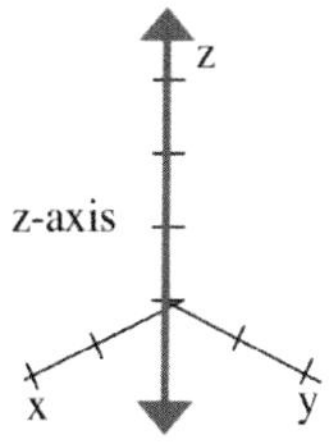

Fig. Problem 55

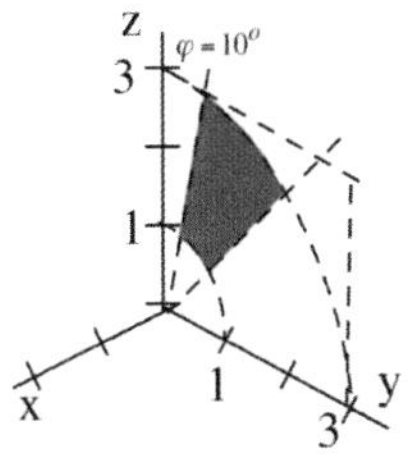

Fig. Problem 57

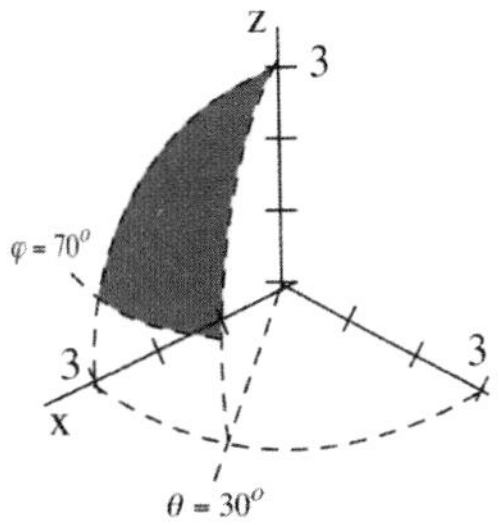

Fig. Problem 59

61. $x = 5\cdot\sin(\pi/3)\cdot\cos(\pi/2) = 0$, $y = 5\cdot\sin(\pi/3)\cdot\sin(\pi/2) = 5\sqrt{3}/2 \approx 4.330$, $z = 5\cdot\cos(\pi/3) = 5/2$

63. $x = 4\cdot\sin(30^o)\cdot\cos(45^o) = (4)(\frac{1}{2})(\frac{\sqrt{2}}{2}) = \sqrt{2} \approx 1.414$

$y = 4\cdot\sin(30^o)\cdot\sin(45^o) = (4)(\frac{1}{2})(\frac{\sqrt{2}}{2}) = \sqrt{2} \approx 1.414$, $z = 4\cdot\cos(30^o) = (4)(\frac{\sqrt{3}}{2}) = 2\sqrt{3} \approx 3.468$

65. $\rho = \sqrt{x^2 + y^2 + z^2} = \sqrt{14} \approx 3.742$, $\theta = \arctan(y/x) = \arctan(2/1) \approx 1.107$ ($\approx 63.4^o$),

$\varphi = \arccos(z/\ ) = \arccos(3/\sqrt{14}) \approx 0.641$ ($\approx 36.7^o$)

67. $\rho = \sqrt{x^2 + y^2 + z^2} = \sqrt{38} \approx 6.164$, $\theta = \arctan(y/x) = \arctan(-3/5) \approx -0.540$ ($\approx -30.9^o$),

$\varphi = \arccos(z/\rho) = \arccos(2/\sqrt{38}) \approx 1.240$ ($\approx 71.0^o$)

69. (a) $5 = \rho = \sqrt{x^2 + y^2 + z^2}$ or $x^2 + y^2 + z^2 = 25$, (b) (Graphically) $x = 0$

71. (a) $\rho = 5\cdot\sin(\varphi)\cdot\cos(\theta)$

$\rho^2 = 5\rho\cdot\sin(\varphi)\cdot\cos(\theta)$

$x^2 + y^2 + z^2 = 5x$

(b) $\rho = 3\cdot\sec(\varphi) = 3\frac{1}{\cos(\varphi)}$

$\rho\cdot\cos(\varphi) = 3$

$z = 3$

73. (a) $\rho^2 = 9$ (b) $\rho\cdot\sin(\varphi)\cdot\cos(\theta) + \rho\cdot\cos(\varphi) = 2$

75. (a) $\rho\cdot\cos(\varphi) = 2\rho^2\cdot\sin^2(\varphi)$ (b) $\rho^2\cdot\cos^2(\varphi) = 25 - \rho^2\cdot\sin^2(\varphi)\cdot\cos^2(\theta)$

13.1 FUNCTIONS OF TWO OR MORE VARIABLES

This section presents many of the precalculus concepts relating to functions of several variables, and the focus is on using formulas and tables of data to create and interpret graphical representations of functions of two variables.

Definition:

> A **function f of two variables** is a rule that assigns to each ordered pair (x, y) in the domain of the function a unique real number z. This can be written $z = f(x,y)$.

As with a function of one variable, a function of two variables is typically given by a table of values, a graph, or a formula.

Tables of Values

If we have data or perform measurements about the elevation of the ground (above sea level) at several locations, it is natural to record the elevation measurements using several columns of numbers as in Fig. 1.

Example 1: Describe the progression of elevations for the data in Fig. 1 if we start at the location (2,1), keep the x–values constant at $x = 2$, and proceed to increase the values of y: list the values of $f(2,1), f(2,2), f(2,3), f(2,4), \ldots$

Solution: $f(2,1) = 4.3, f(2,2) = 5.6, f(2,3) = 6.7, f(2,4) = 7.1, f(2,5) = 6.7$, and $f(2,6) = 5.6$. As y increases from 1 to 4, the z–values increase to a maximum of $z = 7.1$. As the y–values increase from 4 to 6, the z–values then decrease.

Practice 1: Describe the progression of elevations for the data in Fig. 1 if we start at the location (2,1), keep the y–values constant at $y = 1$, and proceed to increase the values of x: list the values of $f(2,1), f(3,1), f(4,1), f(5,1), \ldots$

x	y	z	x	y	z	x	y	z
0	0	1.6	3	0	3.8	6	0	1.6
0	1	1.8	3	1	5.3	6	1	1.8
0	2	2.0	3	2	7.1	6	2	2.0
0	3	2.1	3	3	9.1	6	3	2.1
0	4	2.2	3	4	10.0	6	4	2.2
0	5	2.1	3	5	9.1	6	5	2.1
0	6	2.0	3	6	7.1	6	6	2.0
1	0	2.4	4	0	3.3			
1	1	2.9	4	1	4.3			
1	2	3.3	4	2	5.6			
1	3	3.7	4	3	6.7			
1	4	5.9	4	4	7.1			
1	5	3.7	4	5	6.7			
1	6	3.3	4	6	5.6			
2	0	3.3	5	0	2.4			
2	1	4.3	5	1	2.9			
2	2	5.6	5	2	3.3			
2	3	6.7	5	3	3.7			
2	4	7.1	5	4	3.8			
2	5	6.7	5	5	3.7			
2	6	5.6	5	6	3.3			

Fig. 1: z is the elevation at location (x,y)

Many people find it difficult to recognize patterns and shapes from lists of numbers, and it is common to arrange the information in a table as in Fig. 2. With this table arrangement it is easier to answer the questions in Example 1 and Practice 1 as well as more complicated questions.

Example 2: Describe the progression of elevations for the data in Fig. 2 if we start at the location (2,1) and move "southeast" (x and y both increase at the same rate): list the values of $f(2,1)$, $f(3,2), f(4,3), f(5,4), \ldots$

Solution: $f(2,1) = 4.3, f(3,2) = 7.1, f(4,3) = 6.7,$ $f(5,4) = 3.8,$ and $f(6,5) = 2.1$.

x \ y	0	1	2	3	4	5	6
0	1.6	1.8	2.0	2.1	2.2	2.1	2.0
1	2.4	2.9	3.3	3.7	5.9	3.7	3.3
2	3.3	4.3	5.6	6.7	7.1	6.7	5.6
3	3.8	5.3	7.1	9.1	9.9	9.1	7.1
4	3.3	4.3	5.6	6.7	7.1	6.7	5.6
5	2.4	2.9	3.3	3.7	3.8	3.7	3.3
6	1.6	1.8	2.0	2.1	2.2	2.1	2.0

Fig. 2: Elevations arranged in a table

Graphs

Once the data is arranged in a table, it seems natural to create a partial graph of the function by plotting the elevations as points at the appropriate elevations above the xy–plane, but such a point graph is usually difficult to "read." Instead of a point graph, we could build a surface by plotting a small platform at each point (Fig. 3) or by connecting the points to nearby points (Fig. 4).

For a function f of two variables, the graph of the surface defined by f is the set of points $z = f(x, y)$ where the ordered pairs (x,y) are in the domain of f.

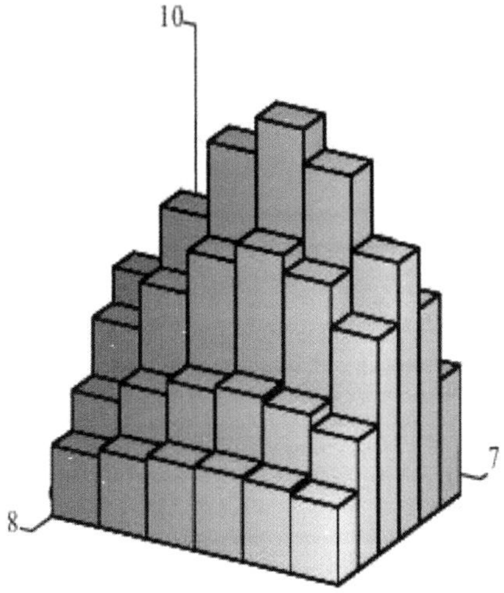

Fig. 3: A "graph" of the data from Fig. 2

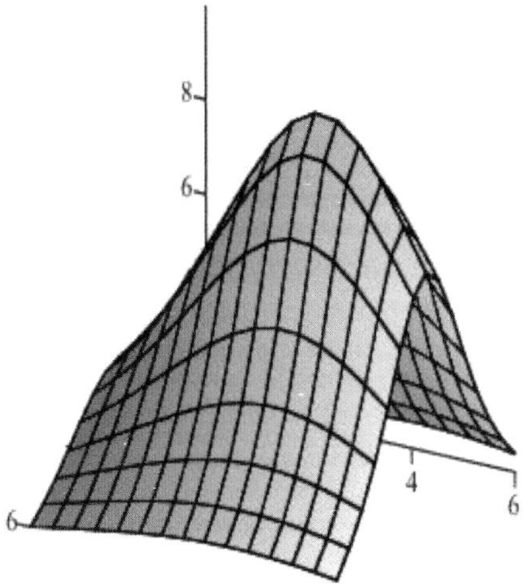

Fig. 4: A surface from the data in Fig. 2

Note: Figures 2 and 3 were drawn by computer using the language Maple. The Maple commands for drawing these and the other 3–dimensional surfaces in this section are given after the problem set.

If we "slice" the surface in Fig. 4 with the plane z = 8, the points where the plane cuts the surface are those points where the elevation of the surface is 8 units above the xy–plane. Fig. 5 shows the surface being sliced by the planes z = 8 and z = 4.

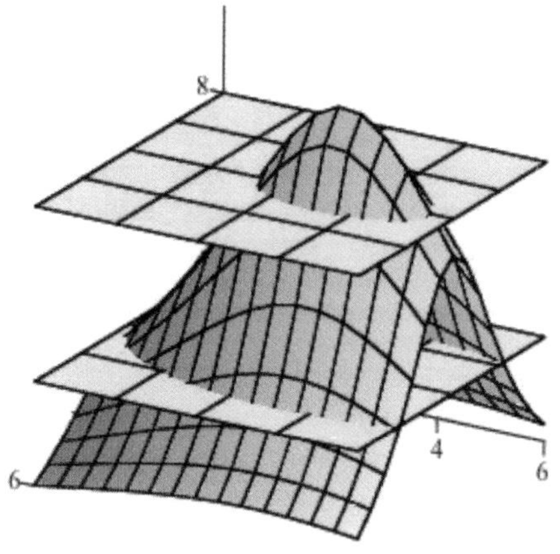

Fig. 5: The planes z=4 and z=8 "slicing" the surface

Fig. 6 shows the results of slicing the surface with several planes at different elevations. The thicker curves are those points where the planes z = 4 and z = 8 intersect the surface.

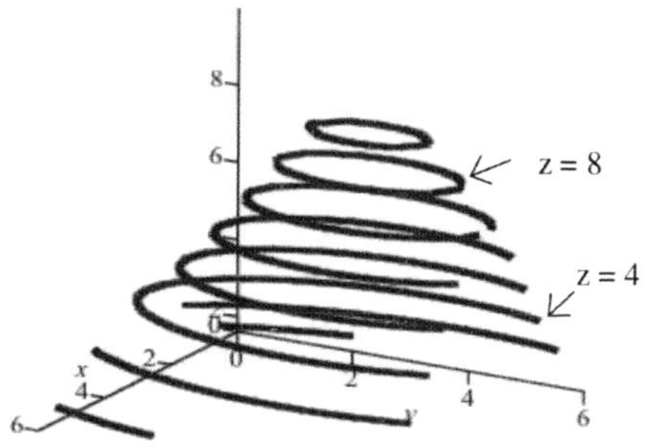

Fig. 6: Intersections of the surface and several planes

Definition:

The **k level curve** of a function f of two variables is the set of points (x,y) that satisfy the equation f(x,y) = k, where k is a constant.

If we move all of the curves in Fig. 6 to the xy–plane (or, equivalently, view them from directly overhead), the result is a 2–dimensional graph of the level curves of the original surface. A graph of several level curves of the surface is shown in Fig. 7. We call such a graph a **level curve graph of f** or a **contour graph of f.**

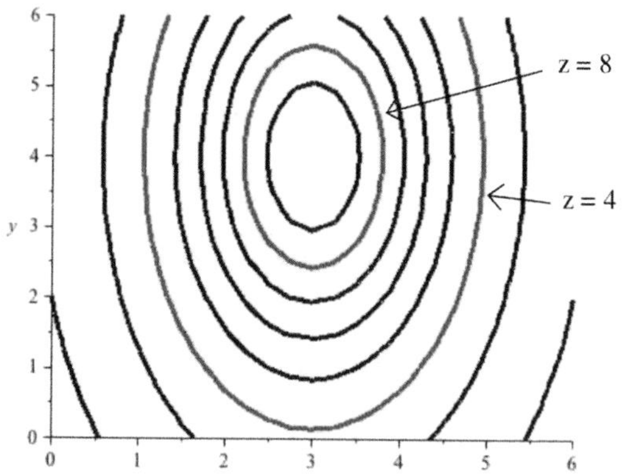

Fig. 7: Several level curves of the surface in Fig. 3

Level curve graphs are an effective and efficient method of presenting a great deal of information about a surface or function of two variables in a 2–dimensional way, and such graphs are very commonly used. Weather maps showing temperatures or barometric pressures over a region use a variation of contour graphs (Fig. 8). Hikers are familiar with topographic maps that show elevations of the terrain over which they plan to hike (Fig. 9). And almost every issue of scientific journals such as *Science* or *Nature* contains a variety of surface and contour graphs.

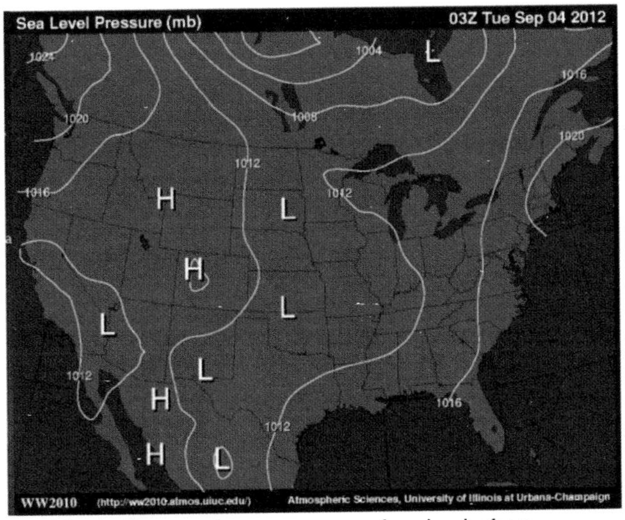

Fig. 8: Baromatric pressure map showing isobars

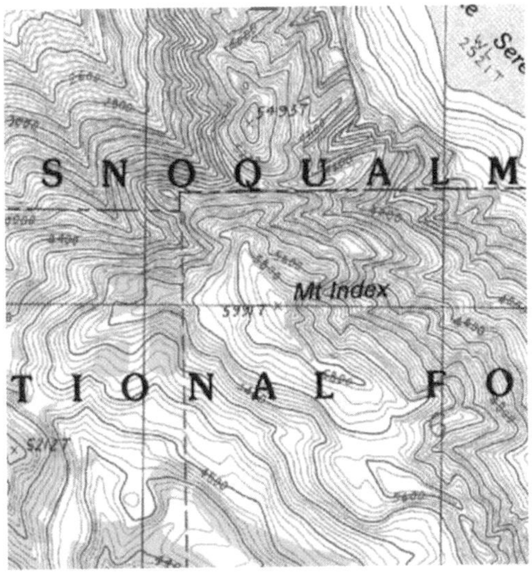

Fig. 9: Topographic map showing level contours

Practice 2: Fig. 10 shows a level curve graph of the temperature at each location on a table. On Fig. 11 sketch the a graph of the temperatures a bug experiences as it moves from point A to point B along a straight line. On Fig. 12 sketch the a graph of the temperatures a bug experiences as it moves from point C clockwise along the ellipse and returns to point C.

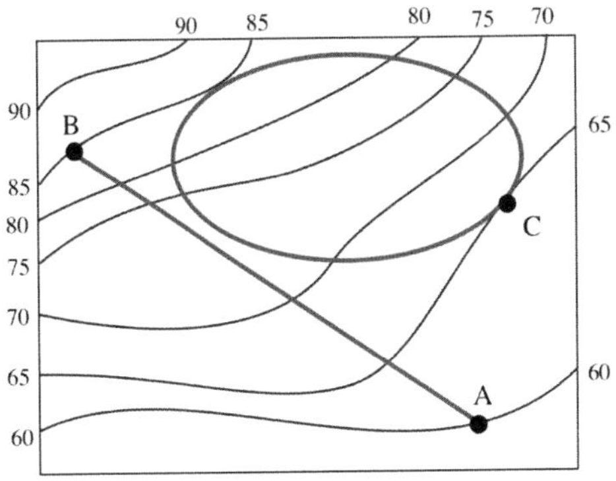

Fig. 10: Temperatures on a table

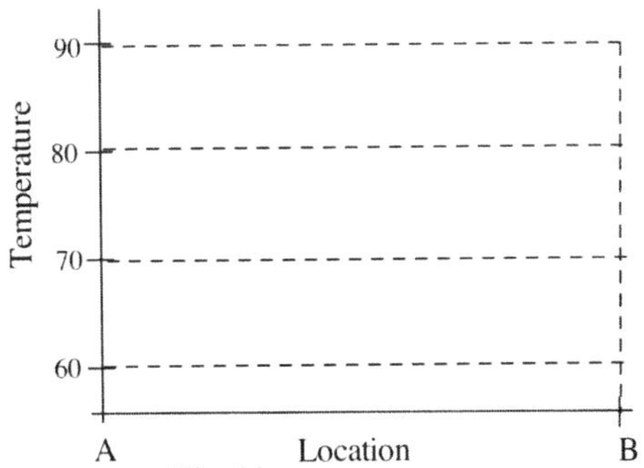

Fig. 11

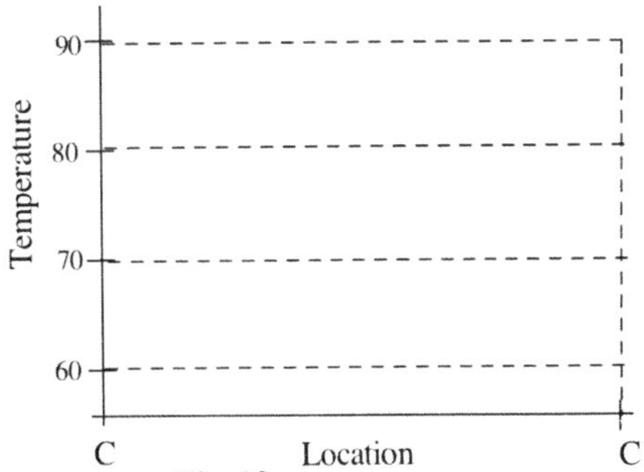

Fig. 12

Sketching Level Curves of a Surface From Data

In general, it is difficult to sketch good surface graphs (such as Fig. 4) from data or even from an equation for a surface, and such surface graphs are typically done by computers. It is much easier, however, to sketch level curves of a surface.

If our surface is described by data (by z–values at given locations (x,y)) and if we can assume that the surface does not have any holes or jumps, then there is a straightforward method for sketching crude level curves for the surface.

"Crude" Level Curve $z = k$ Algorithm for Surface Data (Fig. 13)

Step 1: "Triangulate" the data locations by connecting adjacent locations with light (dotted) lines. The triangulation should be done so no dotted lines cross (but they can meet at the data locations).

Step 2: Plot a point (or small box) on each dotted line segment whose endpoints "surround" the value k (one endpoint is larger than k and one endpoint is smaller).

Step 3: In each triangle, connect the boxes with a line segment.

Step 4: Remove the light (dotted) lines.

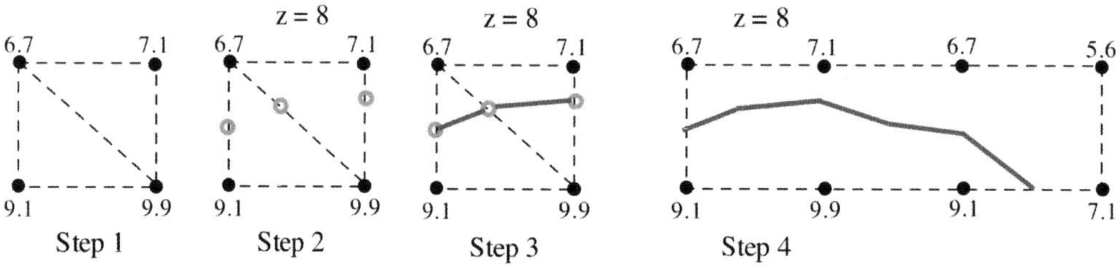

Fig. 13: Steps for drawing a (crude) level curve z=8 from data

Fig. 14 shows the result when this algorithm is used in the data from Fig. 2 to approximate the level curves $z = 4$ and $z = 8$.

x \ y	0	1	2	3	4	5	6
0	1.6	1.8	2.0	2.1	2.2	2.1	2.0
1	2.4	2.9	3.3	3.7	5.9	3.7	3.3
2	3.3	4.3	5.6	6.7	7.1	6.7	5.6
3	3.8	5.3	7.1	9.1	9.9	9.1	7.1
4	3.3	4.3	5.6	6.7	7.1	6.7	5.6
5	2.4	2.9	3.3	3.7	3.8	3.7	3.3
6	1.6	1.8	2.0	2.1	2.2	2.1	2.0

z = 4

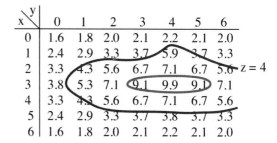

Fig. 14: Crude level curves z=4 and z=8 from data

If the data points are far apart, the resulting level curves may be only crude approximations of the actual level curves of the surface. However, if the data points are relatively close together and if the surface does not change elevation rapidly, then the algorithm results in a good approximation of the actual level curves of the surface. Many computer programs use variations of this algorithm to create level curves of surfaces.

Note: After a bit of practice most people don't bother actually sketching the dotted lines, but they still need to "think triangles."

It is possible to work with functions of three or more variables, but they are difficult (or impossible) to graph. The idea of a level curve for a function of 2 variables generalizes to a **level surface** for a function of 3 variables.

Sketching Level Curves of a Surface From A Formula

If the surface is described by a formula of the form $z = f(x,y)$, then we can often use algebra and our 2–dimensional graphing skills to create level curves for the surface.

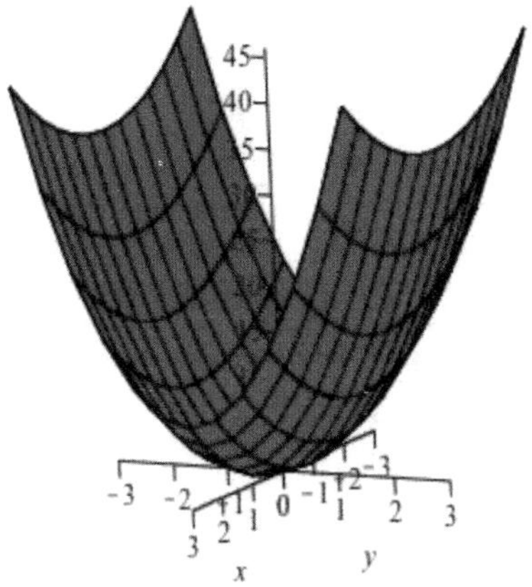
Fig. 15: $z = x^2 + 4y^2$ surface

Example 3: Sketch the level curve $z = 4$ for the surface $z = x^2 + 4y^2$ (Fig. 15).

Solution: The graph of the level curve $z = 4$ is the graph of $4 = x^2 + 4y^2$ or $\frac{x^2}{4} + \frac{y^2}{1} = 1$.

From Chapter 9, we know that the graph of $\frac{x^2}{4} + \frac{y^2}{1} = 1$ is an ellipse. The graph of the level curve $z = 4$ and other level curves are shown in Fig. 16.

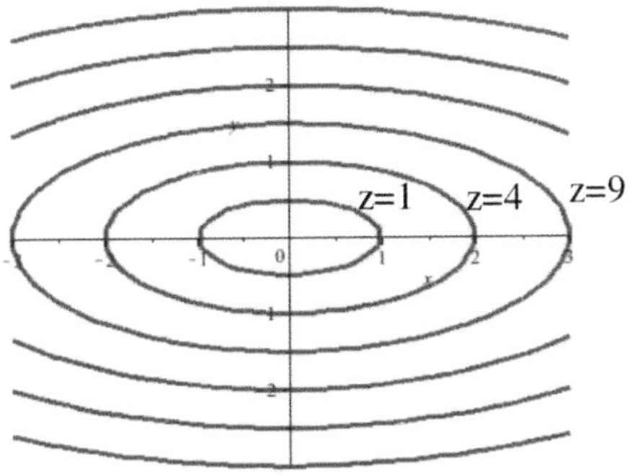

Fig. 16: $z = x^2 + 4y^2$ contours

Practice 3: Sketch the level curves $z = 4$ and $z = -8$ for the surface $z = xy$.

Functions of Three Variables: w = f(x, y, z)

A function f of three variables is a rule that assigns to each ordered triple (x, y, z) in the domain of the function a unique real number w. This can be written $w = f(x,y,z)$. For example, the temperature w at each location (x,y,z) in a classroom is a function of the three variables x, y, and z that specify the location. The cost of a new car is a function of the make of the car, the options included in the car, and the time of year (and probably several more variables). Unfortunately, the graph of $w = f(x,y,z)$ requires four axes and four dimensions. However, a level surface of $w = f(x,y,z)$ is the set of points (x,y,z) such that $w = k$ is constant, and a level surface only requires three dimensions. The $w = 75^o$ level surface in a classroom is the set of locations (x,y,z) at which the temperature is 75°, and sometimes we can graph such level surfaces (Fig. 17).

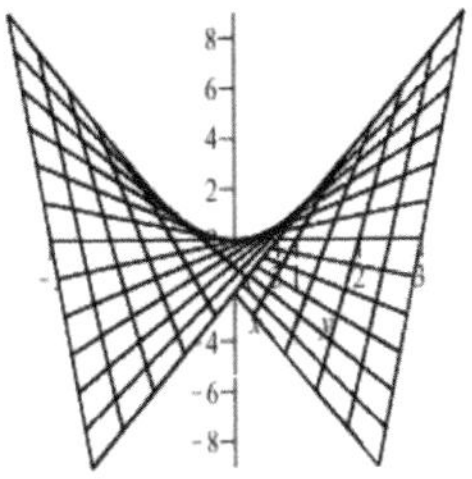
Fig. 17: $z = xy$ surface

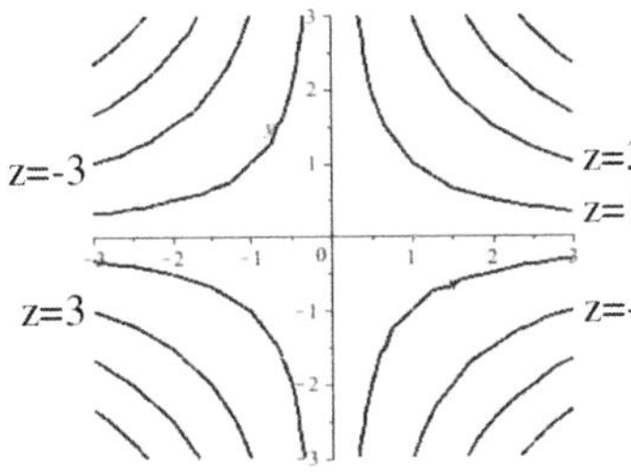

Fig. 18: $z = xy$ contours

We will examine how to differentiate and integrate functions of three or more variables, but we will do very little with their graphs.

Problems

1. The table of data in Fig. 19 shows the number thousands of gallons of drinks sold at a sports stadium as a function of the temperature at the beginning of the game and the number of people attending the game.
   (a) What are the minimum and maximum number of thousands of gallons sold?
   (b) When the attendance is 30,000 people, describe what happens to sales as the temperature increases from 50° to 90°.
   (c) When the temperature is 90°, describe what happens to sales as the attendance increases from 10,000 to 60,000.

Attendance (1000s)	Temperature (°F) 50	60	70	80	90
10	5	2	2	5	5
20	7	5	3	8	10
30	10	8	6	15	20
40	12	10	12	20	25
50	15	12	15	25	30
60	20	15	20	30	30

Fig. 19: Gallons (1000s) of drinks sold

2. The table of data in Fig. 20 shows the usual gas mileage (miles per gallon) for a truck hauling different loads and traveling at different speeds.
   (a) What are the minimum and maximum mileages for the truck?
   (b) When the truck has a load of 15,000 pounds, describe what happens to the mileage as the speed increases from 20 miles per hour (mph) to 60 mph.
   (c) When the truck is traveling at 50 mph, describe what happens to the mileage as the weight of the load varies from 20,000 pounds to 5,000 pounds.

Weight of load (pounds)	Speed (miles per hour) 20	30	40	50	60
5,000	14	18	20	22	20
10,000	11	14	17	19	17
15,000	9	12	16	17	13
20,000	7	11	15	14	10

Fig. 20: Gas Mileage of a Truck (miles per gallon)

3. Fig. 21 shows the depth of a lake at several locations.
   (a) What is the maximum depth of the lake in this region?
   (b) Describe the changing depth a fish swimming along the bottom would experience if the fish started at the location (2,1) and swam east.
   (c) Describe the changing depth a fish swimming along the bottom would experience if the fish started at the location (2,1) and swam south.

x \ y (East →; South ↓)	1	2	3	4	5
1	3	4	5	4	3
2	5	8	4	5	4
3	3	5	6	8	6
4	2	4	7	10	7
5	1	3	5	8	5

Fig. 21: Depth of a lake (meters)

4. Fig. 22 shows a map of several level elevation curves.
   (a) What is the maximum elevation in this region?
   (b) Sketch a graph of the elevation of a hiker moving along the straight path from A to B.
   (c) Sketch a graph of the elevation of a hiker moving along the curved path from C to D.
   (d) Sketch a path from E to F so that the path is never very steep.
   (e) At the point A, which directions can the hiker move (a short distance) without changing elevation?

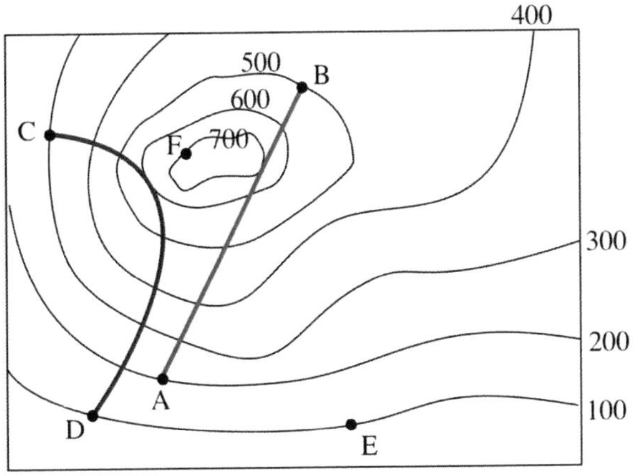

Fig. 22: Level elevation curves (topographic map)

5. Fig. 23 shows the nutrient concentration levels for a region at the bottom of the ocean.
   (a) Label the location of the highest nutrient concentration with an X.
   (b) Sketch a graph of the nutrient concentration level for an animal moving along the path from A to B.
   (c) Sketch a graph of the nutrient concentration level for an animal moving along the curved path from C to D.
   (d) Suppose an animal can sense the nutrient levels in the water near its location and always moves to increase the nutrient level. For an animal that starts at the point P, sketch several nutrient–increasing paths for the animal. Along which path does the nutrient concentration level seem to increase most rapidly?
   (e) Suppose the animal at location Q is at the nutrient concentration level that is best for it. Sketch an approximate path for the animal to move to stay at this nutrient concentration level.

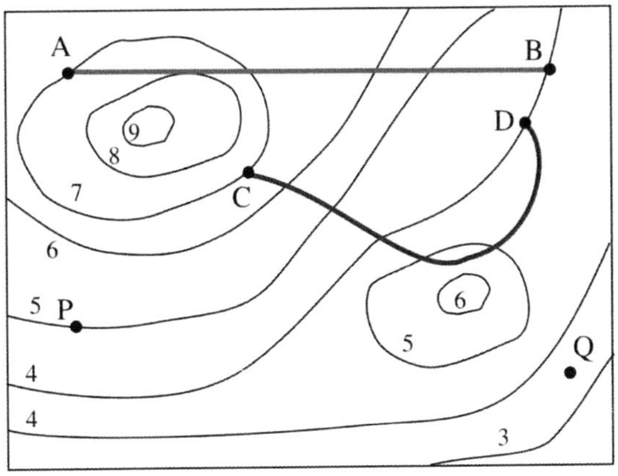

Fig. 23: Nutrient concentrations (parts per million)

6. Fig. 24 shows a level curve elevation graph for a piece of property with a river. Which direction (left–to–right or right–to–left) does the river flow? (The elevations of the level curves are intentionally unlabeled. You should be able to answer the question from the shape and location of the curves.)

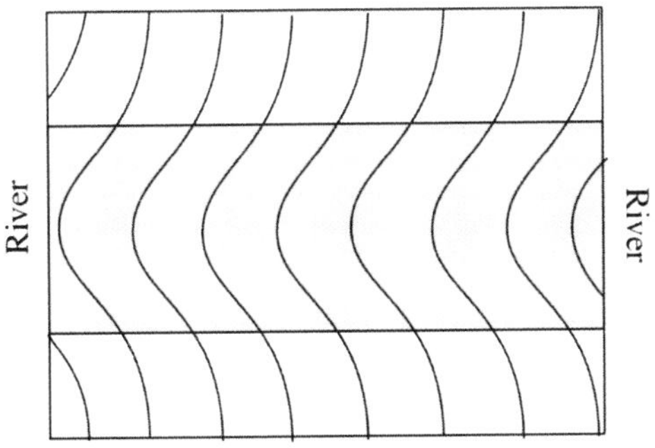

Fig. 24: River property

7. For the data in Fig. 19 sketch the approximate level curves for selling

(a) 4,000 gallons of drinks. (b) 11,000 gallons of drinks (c) 15,000 gallons of drinks.

8. For the depth data in Fig. 23 sketch the approximate level curves

(a) for the depth 9 . (b) for the depth 6.5 . (c) for the depth 4 .

9. Find and sketch the level curves $z = 1, 4,$ and 9 for $z = x^2 + y^2$.

10. Find and sketch the level curves $z = 1, 4,$ and 9 for $z = x^2 - y^2$.

11. Find and sketch the level curves $z = 0, 7, 12$ and 16 for $z = 16 - x^2 - y^2$.

12. Find and sketch the level curves $z = 1, 4$ and 5 for $z = \dfrac{5}{1 + (x-3)^2 + (y-4)^2}$

13. Find and sketch the level curves $z = 0, 1, 4$ and 9 for $z = x^2 - y^2$.

14. Sketch several level curves for a surface with a hill of height 600 feet at the location $(4, 6)$

15. Sketch several level curves for a surface with a hill of height 600 feet at the location $(4, 6)$ so that if we move south or east from the peak the path is very steep, but if we move north or west from the peak the path is not steep.

16. Sketch several level curves for a surface with a hill of height 600 feet at the location $(4, 6)$ and a hill of height 400 feet at the location $(2,2)$.

In problems 17 – 23, a surface is given. Match the surface with one of the level curve graphs A – G.

17. The surface in Fig. 25.

18. The surface in Fig. 26.

19. The surface in Fig. 27.

20. The surface in Fig. 28.

21. The surface in Fig. 29.

22. The surface in Fig. 30.

23. The surface in Fig. 31.

24. A new manufacturing process uses a computer–controlled laser light and a vat of liquid plastic. The part of the plastic exposed to the laser light becomes solid to a depth of 1 cm. The process begins with a platform 1 cm below the liquid level of the plastic, and the laser light moves over a region of plastic which hardens. Then the platform is lowered 1 cm, and the laser light moves over another region of the plastic. This process of hardening a region of plastic and then lowering the platform is repeated until the solid plastic object is complete. How is this process related to the level curves of the object?

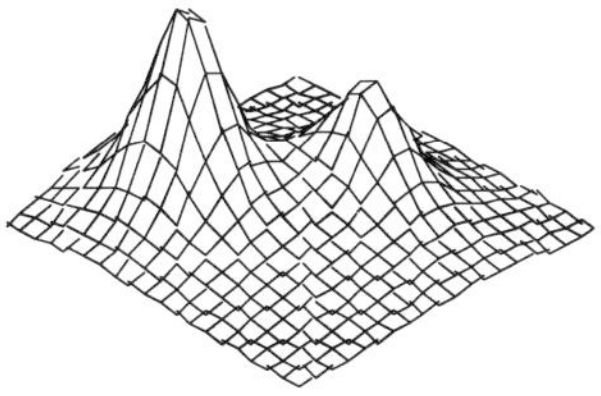
Fig. 25

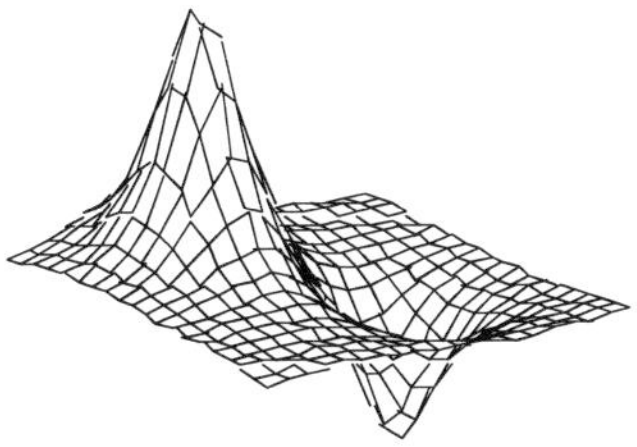
Fig. 26

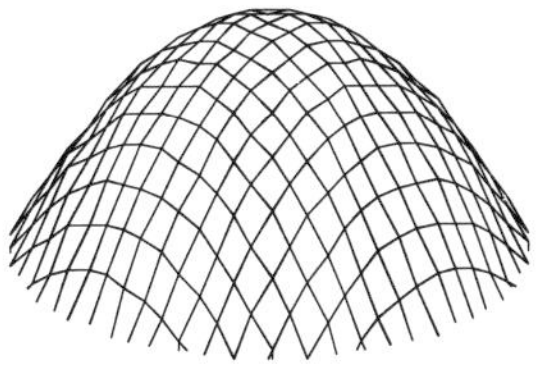
Fig. 27

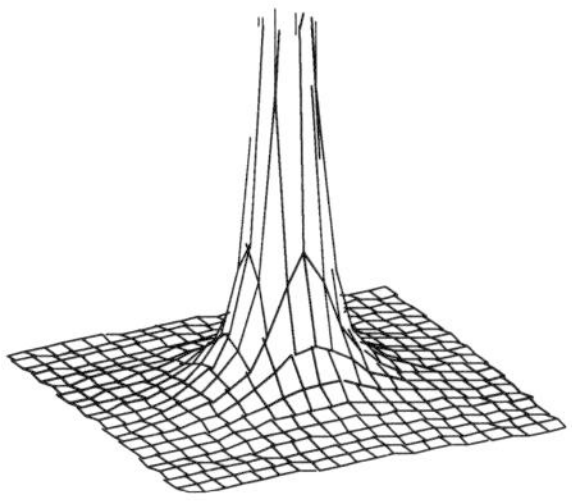
Fig. 28

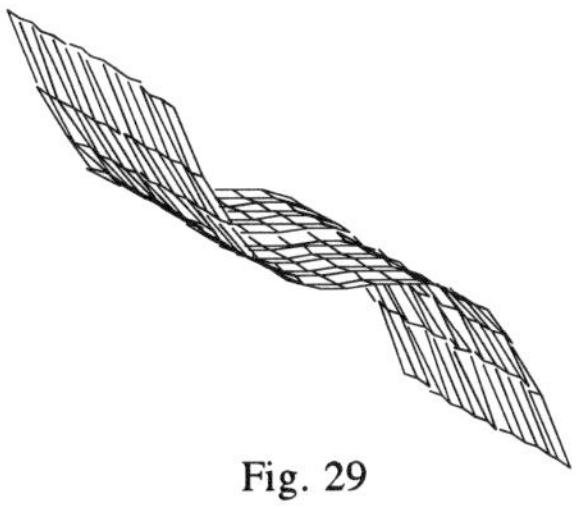
Fig. 29

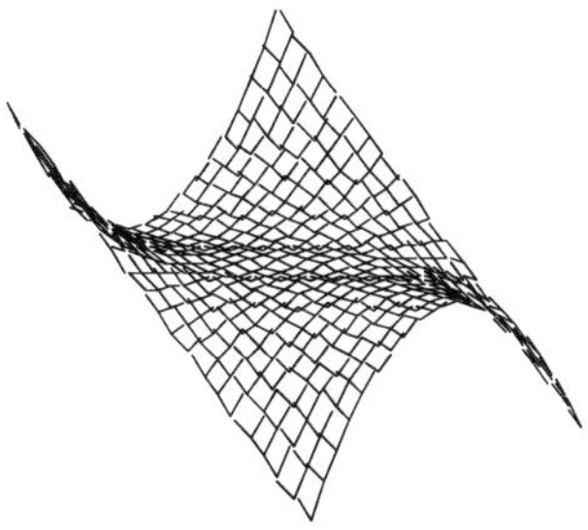
Fig. 30

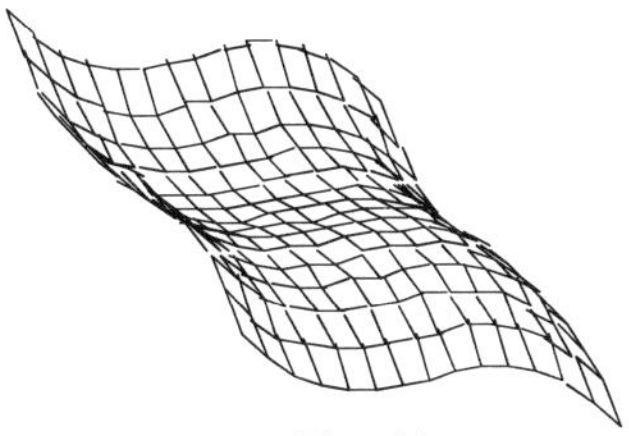
Fig. 31

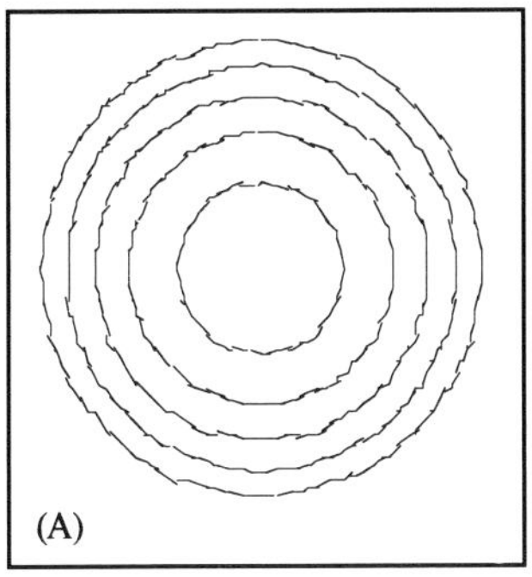
(A)

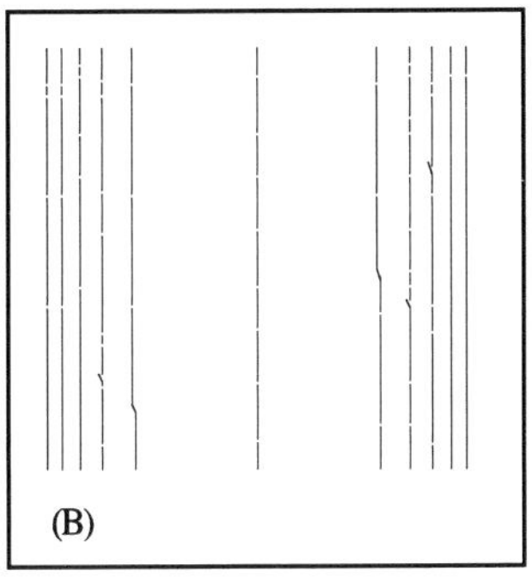
(B)

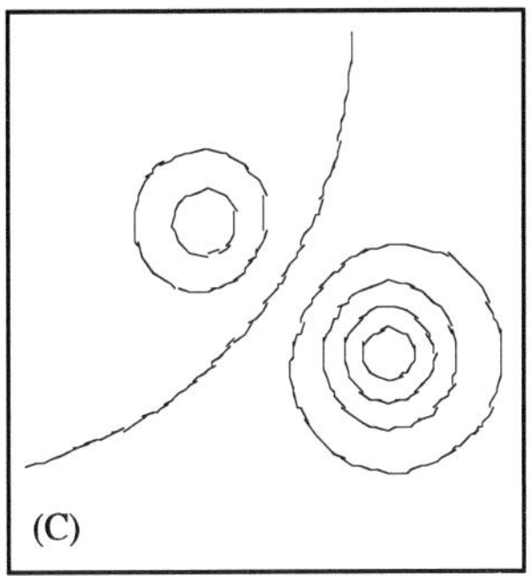
(C)

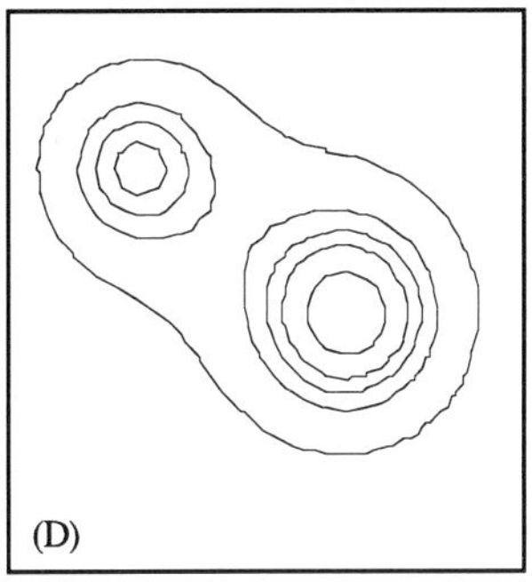
(D)

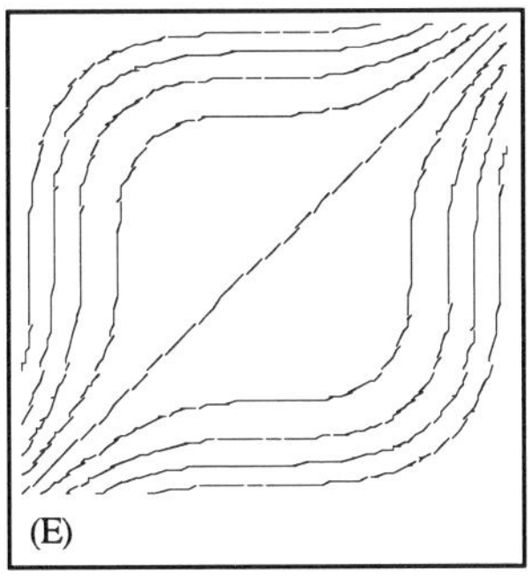
(E)

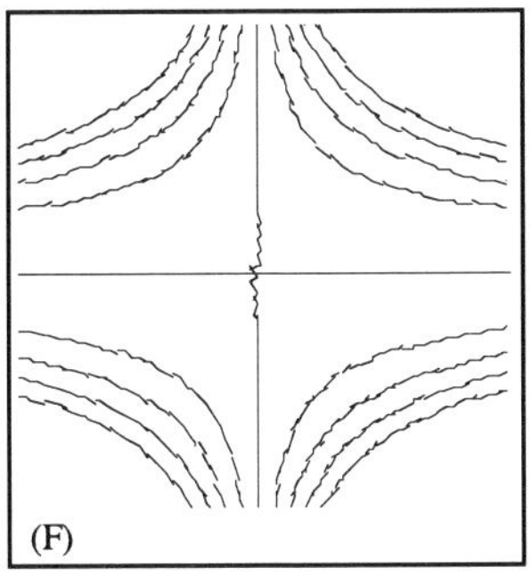
(F)

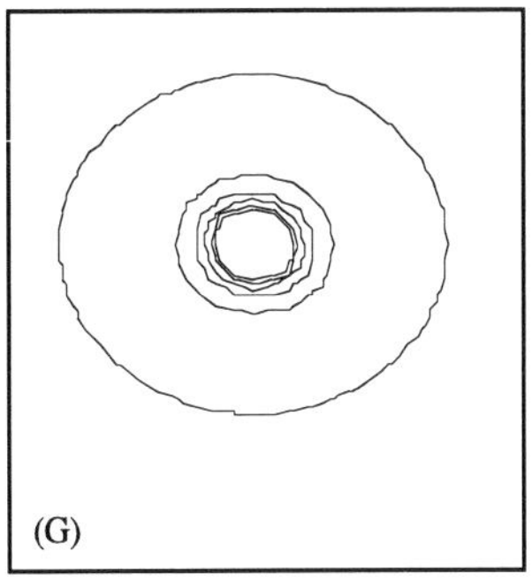
(G)

Practice Answers

Practice 1: f(2,1) = 4.3 , f(3,1) = 5.3 , f(4,1) = 4.3 , f(5,1) = 2.9 , f(6,1) = 1.8

As y is held constant at 1 and the x values increase from 2 to 6, the values of f rise to a maximum of 5.3 (when x = 3) and then decrease to a minimum of 1.8 (when x = 6).

Practice 2: The graphs are given below.

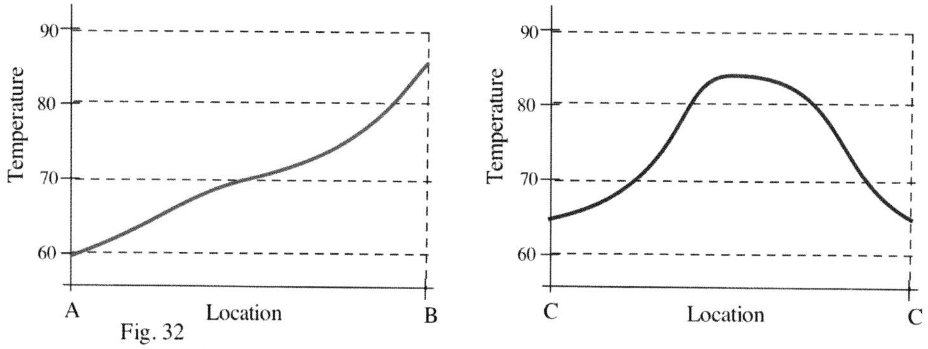

Fig. 32

Practice 3: The level curves are given below.

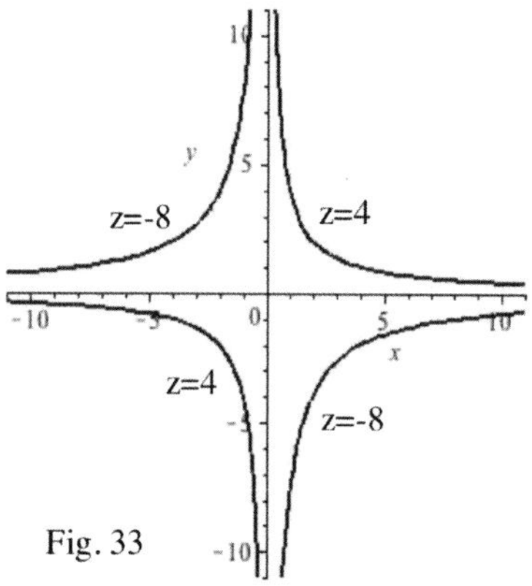

Fig. 33

13.2 LIMITS AND CONTINUITY

Our development of the properties and the calculus of functions $z = f(x,y)$ of two (and more) variables parallels the development for functions $y = f(x)$ of a single variable, but the development for functions of two variables goes much quicker since you already understand the main ideas of limits, derivatives, and integrals. In this section we consider limits of functions of two variables and what it means for a function of two variables to be continuous. In many respects this development is similar to the discussions of limits and continuity in Chapter One and many of the results we state in this section are merely extensions of those results to a new setting. It may be a good idea to spend a little time now in Chapter One rereading the main ideas and results for limits of functions of one variable and reworking a few limit problems.

The main focus of this section is on functions of two variables since it is still possible to visualize these functions and to work geometrically, but the end of this section includes extensions to functions of three and more variables.

Limits of Functions of Two Variables

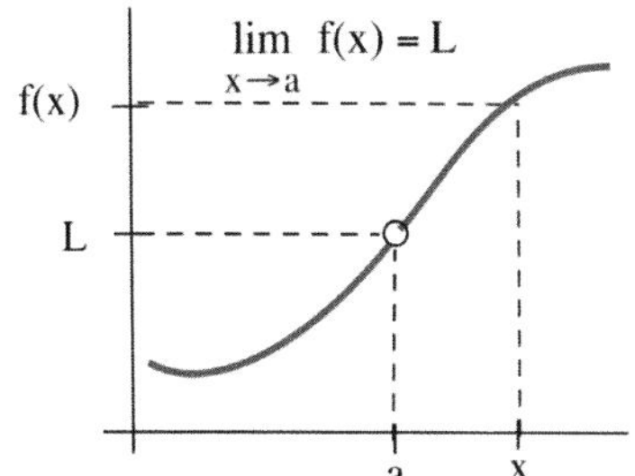

Fig. 1

When we considered limits of functions of one variable, $\lim_{x \to a} f(x)$, we were interested in the values of $f(x)$ when x was close to the point a in the domain of f (Fig. 1), and we often read the symbols "$x \to a$" as "x approaches a."

For the limit of function of two variables, $\lim_{(x,y) \to (a,b)} f(x,y)$, we are interested in the values of $f(x,y)$ when the point (x,y) is close to the point (a,b) in the domain of f (Fig. 2).

> **Definition:**
>
> Let f be a function of two variables defined for all points "near" (a,b) but possibly not defined at the point (a,b). We say the **limit of f(x,y) as (x,y) approaches (a,b) is L**, written as
>
> $$\lim_{(x,y) \to (a,b)} f(x,y) = L ,$$
>
> if the distance from $f(x,y)$ to L, $| f(x,y) - L |$, can be made arbitrarily small by taking (x,y) sufficiently close to (a,b),
> (if $\sqrt{(x-a)^2 + (y-b)^2} = | \langle x-a, y-b \rangle |$ is sufficiently small).

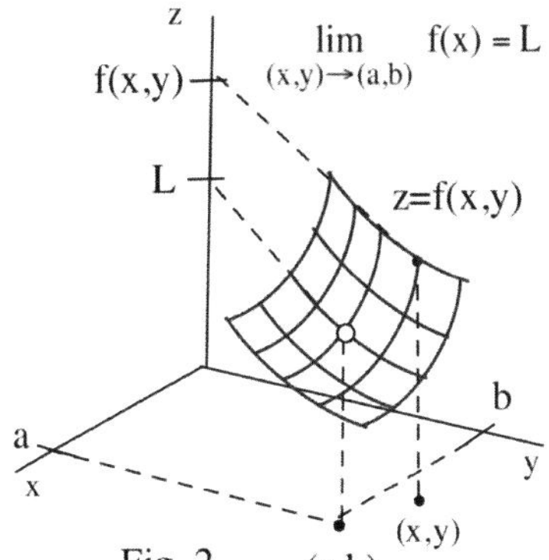

Fig. 2

All of the limit properties in the Main Limit Theorem (Section 1.2) are also true for limits of functions of two variables, and many limits of functions of two variables are easy to calculate.

Example 1: Calculate the following limits: (a) $\lim_{(x,y)\to(1,2)} \frac{xy}{x^2+y^2}$

(b) $\lim_{(x,y)\to(0,2)} \cos(xy^2) + \frac{x+6}{y}$ (c) $\lim_{(x,y)\to(5,3)} \sqrt{x^2-y^3}$

Solution: (a) $\lim_{(x,y)\to(1,2)} \frac{xy}{x^2+y^2} = \frac{1\cdot 2}{1^2+2^2} = \frac{2}{5}$

(b) $\lim_{(x,y)\to(0,2)} \cos(xy^2) + \frac{x+6}{y} = \cos(0\cdot 2) + \frac{0+6}{2} = 4$

(c) $\lim_{(x,y)\to(5,3)} \sqrt{x^2-y^3} = \sqrt{5^2-3^3} = \sqrt{16} = 4$

Practice 1: Calculate the following limits: (a) $\lim_{(x,y)\to(3,1)} \frac{xy}{x^2-y^2}$

(b) $\lim_{(x,y)\to(0,2)} \cos(x^2y) + \frac{x+9}{y+1}$ (c) $\lim_{(x,y)\to(3,2)} \sqrt{x^2-y^3}$

At the end of this section we consider some examples of more complicated situations with functions whose limits do not exist. And we also extend the idea of limits to functions of three or more variables.

Continuity of Functions of Two Variables

A function of one variable is continuous at $x = a$ if $\lim_{x\to a} f(x) = f(a)$. Geometrically that means that the graph of f is connected at the point $(a, f(a))$ and does not have a hole or break there (Fig. 3). The definition and meaning of continous for functions of two variables is quite similar.

> Definition:
>
> A function of two variables defined at the point (a,b) and for all points near (a,b) is
>
> **continuous at (a,b)** if $\lim_{(x,y)\to(a,b)} f(x,y) = f(a,b)$. (Fig. 4)

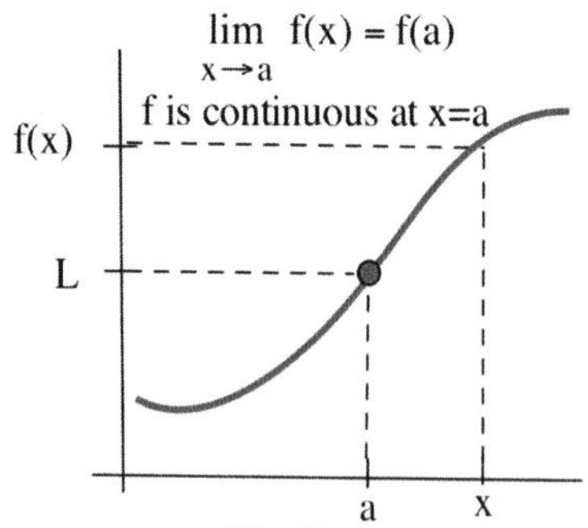

Fig. 3

And just as we talked about a function of one variable being continuous on an interval (or even on the entire real number line), we can talk about a function of two variables being continuous on a set D in the xy–plane or even on the entire xy–plane.

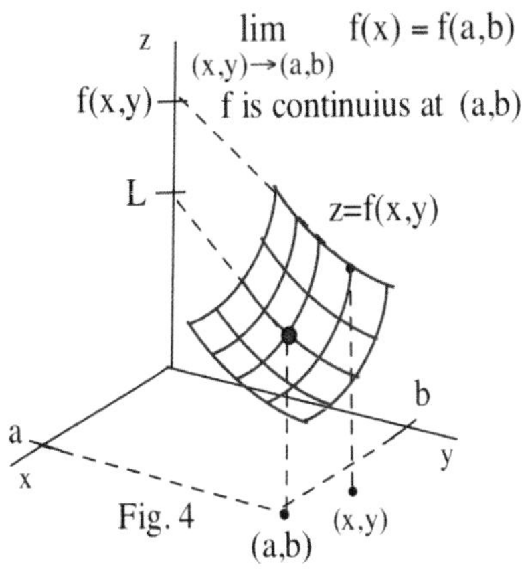

Fig. 4

> Definition:
>
> If a function of two variables is continuous at every point (a,b) in a set D, we say that the function is **continuous on D**.

Most of the functions we will work with are continuous either everywhere (at all points (x,y) in the plane) or continuous everywhere except at a "few" places.

- A **polynomial** function of two variables is **continuous everywhere**, at every point (x,y).
- A **rational** function of two variables is **continuous everywhere in its domain** (everywhere except where division by zero would occur).
- If f(x,y) is continuous at (a,b), then **sin(** f(x,y)), **cos(** f(x,y)), and $\mathbf{e}^{f(x,y)}$ are continuous at (a,b).
- More generally, if f (a function of two variables) is continuous at (a,b) and g (a function of one variable) is continuous at f(a,b), then g(f(x,y)) = gof (x,y) is continuous at (a,b).

Geometrically "f(x,y) is continuous at (a,b)" that means that the surface graph of f is connected at the point (a,b, f(a,b)) and does not have a hole or break there. Fig. 5 shows the surface graphs of several continuous functions of two variables.

Similar definitions and results are used for functions of three or more variables.

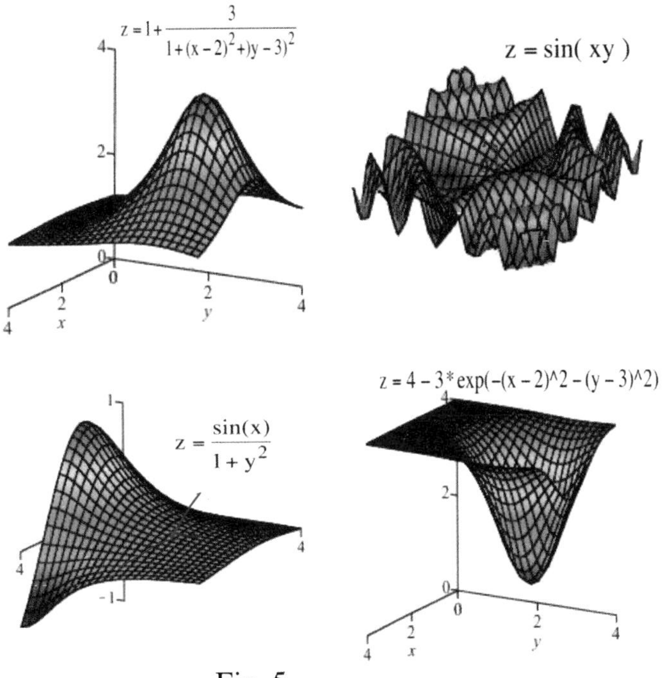

Fig. 5

Limits That Do Not Exist

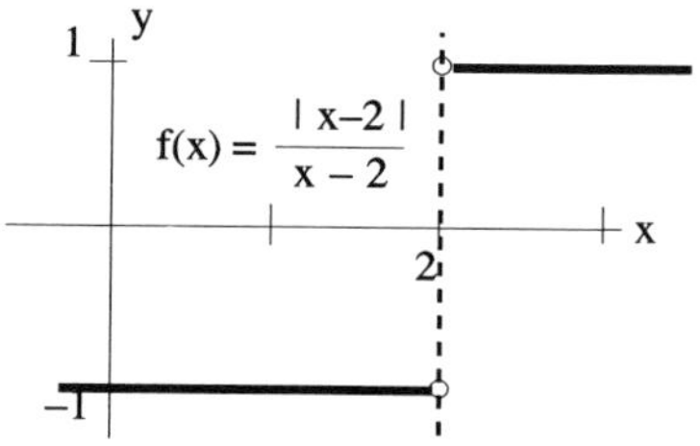

$$\lim_{x \to 2^-} \frac{|x-2|}{x-2} = -1 \qquad \lim_{x \to 2^+} \frac{|x-2|}{x-2} = +1$$

so $\lim_{x \to 2} \frac{|x-2|}{x-2}$ Does not exist

Fig. 6

Most of the functions we work with will have limits and will be continuous, but not all of them. A function of one variable did not have a limit if its left limit and its right limit had different values (Fig. 6). Similar situations can occur with functions of two variable as shown graphically in Fig. 7. For the function $f(x,y)$ in Fig. 7, if (x,y) approaches the point $(1,2)$ along path 1 ($x = 1$ and $y \to 2^-$) then the values of $f(x,y)$ approach 2. But if (x,y) approaches the point $(1,2)$ along path 2 ($x = 1$ and $y \to 2^+$) then the values of $f(x,y)$ approach 1. Since two paths to the point $(1,2)$ result in two different limiting values for f, we say that the limit of $f(x,y)$ as (x,y) approaches $(1,2)$ does not exist.

Showing a limit does not exist:

If there are two paths so that $f(x,y) \to L_1$ as $(x,y) \to (a,b)$ along path 1, and $f(x,y) \to L_2$ as $(x,y) \to (a,b)$ along path 2, and $L_1 \neq L_2$,

then $\lim_{(x,y) \to (a,b)} f(x,y)$ does not exist.

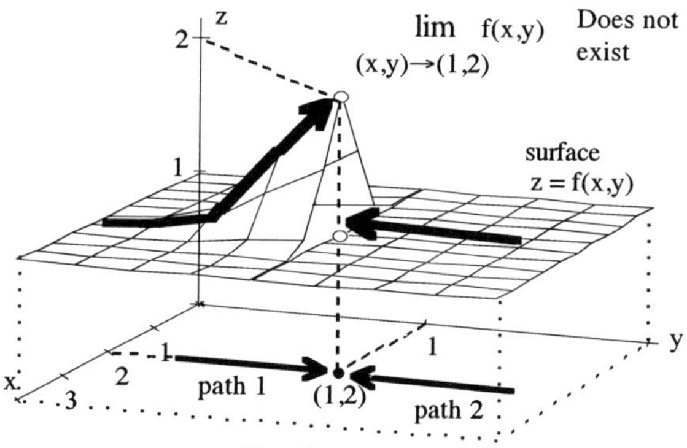

Fig. 7

Example 2: Find $\lim_{(x,y) \to (0,0)} \frac{xy}{x^2 + y^2}$

Solution: Let path 1 be the x–axis, so y = 0.

Then $\frac{xy}{x^2+y^2} = \frac{0}{x^2+0} = 0$ and the limit of $\frac{xy}{x^2+y^2}$ as $(x,y) \to (0,0)$ along path 1 is 0.

However, if we take the path 2 to be the line $y = x$,

then $\frac{xy}{x^2+y^2} = \frac{x^2}{x^2+x^2} = \frac{1}{2}$ so the limit of $\frac{xy}{x^2+y^2}$ as $(x,y) \to (0,0)$ along path 2 is $\frac{1}{2}$.

Since the limits of f as $(x,y) \to (0,0)$ along two different paths is two different numbers, the limit of this $f(x,y)$ as $(x,y) \to (0,0)$ **does not exist.**

Fig 8 shows these two paths and the different limits of f along them.

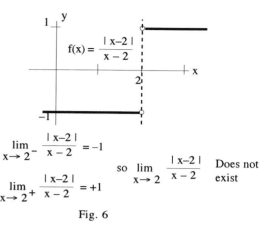

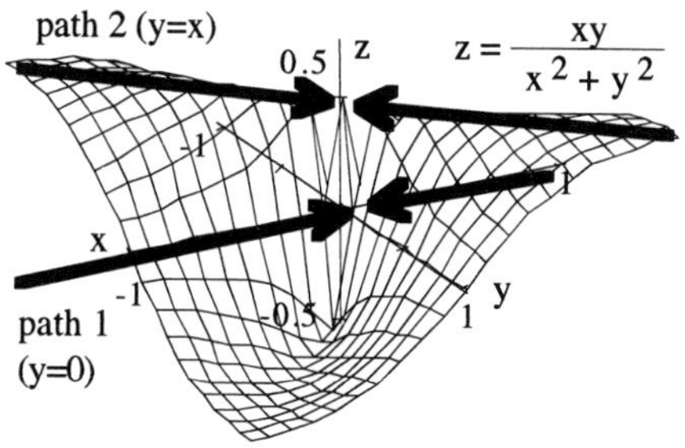

Fig. 8

Practice 2: Find the limit of $f(x,y) = \frac{xy}{x^2+y^2}$ as $(x,y) \to (0,0)$ along the path $y = 3x$.

Since the limit of $f(x,y) = \frac{xy}{x^2+y^2}$ as $(x,y) \to (0,0)$ does not exist, this function is not continuous at $(0,0)$.

Practice 3: Show that $\lim_{(x,y)\to(0,0)} \frac{x^2 - y^2}{x^2 + y^2}$ does not exist.

Note: The "path method" only shows that a limit does not exist. Even if the limit of a function as (x,y)→(a,b) is the same value along two or three paths (or even along an infinite number of paths) we still cannot validly conclude that the limit exists.

Example 3: Evaluate $\lim_{(x,y)\to(0,0)} \frac{x^2 y}{x^4 + y^2}$ along the paths (a) the y–axis (x=0), (b) the x–axis (y=0), (c) the lines $y = mx$ for all values of $m \neq 0$, and (d) along the parabola $y = x^2$.

Solution: (a) Since x=0, $\lim_{(x,y)\to(0,0)} \frac{x^2 y}{x^4 + y^2} = \lim_{y\to 0} \frac{0}{0 + y^2} = 0$.

(b) Since y=0, $\lim_{(x,y)\to(0,0)} \frac{x^2 y}{x^4 + y^2} = \lim_{x\to 0} \frac{0}{x^4 + 0} = 0$.

(c) Since $y = mx$ ($m \neq 0$), $\lim_{(x,y)\to(0,0)} \frac{x^2 y}{x^4 + y^2} = \lim_{x\to 0} \frac{x^2(mx)}{x^4 + (mx)^2} = \lim_{x\to 0} \frac{x^2}{x^2} \cdot \frac{mx}{x^2 + m^2} = 0.$

From parts (a), (b), and (c) we know that the limit of $\frac{x^2 y}{x^4 + y^2}$ is 0 as (x,y)→(0,0) along **every straight line path**. But that is not enough to conclude that the limit along every path is 0.

(d) Along the parabolic path $y = x^2$, $\lim_{(x,y)\to(0,0)} \frac{x^2 y}{x^4 + y^2} = \lim_{x\to 0} \frac{x^2(x^2)}{x^4 + (x^2)^2} = \lim_{x\to 0} \frac{x^4}{2x^4} = \frac{1}{2}$.

Part (d) together with any one of parts (a), (b), or (c) lets us conclude that

$\lim_{(x,y)\to(0,0)} \frac{x^2 y}{x^4 + y^2}$ does not exist.

Fig. 9 shows this surface.

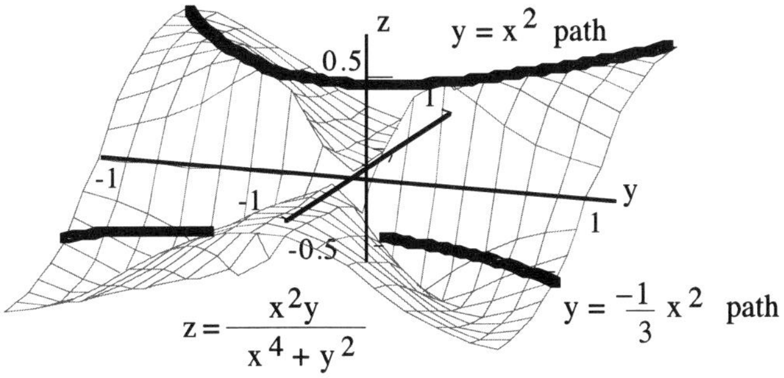

Fig. 9

Functions of More Than Two Variables: Limits and Continuity

Once we have made the adjustments to extend the ideas and definitions of limits and continuity to functions of two variables, it is straightforward to extend them to functions of three or more variables.

Definition:

Let f be a function of three variables defined for all points "near" (a,b,c) but possibly not defined at the point (a,b,c). We say the

limit of f(x,y,z) as (x,y,z) approaches (a,b,c) is L, written as

$$\lim_{(x,y,z)\to(a,b,c)} f(x,y,z) = L,$$

if the distance from f(x,y,z) to L, $| f(x,y,z) - L |$, can be made arbitrarily small by taking (x,y,z) sufficiently close to (a,b,c),

(if $\sqrt{(x-a)^2 + (y-b)^2 + (z-c)^2} = | \langle x-a, y-b, z-c \rangle |$ is sufficiently small).

Definition:

A function of three variables defined at the point (a,b,c) and for all points near (a,b,c) is **continuous at (a,b,c)** if $\lim_{(x,y,z)\to(a,b,c)} f(x,y,z) = f(a,b,c)$.

PROBLEMS

In Problems 1 – 4, the level curves of functions are given. Use the information from these level curves to determine the limits. (Since only a few level curves are shown, you need to make reasonable assumptions about the behavior of the functions.)

1. The level curves of z = f(x,y) are shown in Fig. 10.

(a) $\lim_{(x,y)\to(1,2)} f(x,y)$ (b) $\lim_{(x,y)\to(1,1)} f(x,y)$

(c) $\lim_{(x,y)\to(2,1)} f(x,y)$ (d) $\lim_{(x,y)\to(3,2)} f(x,y)$

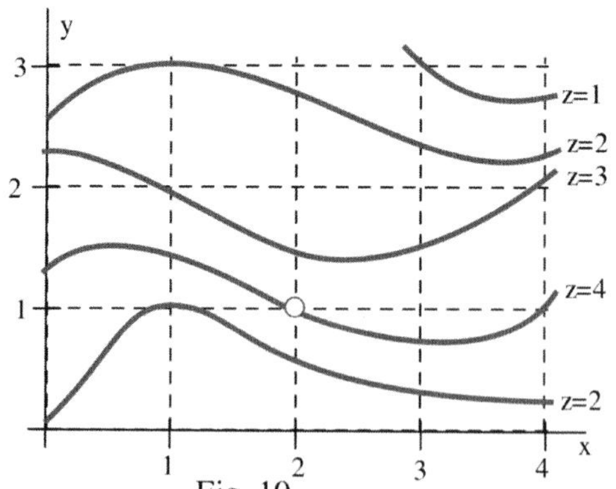

Fig. 10

2. The level curves of $z = g(x,y)$ are shown in Fig. 11.

(a) $\lim_{(x,y)\to(2,2)} g(x,y)$ (b) $\lim_{(x,y)\to(2,1)} g(x,y)$

(c) $\lim_{(x,y)\to(1,2)} g(x,y)$ (d) $\lim_{(x,y)\to(3,2)} g(x,y)$

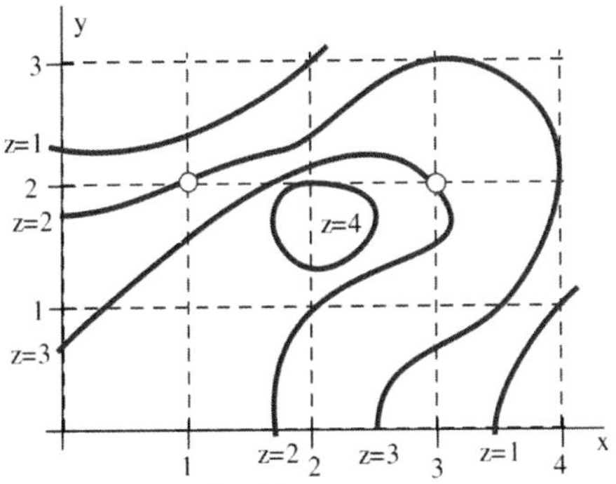

Fig. 11

3. The level curves of $z = S(x,y)$ are shown in Fig. 12.

(a) $\lim_{(x,y)\to(1,2)} S(x,y)$ (b) $\lim_{(x,y)\to(2,1)} S(x,y)$

(c) $\lim_{(x,y)\to(1,1)} S(x,y)$ (d) $\lim_{(x,y)\to(3,2)} S(x,y)$

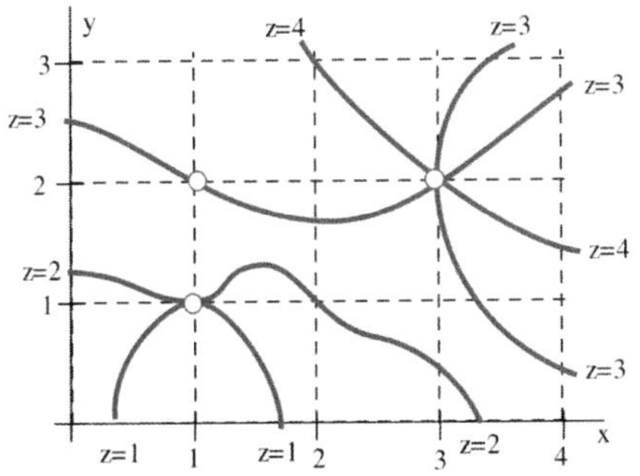

Fig. 12

4. The level curves of $z = T(x,y)$ are shown in Fig. 13.

(a) $\lim_{(x,y)\to(3,3)} T(x,y)$ (b) $\lim_{(x,y)\to(2,2)} T(x,y)$

(c) $\lim_{(x,y)\to(1,2)} T(x,y)$ (d) $\lim_{(x,y)\to(4,1)} T(x,y)$

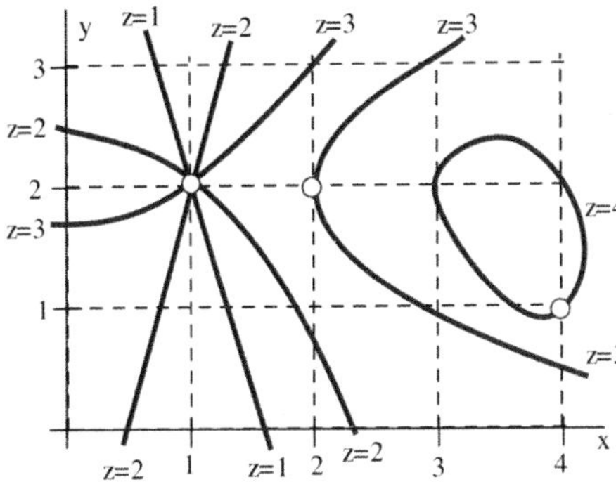

Fig. 13

In Problems 5 – 26, determine the limit if it exists or show that the limit does not exist.

5. $\lim_{(x,y)\to(2,3)} (x^2y^2 - 2xy^5 + 3y)$

6. $\lim_{(x,y)\to(-3,4)} (x^3 + 3x^2y^2 - 5y^3 + 1)$

7. $\lim_{(x,y)\to(0,0)} \dfrac{x^2y^3 + x^3y^2 - 5}{2 - xy}$

8. $\lim_{(x,y)\to(-2,1)} \dfrac{x^2 + xy + y^2}{x^2 - y^2}$

9. $\lim_{(x,y)\to(\pi,\pi)} x\cdot\sin\left(\dfrac{x+y}{4}\right)$

10. $\lim_{(x,y)\to(1,4)} e^{(\sqrt{x+2y}\,)}$

11. $\lim_{(x,y)\to(0,0)} \dfrac{\sin(x+y)}{x+y}$

12. $\lim_{(x,y)\to(0,0)} \dfrac{x^2 - y^2}{x+y}$

13. $\lim_{(x,y)\to(0,0)} \dfrac{x-y}{x^2+y^2}$

14. $\lim_{(x,y)\to(0,0)} \dfrac{x^2}{x^2+y^2}$

15. $\lim_{(x,y)\to(0,0)} \dfrac{8x^2y^2}{x^4+y^4}$

16. $\lim_{(x,y)\to(0,0)} \dfrac{x^3+xy^2}{x^2+y^2}$

17. $\lim_{(x,y)\to(0,0)} \dfrac{2xy}{x^2+y^2}$

18. $\lim_{(x,y)\to(0,0)} \dfrac{(x+y)^2}{x^2+y^2}$

19. $\lim_{(x,y)\to(0,0)} \dfrac{\sqrt{xy}}{\sqrt{x^2+y^2}}$

20. $\lim_{(x,y)\to(0,0)} \dfrac{2x^2+3xy+4y^2}{3x^2+5y^2}$

21. $\lim_{(x,y)\to(0,0)} \dfrac{xy+1}{x^2+y^2+1}$

22. $\lim_{(x,y)\to(0,0)} \dfrac{xy^3}{x^2+y^6}$

23. $\lim_{(x,y)\to(0,0)} \dfrac{2x^2y}{x^4+y^2}$

24. $\lim_{(x,y)\to(0,0)} \dfrac{x^3y^2}{x^2+y^2}$

25. $\lim_{(x,y)\to(0,0)} \dfrac{x^2+y^2}{\sqrt{x^2+y^2+1}-1}$

26. $\lim_{(x,y)\to(0,0)} \dfrac{\sqrt{x^2+y^2+1}-1}{x^2+y^2}$

27. $\lim_{(x,y)\to(0,1)} \dfrac{xy-x}{x^2+y^2-2x+2y+2}$

28. $\lim_{(x,y)\to(1,-1)} \dfrac{x^2+y^2-2x-2y}{x^2+y^2-2x+2y+2}$

29. $\lim_{(x,y,z)\to(1,2,3)} \dfrac{xz^2-y^2z}{xyz-1}$

30. $\lim_{(x,y,z)\to(2,3,0)} \{ xe^x + \ln(2x-y) \}$

31. $\lim_{(x,y,z)\to(0,0,0)} \dfrac{x^2-y^2-z^2}{x^2+y^2+z^2}$

32. $\lim_{(x,y,z)\to(0,0,0)} \dfrac{xy+yz+zx}{x^2+y^2+z^2}$

33. $\lim_{(x,y,z)\to(0,0,0)} \dfrac{xy+yz^2+xz^2}{x^2+y^2+z^2}$

34. $\lim_{(x,y,z)\to(0,0,0)} \dfrac{x^2y^2z^2}{x^2+y^2+z^2}$

35. The function f whose level curves are shown in Fig. 10 is not defined at (2,1). Define a value for f(2,1) so f is continuous at (2,1).

36. The function g whose level curves are shown in Fig. 11 is not defined at (1,2) and (3,2). Can we define values for g(1,2) and g(3,2) so g is continuous at (1,2) and (3,2)?

37. The function S whose level curves are shown in Fig. 12 is not defined at (1,1), (1,2) and (3,2). Can we define values for S(1,1), S(1,2) and S(3,2) so S is continuous at each of those points?

38. The function T whose level curves are shown in Fig. 13 is not defined at (1,2), (2,2) and (4,1). Can we define values for T(1,2), T(2,2) and T(4,1) so T is continuous at eaach of those points?

In Problems 39 – 51, determine where the given function is not continuous.

39. $f(x,y) = \dfrac{x^2 + y^2 + 1}{x^2 + y^2 - 1}$

40. $f(x,y) = \dfrac{x^6 + x^3y^3 + y^6}{x^3 + y^3}$

41. $g(x,y) = \ln(2x + 3y)$

42. $S(x,y) = e^{xy}\sin(x + y)$

43. $T(x,y) = \sqrt{x + y} - \sqrt{x - y}$

44. $T(x,y) = 2^{x\tan(y)}$

45. $F(x,y) = x\ln(yz)$

46. $F(x,y) = x + y\sqrt{x + z}$

Practice Answers

Practice 1: (a) $\displaystyle\lim_{(x,y)\to(3,1)} \frac{xy}{x^2 - y^2} = \frac{3\cdot1}{3^2 - 1^2} = \frac{3}{8}$

(b) $\displaystyle\lim_{(x,y)\to(0,2)} \cos(x^2y) + \frac{x+9}{y+1} = \cos(0^2\cdot2) + \frac{0+9}{2+1} = 1 + \frac{9}{3} = 4$

(c) $\displaystyle\lim_{(x,y)\to(3,2)} \sqrt{x^2 - y^3} = \sqrt{3^2 - 2^3} = \sqrt{1} = 1$

Practice 2: Along the path $y = 3x$, $\dfrac{xy}{x^2 + y^2} = \dfrac{x(3x)}{x^2 + (3x)^2} = \dfrac{3x^2}{10x^2} .= \dfrac{3}{10}$ for $x \neq 0$.

Then the limit of $\dfrac{xy}{x^2 + y^2}$ as $(x,y) \to (0,0)$ along the path $y = 3x$ is $\dfrac{3}{10}$.

Practice 3:

Along the path $y = x$, $\dfrac{x^2 - y^2}{x^2 + y^2} = \dfrac{x^2 - x^2}{x^2 + x^2} = \dfrac{0}{2x^2} = 0$ for $x \neq 0$, so the limit along this path is 0.

Along the x–axis $y = 0$, so $\dfrac{x^2 - y^2}{x^2 + y^2} = \dfrac{x^2}{x^2} = 1$ for $x \neq 0$, so the limit along this path is 1.

(Also, along the y–axis $x = 0$, so $\dfrac{x^2 - y^2}{x^2 + y^2} = \dfrac{-y^2}{y^2} = -1$ for $y \neq 0$, so the limit along this path is –1.)

13.3 PARTIAL DERIVATIVES

For a function $y = f(x)$ of one variable, the derivative $\frac{dy}{dx}$ measured the rate of change of the variable y with respect to the variable x. For a function $z = f(x,y)$ of two variables we can ask about the rate of change of z with respect to the variable x or the variable y: how do changes in x effect z, and how do changes in y effect z? In scientific and economic settings with many variables, it is common to try to determine the effect of each variable by holding all of the other variables constant and then measuring the outcomes as that single variable in allowed to vary.

Example 1: The table of data in Fig. 1 shows the number thousands of gallons of drinks sold at a sports stadium as a function of the temperature at the beginning of the game and the number of people attending the game. At a game with 30,000 people on a 70° day,

(a) what is the average rate of change of drink sales as the temperature rises to 80°?

(b) what is the average rate of change of drink sales as the attendance increases to 40,000 people?

Attendance (1000s)	Temperature (°F) 50	60	70	80	90
10	5	2	2	5	5
20	7	5	3	8	10
30	10	8	6	15	20
40	12	10	12	20	25
50	15	12	15	25	30
60	20	15	20	30	30

Fig. 1: Gallons (1000s) of drinks sold

Solution: (a) In this situation the attendance is constant at 30,000 people, and the temperature changes from 70° to 80°. The average rate of change is

$$\frac{f(30,80^{o}) - f(30,70^{o})}{80^{o} - 70^{o}} = \frac{15000 - 6000 \text{ gallons}}{10^{o}} = 900 \text{ gallons per degree rise in temperature.}$$

(b) In this case the temperature is constant at 70°, and the attendance changes from 30,000 people to 40,000 people. The average rate of change is

$$\frac{f(40,70) - f(30,70)}{40000 - 30000} = \frac{12000 - 6000 \text{ gallons}}{10000 \text{ people}} = 0.6 \text{ gallons per additional person in attendance.}$$

Note that these rates of change depend on the starting attendance and temperature as well as on the variable that is allowed to change. You should also notice that the units of the two answers are different — one is "gallons/degree" and the other is "gallons/person."

Practice 1: Using the data in Fig. 1 and at a game with 20,000 people on a 80° day,

(a) what is the average rate of change of drink sales as the temperature rises to 90°?

(b) what is the average rate of change of drink sales as the attendance increases to 30,000 people?

The definition of a partial derivative follows from this idea of holding one variable constant and measuring the rate of change as the other variable changes.

Definition:

The **partial derivative of f(x,y) with respect to x at the point (a,b)** is

$$\mathbf{f_x(a,b)} = \lim_{h\to 0} \frac{f(a+h,b)-f(a,b)}{h} \quad \text{(if the limit exists and is finite).}$$

Meaning: $f_x(x,y)$ measures the ***instantaneous rate of change*** of f at the point (x,y) in the direction of increasing x values.

To calculate $f_x(x,y)$ when $z = f(x,y)$ is given by a formula,

treat y as a constant and differentiate with respect to x.

Example 2: (a) For $f(x,y) = 3x^2 + 7y^2 - 10xy$, find $f_x(x,y)$, $f_x(1,2)$ and $f_x(3,1)$.

(b) For $g(x,y) = \sin(3xy) + \ln(5y) + x^3y^5$, find $g_x(x,y)$ and $g_x(1,2)$.

Solution: (a) $f_x(x,y) = 6x + 0 - 10y = 6x - 10y$ and $f_x(1,2) = 6(1) - 10(2) = -14$. $f_x(3,1) = 6(3) - 10(1) = 8$.

(b) $g_x(x,y) = 3y\cdot\cos(3xy) + 0 + 3x^2y^5$ and

$g_x(1,2) = 3(2)\cos(3(1)(2)) + 3(1)^2(2)^5 = 6\cos(6) + 96 \approx 101.8$.

Practice 2: (a) For $f(x,y) = x^3 + 4y^2 + 5x^2y$, find $f_x(x,y)$ and $f_x(2,5)$.

(b) For $g(x,y) = e^{xy} + \frac{x}{y}$, find $g_x(x,y)$ and $g_x(0,2)$.

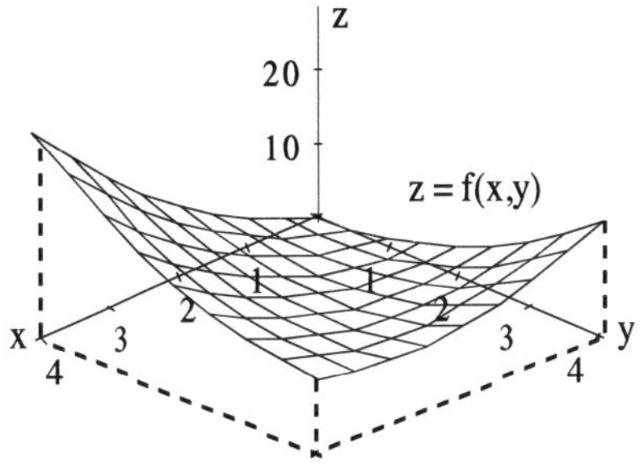

Fig. 2

We can also interpret the partial derivatives graphically. The graph of $z = f(x,y)$ is typically a surface (Fig. 2) and the graph of "y = a constant" is a plane, so the graph of "z = f(x,y) with y held constant" is the curve resulting from the intersection of the surface and the plane. Fig. 3 shows such a surface and plane and their curve of intersection when $y = 2$. $f_x(1,2)$ is the slope of the line tangent to this curve at the point (1,2) as shown in Fig. 4.

Example 3: Use the information in Fig. 4 to estimate the value of $f_x(1,2)$.

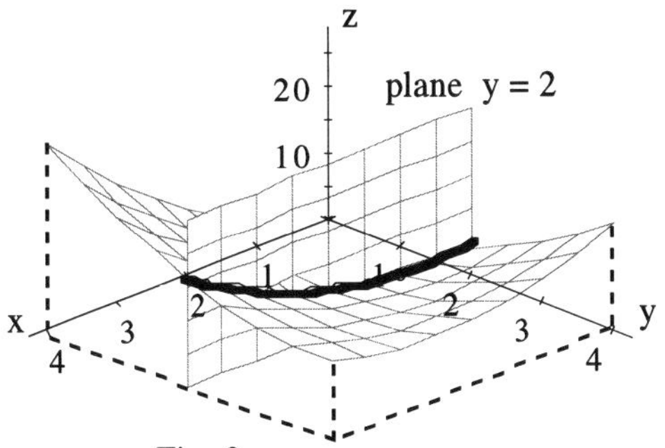

Fig. 3

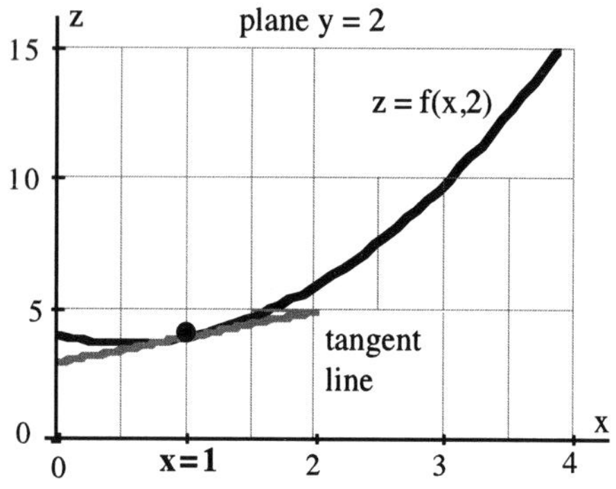

Fig. 4

Solution: We can estimate the slope of the tangent line in Fig. 4 by picking two points on the line and calculating the slope of the line connecting the two points. It looks like (0,3) and (2,5) are points on the tangent line, and the slope of the segment between those two points is $\frac{5-3}{2-0} = 1$. Then we estimate $f_x(1,2) \approx 1$.

Practice 3: Use the information in Fig. 4 to estimate the values of $f(3,2)$ and $f_x(3,2)$.

The partial derivative with respect to y is similar, but now we treat x as the constant.

> Definition:
>
> The **partial derivative of f(x,y) with respect to y at the point (a,b)** is
>
> $$\mathbf{f_y(a,b)} = \lim_{h\to 0} \frac{f(a,b+h) - f(a,b)}{h} \quad \text{(if the limit exists and is finite).}$$
>
> Meaning: $f_y(x,y)$ measures the *instantaneous rate of change* of f at the point (x,y) in the direction of increasing y values.
>
> To calculate $f_y(x,y)$ when $z = f(x,y)$ is given by a formula,
>
> **treat x as a constant and differentiate with respect to y.**

Example 4: (a) For $f(x,y) = 3x^2 + 7y^2 - 10xy$, find $f_y(x,y)$, $f_y(1,2)$, and $f_y(3,1)$.

(b) For $g(x,y) = \sin(3xy) + \ln(5y) + x^3y^5$, find $g_y(x,y)$ and $g_y(1,2)$.

Solution: (a) $f_y(x,y) = 0 + 14y - 10x = 14y - 10x$. Then $f_y(1,2) = 14(2) - 10(1) = 18$ and $f_y(3,1) = -16$.

(b) $g_y(x,y) = 3x\cdot\cos(3xy) + \frac{5}{5y} + 5x^3y^4$. Then

$$g_y(1,2) = 3(1)\cdot\cos(3(1)(2)) + \frac{5}{5(2)} + 5(1)^3(2)^4 \approx 2.88 + 0.5 + 80 = 83.38 .$$

Practice 4: (a) For $f(x,y) = x^3 + 4y^2 + 5x^2y$, find $f_y(x,y)$ and $f_y(2,5)$.

(b) For $g(x,y) = e^{xy} + \frac{x}{y}$, find $g_y(x,y)$ and $g_y(0,2)$.

> **Notations**: The following notations are all commonly used to represent partial derivatives of $z = f(x,y)$
>
> $$f_x(x,y) = f_x = \frac{\partial f}{\partial x} = \frac{\partial}{\partial x} f(x,y) = \frac{\partial z}{\partial x} = D_x f$$ Partial derivative of f with respect to x
>
> $$f_y(x,y) = f_y = \frac{\partial f}{\partial y} = \frac{\partial}{\partial y} f(x,y) = \frac{\partial z}{\partial y} = D_y f$$ Partial derivative of f with respect to y

Example 5: Use the information in Figures 5 and 6 to estimate the value of $f_y(1,2)$.

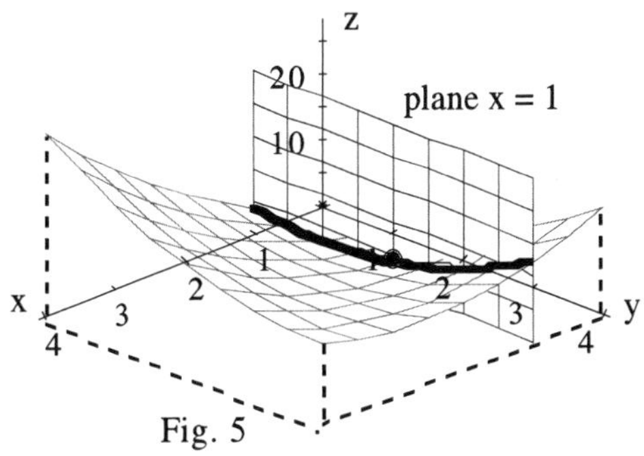

Fig. 5

Solution: Fig. 5 shows the surface $z = f(x,y)$ and the plane $x = 1$, but it is difficult to estimate the value of $f_y(1,2)$ from it. Fig. 6 shows the intersection of the surface graph with the plane, and the tangent line at the point (1,2) is included. We can estimate the value of $f_y(1,2)$ by picking two points on the tangent line and calculating the slope between them. It looks like (1,2) and (3,6) are points on the tangent line, and the slope of the segment is $\frac{6-2}{3-1} = 2$. Then we estimate $f_y(1,2) \approx 2$.

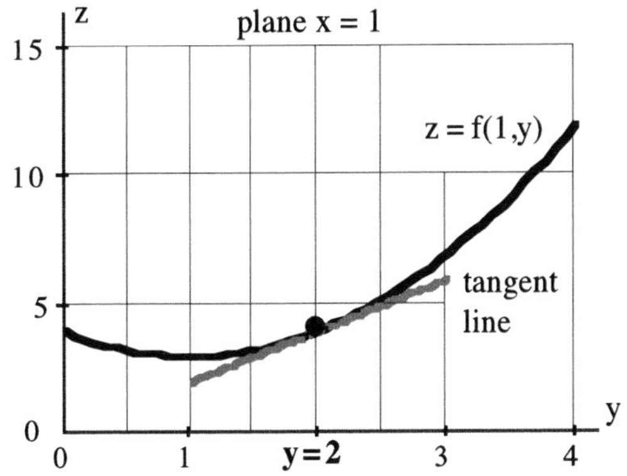

Fig. 6

Practice 5: Use the information in Figures 5 and 6 to estimate the signs (positive, negative or zero) of (a) $f_y(1,3)$, (b) $f_y(1,1)$, and (c) $f_y(1,4)$.

Partial Derivatives in Context

Of course it is very important to be able to calculate partial derivatives, but it also important to understand and to be able to communicate what they mean and measure. And you need to be able to attach the correct units to your answers.

Example 6: The surface area A (square inches) of a small child is a function of the length L (inches) and the weight W (pounds) of the child: $A = A(L, W)$. Explain (in clear English sentences) the meaning of the following. Be sure to include units.

(a) $A(26, 46) = 164$ (b) $\frac{\partial A(26,46)}{\partial W} = 7$ (c) $\frac{\partial A(26,46)}{\partial L} = 5$

Solution: (a) $A(26, 46) = 164$ square inches. A child who is 26 inches long and weighs 46 pounds will have a surface area of 164 square inches.

(b) $\frac{\partial A(26,46)}{\partial W} = 7$ square inches per pound . The surface area of this child (length 26 inches, weight 46 pounds, area 164 square inches) is **increasing at an INSTANTANEOUS RATE of** 7 square inches per each additional pound of weight if the length stays constant. Units: (square inches)/pound

(c) $\frac{\partial A(26,46)}{\partial L} = 5$ square inches per inch. The surface area of this child (length 26 inches, weight 46 pounds, area 164 square inches) is **increasing at an INSTANTANEOUS RATE of** 5 square inches per each additional inch of length if the weight stays constant. Units: (square inches)/inch

Practice 6: A certain biotech process using bacteria to produce a vaccine V (in grams) depends on the number B of bacteria and the temperature T (in ^oC) of the laboratory: V = V(B, T). Explain (in clear English sentences) the meaning of the following. Be sure to include units.

(a) $V(2000,40) = 8.9$ (b) $\dfrac{\partial V(2000,40)}{\partial B} = 0.003$ (c) $\dfrac{\partial V(2000,40)}{\partial T} = -1.4$

Partial Derivatives and Level Curves

Level curves of a surface z = f(x,y) give us information about f and also about the rate of change of f as x and y increase, the partial derivatives f_x and f_y.

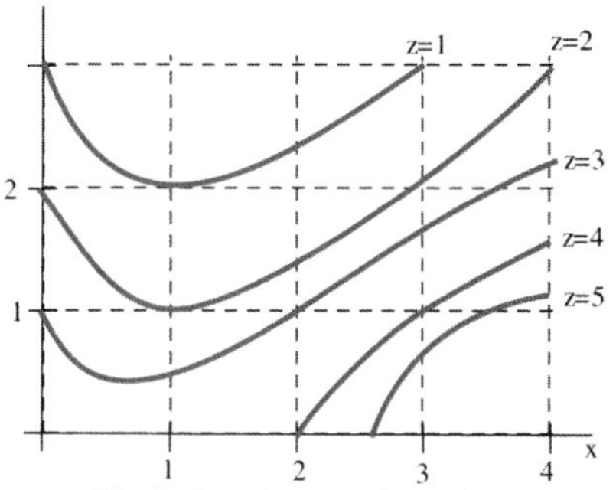

Fig. 7: Level curves for z=f(x,y)

Example 7: Use the information in Fig. 7 to estimate the signs (positive, negative or zero) of (a) $f_x(3,2)$ and (b) $f_y(3,2)$.

Solution: (a) As we move through the point (3,2) in the increasing x direction (Fig. 8a), the level curves are increasing in value so $f_x(3,2)$ is positive. Fig. 8b shows the graph of z along the line segment of increasing x–values (y is constantly 2), and the slope of the tangent line to this graph is positive when x = 3 so $f_x(3,2)$ is positive.

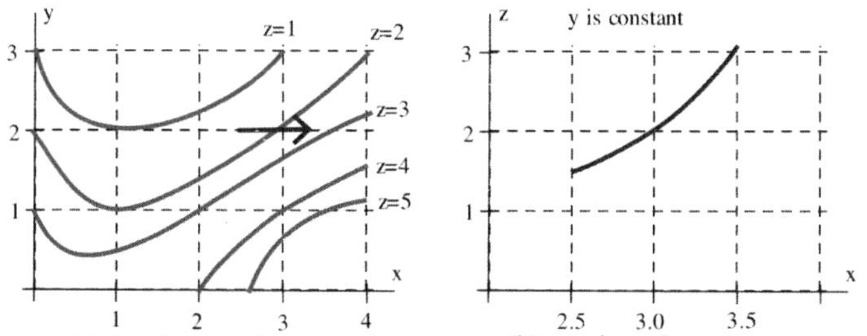

(a) level curves for z=f(x,y) (b) z-values along the arrow

Fig. 8: the sign of $f_x(3,2)$

(b) As we move through the point (3,2) in the increasing y direction (Fig. 9a), the level curves are decreasing in value so $f_y(3,2)$ is negative. Fig. 9b shows the graph of z along the line segment of increasing y–values (x is constantly 3), and the slope of the tangent line to this graph is negative when y = 2 so $f_y(3,2)$ is negative.

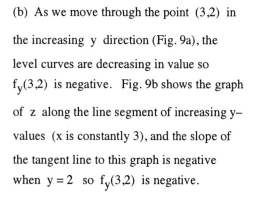

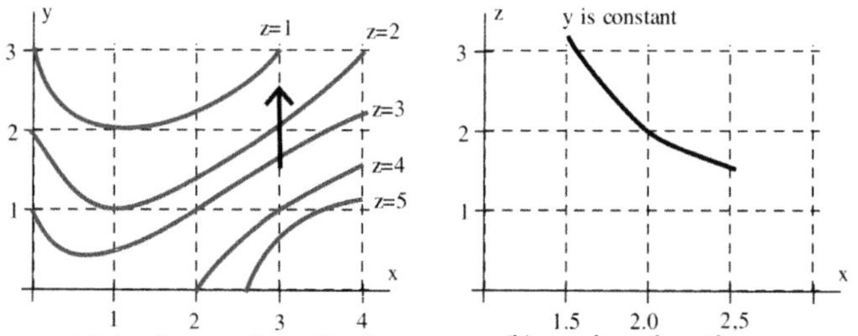

(a) level curves for z=f(x,y) (b) z-values along the arrow

Fig. 9: the sign of $f_y(3,2)$

Note: If the level curves are close together in our direction of movement, then the z–values are changing rapidly in that direction and the magnitude (the absolute value of the magnitude) of the rate of change in that direction is large. For the function described by the level curves in Fig. 7, $| f_x(3,2) | > | f_y(3,2) |$ because the level curve lines are closer together as when we move from (3,2) in the x–direction than when we move in the y–direction.

Practice 7: Use the information in Fig. 7 to estimate the signs (positive, negative or zero) of $f_x(3,1)$, $f_y(3,1)$, $f_x(1,1)$ and $f_y(1,1)$. Which of these partial derivatives has the largest absolute value?

Second Partial Derivatives

For a function $y = f(x)$ of one variable, the second derivative $f''(x) = \frac{d}{dx}\left(\frac{df}{dx}\right) = \frac{d^2 f}{dx^2}$ is the rate of change of the rate of change of f, and it measures the concavity of the graph of f. The situation for $z = f(x,y)$ is similar. The second derivative of a function of one variable was used to determine whether a critical point was a local maximum or minimum, and the second partial derivatives will be used to help determine whether critical points of functions of two variables are local maximums or minimums.

> Definition: Second Partial Derivatives of $z = f(x,y)$:
>
> $f_{xx}(x,y) = f_{xx} = \frac{\partial}{\partial x}\left(\frac{\partial f}{\partial x}\right) = \frac{\partial^2 f}{\partial x^2} = \frac{\partial^2 z}{\partial x^2}$ differentiate twice with respect to x
>
> $f_{yy}(x,y) = f_{yy} = \frac{\partial}{\partial y}\left(\frac{\partial f}{\partial y}\right) = \frac{\partial^2 f}{\partial y^2} = \frac{\partial^2 z}{\partial y^2}$ differentiate twice with respect to y

$f_{xx}(x,y)$ measures the concavity of the graph of f in the x–direction. $f_{yy}(x,y)$ measures the concavity in the y–direction.

We can also differentiate first with respect to one variable and then differentiate the result with respect to the other variable.

> Definition: Second Mixed Partial Derivatives of $z = f(x,y)$:
>
> $f_{xy} = (f_x)_y = \frac{\partial}{\partial y}\left(\frac{\partial f}{\partial x}\right) = \frac{\partial^2 f}{\partial y \partial x}$ differentiate first with respect to x, then with respect to y
>
> $f_{yx} = (f_y)_x = \frac{\partial}{\partial x}\left(\frac{\partial f}{\partial y}\right) = \frac{\partial^2 f}{\partial x \partial y}$ differentiate first with respect to y, then with respect to x

$f_{xy}(x,y)$ measures the rate of change in the y–direction of the rate of change in the x–direction. This is more difficult to interpret graphically.

Note how order of the x and y changes depending on the notation: $f_{xy} = \frac{\partial}{\partial y}\left(\frac{\partial f}{\partial x}\right) = \frac{\partial^2 f}{\partial y \partial x}$ and $f_{yx} = \frac{\partial^2 f}{\partial x \partial y}$.

Example 8: For $f(x,y) = 3x^3 + 7y^4 - 10x^2y$, calculate f_{xx}, f_{yy}, f_{xy} and f_{yx} .

Solution: $f_x = 9x^2 - 20xy$ and $f_y = 28y^3 - 10x^2$. Then

$$f_{xx} = \frac{\partial}{\partial x}(9x^2 - 20xy) = 18x - 20y. \qquad f_{yy} = \frac{\partial}{\partial y}(28y^3 - 10x^2) = 84y^2 .$$

$$f_{xy} = \frac{\partial}{\partial y}(9x^2 - 20xy) = -20x . \qquad f_{yx} = \frac{\partial}{\partial x}(28y^3 - 10x^2) = -20x .$$

Practice 8: For $g(x,y) = e^{xy} + \frac{x}{y}$, calculate g_{xx}, g_{yy}, g_{xy} and g_{yx} .

In the previous Example and Practice it turned out that the mixed partials were equal: $f_{xy} = f_{yx}$ and $g_{xy} = g_{yx}$. The next theorem says this is always the case for "nice" (sufficiently smooth) surfaces.

Clairaut's Theorem:

If $f(x,y)$ is defined and continuous at (a,b) and for all points near (a,b) and f_{xy} and f_{yx} are both continuous at all points near (a,b),

then $f_{xy}(a,b) = f_{yx}(a,b)$.

We can also define higher partial derivatives in a natural way such as $f_{xyy} = (f_{xy})_y = \frac{\partial}{\partial y}\left(\frac{\partial^2 f}{\partial y \partial x}\right) = \frac{\partial^3 f}{\partial y \partial y \partial x}$. These higher partial derivatives are sometimes useful in physics and other areas , but we will not use them.

Partial Derivatives Implicitly

In all of the previous examples we knew z explicitly as a function of x and y. But sometimes it is not possible to algebraically isolate z in order to calculate a partial derivative. In that case we can still determine the partial derivatives, but we need to do so implicitly.

Example 8: $xy + yz = xz$. Determine $\frac{\partial z}{\partial x}$ and $\frac{\partial z}{\partial y}$ in general and at the point (3, 2, 6).

Solution: In this case we can calculate the partial derivatives both explicitly and implicitly.

Explicitly: Solving for z we get $xy = xz - yz$ so $z = \frac{xy}{x - y}$. Then, using the quotient rule,

$$\frac{\partial z}{\partial x} = \frac{(x-y)\cdot\frac{\partial(xy)}{\partial x} - xy\cdot\frac{\partial(x-y)}{\partial x}}{(x-y)^2} = \frac{(x-y)\cdot y - xy\cdot 1}{(x-y)^2} = \frac{-y^2}{(x-y)^2} .$$

At (3, 2, 6), $\frac{\partial z}{\partial x} = -4$. Similarly, $\frac{\partial z}{\partial y} = \frac{x^2}{(x-y)^2}$ which equals 9 at (3, 2, 6).

Implicitly: Taking the partial derivative of each side, $\frac{\partial}{\partial x}(xy + yz) = \frac{\partial}{\partial x}(xz)$, we get

$\left[x \cdot \frac{\partial y}{\partial x} + y \cdot \frac{\partial x}{\partial x}\right] + \left[y \cdot \frac{\partial z}{\partial x} + z \cdot \frac{\partial y}{\partial x}\right] = x \cdot \frac{\partial z}{\partial x} + z \cdot \frac{\partial x}{\partial x}$. But $\frac{\partial y}{\partial x} = 0$ (why?) and

$\frac{\partial x}{\partial x} = 1$ so the previous equation simplifies to $\left[0 + y\right] + \left[y \cdot \frac{\partial z}{\partial x} + 0\right] = x \cdot \frac{\partial z}{\partial x} + z$.

Then $\frac{\partial z}{\partial x} = \frac{z - y}{y - x}$ which equals -4 at (3, 2, 6), the same result we got differentiating explicitly. Similarly, $\frac{\partial z}{\partial u} = \frac{x + z}{x - y} = \frac{9}{1}$ at (3, 2, 6).

Practice 9: $xy^2 + \sin(z) + 3 = 2x + 3z$. Determine $\frac{\partial z}{\partial x}$ in general and at the point (3, 1, 0).

A Final Comment!

Partial derivatives are used extensively in the remaining sections on multivariate calculus, and it is vital that you understand what they measure and that you become able to calculate partial derivatives quickly and accurately. Extra practice now will save you time (and points) in the rest of the course.

PROBLEMS

1. For $f(x,y) = 16 - 4x^2 - y^2$, find $f_x(1,2)$ and $f_y(1,2)$ and interpret these numbers as slopes. Illustrate with sketches.

2. For $f(x,y) = \sqrt{4 - x^2 - 4y^2}$, find $f_x(1,0)$ and $f_y(1,0)$ and interpret these numbers as slopes. Illustrate with sketches.

In problems 3 – 11, find the indicated partial derivatives.

3. $f(x,y) = x^3y^5$; $f_x(3,-1)$

4. $f(x,y) = xe^{-y} + 3y$; $\frac{\partial}{\partial y}(1,0)$

5. $z = \frac{x^3 + y^3}{x^2 + y^2}$; $\frac{\partial z}{\partial x}$, $\frac{\partial z}{\partial y}$

6. $z = \frac{x}{y} + \frac{y}{x}$; $\frac{\partial z}{\partial x}$

7. $xy + yz = xz$; $\frac{\partial z}{\partial x}$, $\frac{\partial z}{\partial y}$

8. $\sin(x) + y \cdot e^z = z$; $\frac{\partial z}{\partial x}$, $\frac{\partial z}{\partial y}$

9. $y^2 + yz^2 = zx^2$; $\frac{\partial z}{\partial x}$, $\frac{\partial z}{\partial y}$

10. $x^2 + y^2 - z^2 = 2x(y + z)$; $\frac{\partial z}{\partial x}$, $\frac{\partial z}{\partial y}$

11. $u = xy \sec(xy)$; $\frac{\partial u}{\partial x}$

12. $f(x,y,z) = xyz$; $f_y(0,1,2)$

13. $u = xy + yz + zx$; u_x , u_y , u_z

In problems 14 – 29, find the first partial derivatives of the given functions.

14. $f(x,y) = x^3y^5 - 2x^2y + x$ 15. $f(x,y) = x^4 + x^2y^2 + y^4$ 16. $f(x,y) = \frac{x-y}{x+y}$

17. $f(x,y) = e^x \tan(x - y)$ 18. $f(s,t) = \sqrt{2 - 3s^2 - 5t^2}$ 19. $f(u,v) = \arctan(u/v)$

20. $g(x,y) = y \tan(x^2y^3)$ 21. $z = \ln(x + \sqrt{x^2 + y^2})$ 22. $z = \sinh(\sqrt{3x + 4y})$

23. $f(x,y) = \int_x^y e^{(t^2)}\, dt$ 24. $f(x,y,z) = x^2yz^3 + xy - z$ 25. $f(x,y,z) = x^{yz}$

26. $u = z \sin\left(\frac{y}{x+z}\right)$ 27. $u = xy^2z^3 \ln(x + 2y + 3z)$ 28. $f(x,y,z,t) = \frac{x-y}{z-t}$

29. $u = \sqrt{x_1^2 + x_2^2 + \ldots + x_n^2}$

30. Use the definition of partial derivatives as limits to find $f_x(x,y)$ and $f_y(x,y)$ when $f(x,y) = x^2 - xy + 2y^2$.

In problems 31 – 33, find $\partial z/\partial x$ and $\partial z/\partial y$.

31. $z = f(x) + g(y)$ 32. $z = f(x + y)$ 33. $z = f(x/y)$

In problems 34 – 36, find all of the second partial derivatives.

34. $f(x,y) = x^2y + x\sqrt{y}$ 35. $z = (x^2 + y^2)^{3/2}$ 36. $z = t \cdot \arcsin(\sqrt{x})$

In problems 37 and 38, verify that the conclusion of Clairaut's Theorem holds, that is, $u_{xy} = u_{yx}$.

37. $u = x^5y^4 - 3x^2y^3 + 2x^2$ 38. $u = \arcsin(xy^2)$

39. Verify that the function $u = e^{-a^2k^2t} \sin(kx)$ is a solution of the heat equation $u_t = a^2 u_{xx}$.

40. The total resistance R produced by three conductors with resistances R_1, R_2, R_3 connected in a parallel electrical circuit is given by the formula $\frac{1}{R} = \frac{1}{R_1} + \frac{1}{R_2} + \frac{1}{R_3}$. Find $\partial R/\partial R_1$.

Practice Answers

Practice 1: (a) (2000 gallons)/(10 degrees) = 200 gallons/degree

(b) (7000 gallons)/(10,000 people) = 0.7 gallons/person

Practice 2: (a) $f_x(x,y) = 3x^2 + 8y + 10xy$, $f_x(2,5) = 152$

(b) $g_x(x,y) = y \cdot e^{xy} + \frac{1}{y}$, $g_x(0,2) = 2.5$

Practice 3: $f(3,2) \approx 9$, $f_x(3,2) \approx 9$

Practice 4: (a) $f_y(x,y) = 8y + 5x^2$, $f_y(2,5) = 60$

(b) $g_y(x,y) = x\, e^{xy} - x/y^2$, $g_y(0,2) = 0$

Practice 5: (a) $f_y(1,3)$ is positive (b) $f_y(1,1)$ is approximately 0 (c) $f_y(1,4)$ is positive

Practice 6: (a) This process will produce 8.9 grams of vaccine when we have 2000 bacteria and the laboratory temperature is 40 °C.

(b) The amount of vaccine produced (when we have 2000 bacteria at a temperature of 40 °C) will **increase at an INSTANTANEOUS RATE of** 0.003 grams for each additional bacteria when the temperature stays constant. Units: grams/bacteria

(c) The amount of vaccine produced (when we have 2000 bacteria at a temperature of 40 °C) will **decrease at an INSTANTANEOUS RATE of** 1.4 grams for each degree increase in temperature when the number of bacteria stays constant. Units: grams/°C

Practice 7: $f_x(3,1)$ is positive , $f_y(3,1)$ is negative ,

$f_x(1,1)$ is zero (z has alocal max for increasing x values) , and $f_y(1,1)$ is negative .

I estimate that $f_y(3,1)$ has the largest absolute value (contour lines are closest together).

Practice 8: If $g(x,y) = e^{xy} + \frac{x}{y}$, then

$g_x = y \cdot e^{xy} + \frac{1}{y}$ $\qquad$ $g_y = x \cdot e^{xy} - \frac{x}{y^2}$.

$g_{xx} = \frac{\partial}{\partial x}\left(y \cdot e^{xy} + \frac{1}{y}\right) = y^2 \cdot e^{xy}$ $\qquad$ $g_{yy} = \frac{\partial}{\partial y}\left(x \cdot e^{xy} - \frac{x}{y^2}\right) = x^2 \cdot e^{xy} + \frac{2x}{y^3}$

$$g_{xy} = \frac{\partial}{\partial y}\left(y \cdot e^{xy} + \frac{1}{y}\right) = xy \cdot e^{xy} + e^{xy} - \frac{1}{y^2}$$

$$g_{yx} = \frac{\partial}{\partial x}\left(x \cdot e^{xy} - \frac{x}{y^2}\right) = xy \cdot e^{xy} + e^{xy} - \frac{1}{y^2}$$

Practice 9: $\frac{\partial}{\partial x}(x \cdot y^2 + \sin(z) + 3) = \frac{\partial}{\partial x}(2x + 3z)$ so

$\left[x \cdot 2y \cdot \frac{\partial y}{\partial x} + y^2 \cdot \frac{\partial x}{\partial x}\right] + \cos(z) \cdot \frac{\partial z}{\partial x} + 0 = 2 \cdot \frac{\partial x}{\partial x} + 3 \cdot \frac{\partial z}{\partial x}$ which simplifies to

$\left[y^2 \right] + \cos(z) \cdot \frac{\partial z}{\partial x} = 2 + 3 \cdot \frac{\partial z}{\partial x}$ so $\frac{\partial z}{\partial x} = \frac{2 - y^2}{\cos(z) - 3}$ which equals $\frac{1}{-2}$ at (3, 1, 0).

13.4 TANGENT PLANES and DIFFERENTIALS

In Section 2.8 we were able to use the derivative f' of a function $y = f(x)$ of one variable to find the equation of the line tangent to the graph of f at a point $(a, f(a))$ (Fig. 1): $y = f(a) + f'(a)\cdot(x - a)$. And then we used this tangent line to approximate values of f near the point $(a, f(a))$, and we introduced the idea if the differential $df = f'(a)\cdot dx$ of the function f. In this section we extend these ideas to functions $z = f(x,y)$ of two variables. But here we will find tangent planes (Fig. 2) rather than tangent lines, and we will use the tangent plane to approximate values of $f(x,y)$. Finally, we will extend the concept of a differential to functions of two variables.

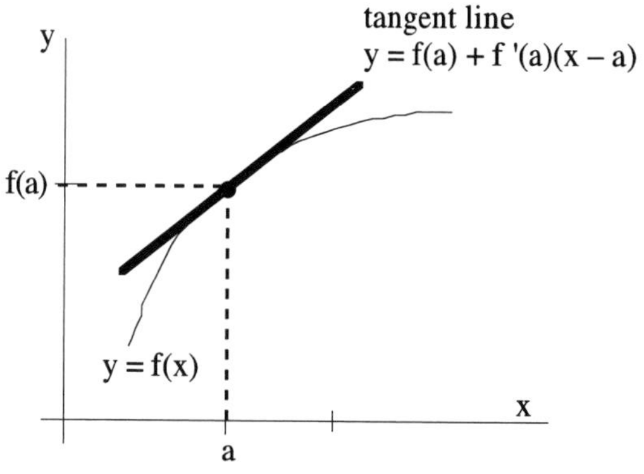

Fig. 1: Tangent line to $y = f(x)$ at $x = a$

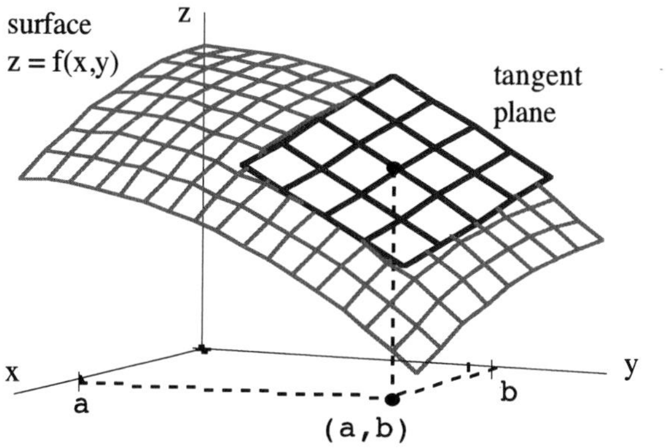

Fig. 2: Tangent plane to $z = f(x,y)$ at (a,b)

Tangent Planes

In Section 11.6 we saw how to use a point (a, b, c) and two (nonparallel) vectors to determine the equation of the plane through the point and containing lines parallel to the given vectors (Fig. 3):

(1) we used the cross product of the two given vectors to find a normal vector $\mathbf{N} = \langle n_1, n_2, n_3 \rangle$ to the plane, and then

(2) we used the normal vector $\mathbf{N}$ and the point to write the equation of the plane as $n_1(x - a) + n_1(y - b) + n_1(z - c) = 0$.

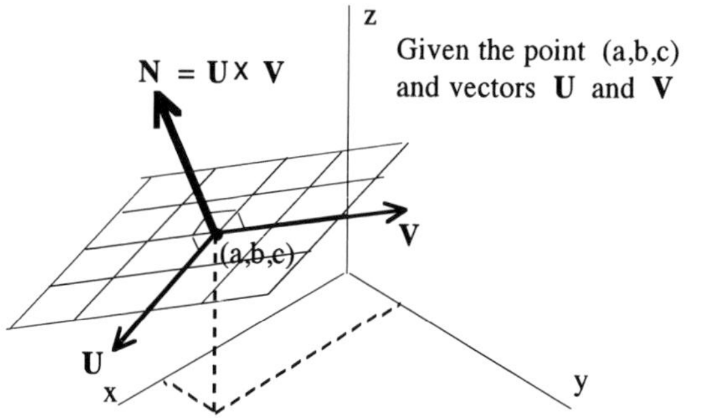

Equation of the tangent plane is
$n_1(x-a) + n_2(y-b) + n_3(z-c) = 0$

Fig. 3

This approach also works when we need the equation of a plane tangent to a surface $z = f(x,y)$, but we will use the formula for the surface to find the two needed vectors.

Example 1: Find the equation of the plane tangent to the surface $f(x,y) = 2x^3 + y^2 + 3$ at the point $P = (1, 2, f(1,2)) = (1, 2, 9)$ on the surface (Fig. 4).

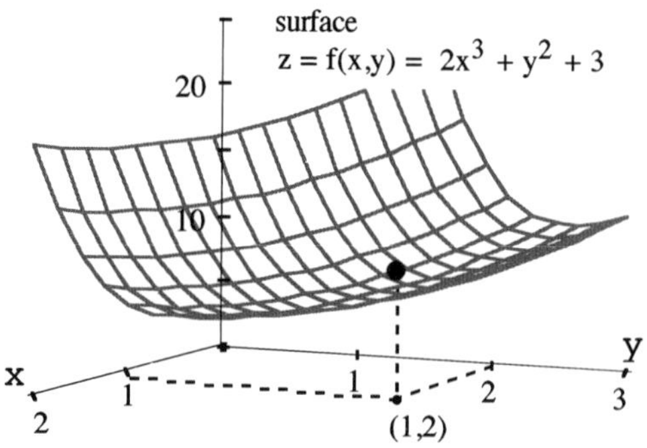

Fig. 4

Solution: We are given a point $(1, 2, 9)$ on the plane, but we need two vectors. These vectors are the rates of change of the surface $f(x,y)$ in the x and y directions. The rate of change of $f(x,y)$ in the x–direction is $f_x(x,y) = 6x^2$, and at the point $(1,2,9)$ we have $f_x(1,2) = 6(1)^2 = 6$. Similarly, the rate of change of $f(x,y)$ in the y–direction is $f_y(x,y) = 2y$, and at the point $(1,2,9)$ we have $f_y(1,2) = 2(2) = 4$.

Then a "rate of change vector in the x–direction" is $\mathbf{U} = \langle 1, 0, 6 \rangle$ formed by taking 1 "step" in the x–direction, taking 0 "steps" in the y–direction (y is constant), and then taking 6 "steps" in the z–direction ($6 = f_x(1,2)$ = rate of change of z with respect to increasing x–values). Similarly, a "rate of change" vector in the y–direction is $\mathbf{V} = \langle 0, 1, 4 \rangle$. These vectors are shown in Fig. 5.

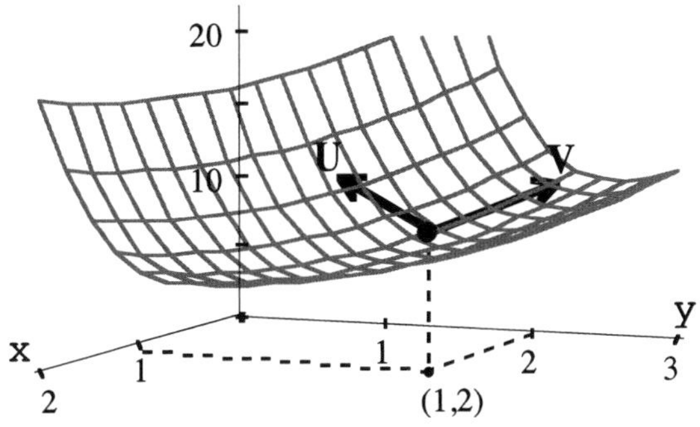

Fig. 5: Surface and tangent vectors **U** and **V**

Now a normal vector **N** to the plane we want is formed by taking

$$\mathbf{N} = \mathbf{V} \times \mathbf{U} = \begin{vmatrix} \mathbf{i} & \mathbf{j} & \mathbf{k} \\ 0 & 1 & 4 \\ 1 & 0 & 6 \end{vmatrix} = (6)\mathbf{i} - (-4)\mathbf{j} + (-1)\mathbf{k} = 6\mathbf{i} + 4\mathbf{j} - 1\mathbf{k}$$

(Note: taking $\mathbf{N} = \mathbf{U} \times \mathbf{V} = -6\mathbf{i} - 4\mathbf{j} + 1\mathbf{k}$ also works.)

Finally, using the point $P = (1, 2, 9)$ and the normal vector $\mathbf{N} = \mathbf{V} \times \mathbf{U} = 6\mathbf{i} + 4\mathbf{j} - 1\mathbf{k}$, we know that the equation of the plane is

$$\mathbf{6(x - 1) + 4(y - 2) - 1(z - 9) = 0} \quad \text{or} \quad \mathbf{z = 9 + 6(x - 1) + 4(y - 2)}$$

Looking at the equation $z = 9 + 6(x - 1) + 4(y - 2)$ of the plane, you should notice that the 9 is the z–coordinate of our original point, that the coefficient of the x variable, 6, is $f_x(1,2)$, and that the coefficient of the y variable, 4, is $f_y(1,2)$. Fig. 6 shows the surface and the tangent plane.

Fortunately we do not need to go through all of those calculations every time we need the equation of a plane tangent to a surface at a point: the pattern that we noted about the coefficients of the variables in the tangent plane equation and the values of the partial derivatives holds for every differentiable function.

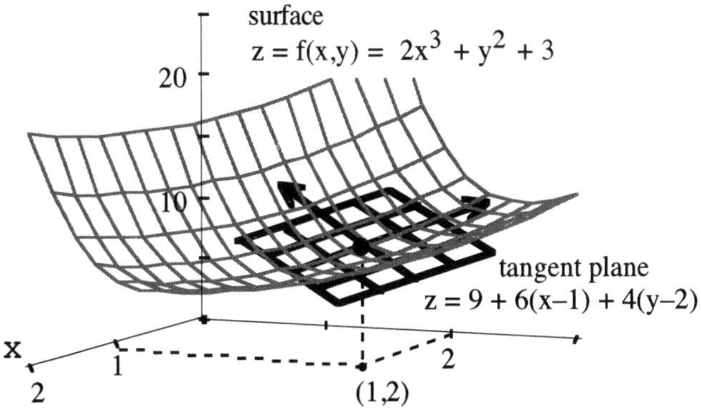

Fig. 6: Surface and tangent plane

Equation for a Tangent Plane

If $f(x,y)$ is differentiable at the point $(a, b, f(a,b))$,

then the equation of the plane tangent to the surface $z = f(x, y)$ at the point $P(a, b, f(a,b))$

is $\mathbf{z = f(a,b) + f_x(a,b)(x - a) + f_y(a,b)(y - b)}$.

Proof: The proof simply involves the steps we went through for Example 1. $\mathbf{U} = \langle 1, 0, f_x(a,b) \rangle$ is formed by taking 1 "step" in the x–direction, taking 0 "steps" in the y–direction (y is constant), and then taking $f_x(a,b)$ "steps" in the z–direction . SImilarly, $\mathbf{V} = \langle 0, 1, f_y(a,b) \rangle$. Then

$$\mathbf{N} = \mathbf{V} \times \mathbf{U} = \begin{vmatrix} \mathbf{i} & \mathbf{j} & \mathbf{k} \\ 0 & 1 & f_y(a,b) \\ 1 & 0 & f_x(a,b) \end{vmatrix} = (f_x(a,b))\mathbf{i} - (-f_y(a,b))\mathbf{j} + (-1)\mathbf{k} = f_x(a,b)\,\mathbf{i} + f_y(a,b)\,\mathbf{j} - 1\mathbf{k} .$$

Finally, using the point $(a, b, f(a,b))$ and $\mathbf{N} = \mathbf{V} \times \mathbf{U} = f_x(a,b)\,\mathbf{i} + f_y(a,b)\mathbf{j} - 1\mathbf{k}$, we have that the equation of the plane is

$$f_x(a,b)(x - a) + f_y(a,b)(y - b) - 1(z - f(a,b)) = 0 \quad \text{or} \quad z = f(a,b) + f_x(a,b)(x - a) + f_y(a,b)(y - b) .$$

Example 2: Find the plane tangent to the surface $z = 2x^2y^3 + \ln(xy) + 7$ at the point $(1, 1, 9)$.

Solution: $f_x(x, y) = 4xy^3 + \frac{1}{xy}(y) = 4xy^3 + \frac{1}{x}$ and $f_y(x, y) = 6x^2y^2 + \frac{1}{xy}(x) = 6x^2y^2 + \frac{1}{y}$

so $f_x(1, 1) = 5$ and $f_y(1, 1) = 7$.

Then the equation of the tangent plane is $z = 9 + 5(x - 1) + 7(y - 1)$ or $\mathbf{z = 5x + 7y - 3}$.

(This is much quicker than the "from scratch" method of Example 1.)

Fig. 7 shows two views of this surface and the tangent plane — notice that in this case the tangent plane cuts through the surface.

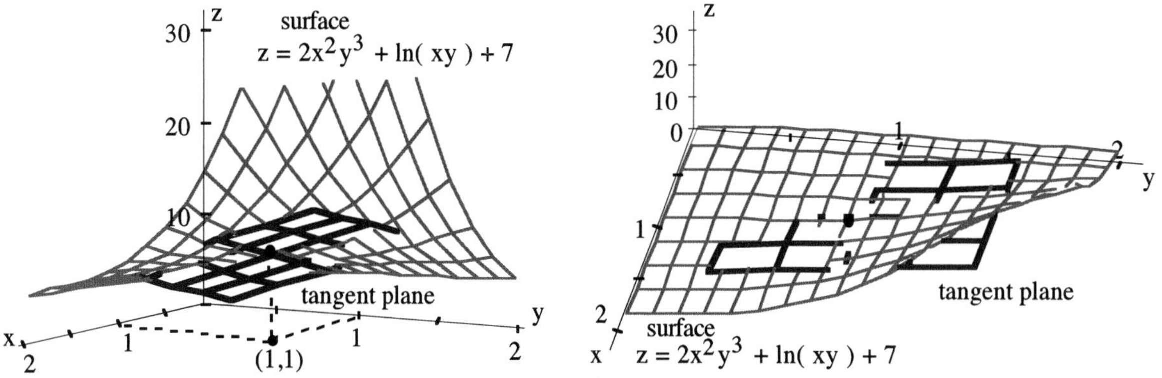

Fig. 7: Two views of the surface and tangent plane

Practice 1: Find the plane tangent to the surface $z = 5xy^2 + 7y + \sin(xy) - 2$ at the point $(0, 1, 5)$.

Differentials

The following "boxed" material summarizes, from Section 2.8, the definition and results about the differential dy of a function $y = f(x)$ of one variable.

Definition for $y = f(x)$: The **differential** of $y = f(x)$ is $\mathbf{dy = f'(x)\, dx} = \frac{df}{dx}\, dx$.

Meaning of dy: **dy** is the change in the y–value, **along the tangent line** to f obtained by a step of dx in the x–value.

Result: If f is differentiable at $x = a$ and dx is "small"
then $f(a + dx) - f(a) \approx \mathbf{dy}$ or $f(a + dx) \approx f(a) + \mathbf{dy}$.

Meaning of the Result: For a small step dx, the actual change in f is approximately equal to the change along the tangent line: $f(a + dx) \approx f(a) + f'(a)\, dx$.

The extension to functions of two variables is given in the next "box."

Definition for $z = f(x,y)$: The **differential** (or **total differential**) of $z = f(x,y)$ is

$$\mathbf{dz = f_x(x, y)\, dx + f_y(x, y)\, dy} = \frac{\partial f}{\partial x}\, dx + \frac{\partial f}{\partial y}\, dy\,.$$

Meaning of dz: **dz** is the change in the z–value, **along the tangent plane** to f, obtained by a step of dx in the x direction and a step of dy in the y direction. (Fig. 8)

Result: If $z = f(x,y)$ is differentiable at the point (a, b),
then $f(a + dx, b + dy) - f(a,b) \approx \mathbf{dz} = f_x(a, b)\, dx + f_y(a, b)\, dy = \frac{\partial f}{\partial x}\, dx + \frac{\partial f}{\partial y}\, dy$.

Meaning of the Result: For a small step of dx in the x direction and a small step dy in the y direction, the change in f is approximately equal to the change along the tangent plane:
$$f(a + dx, b + dy) \approx f(a,b) + dz = f(a,b) + f_x(a, b)\, dx + f_y(a,b)\, dy$$

Example 3: Find the differential of $z = 5 + 3x^2y^3$ (a) in general, and (b) at the point $(x,y) = (2, 1)$.

Solution: (a) $\mathbf{dz} = f_x(a, b)\, dx + f_y(a, b)\, dy = \{ 6xy^3 \}\, dx + \{ 9x^2y^2 \}\, dy$

(b) at $(2, 1)$, $\mathbf{dz} = \{ 6(2)(1)^3 \}\, dx + \{ 9(2)^2(1)^2 \}\, dy = (12)\, dx + (36)\, dy$.

Example 4: For $z = 5 + 3x^2y^3$ and the point (2,1), use the result of Example 3 that $dz = (12)\, dx + (36)\, dy$ to approximate $f(2.02, 1.01)$ and $f(2.01, 0.98)$.
Compare these approximate values with the exact values of $f(2.02, 1.01)$ and $f(2.01, 0.97)$.

Solution: For f(2.02, 1.01) , $dx = 0.02$ and $dy = 0.01$ so $dz = (12)(0.02) + (36)(0.01) = 0.6$.

Then $f(2.02, 1.01) \approx f(2, 1) + dz = 17 + 0.6 = 17.6$.

Actually, $f(2.02, 1.01) = 17.6121206012$ so the approximation "error" using the differential is 0.012 .

For f(2.01, 0.97) , $dx = 0.01$ and $dy = -0.07$, so $dz = (12)(0.01) + (36)(-0.03) = -0.96$.

Then $f(2.01, 0.97) \approx f(2,1) + dz = 17 + (-0.96) = 16.04$.

Actually, $f(2.01, 0.97) = 16.0618705619$ so the approximation "error" using the differential is 0.022 .

Practice 2: Find the differential of $z = 3 + x \cdot \sin(2xy)$ (a) in general, and (b) at the point $(x,y) = (1, \pi/2)$.

(c) Use the result of part (b) to approximate $f(1.3, \frac{\pi}{2} - 0.1)$ and $f(0.99, \frac{\pi}{2} + 0.2)$.

(d) Compare the results of (c) with the exact values of $f(1.3, \frac{\pi}{2} - 0.1)$ and $f(0.99, \frac{\pi}{2} + 0.2)$.

Examples 4 and Practice 2(c and d) compare the value of f found by moving **along the tangent plane** to the actual value of f found **on the surface**. When the sideways movement is "small" (when dx and dy are both small), then the "along the tangent plane" value of z is close to the "on the surface" value of z, the actual value of f.

PROBLEMS

In problems 1 – 8, find an equation for the tangent plane to the given surface at the given point.

1. $z = x^2 + 4y^2$ at (2, 1, 8)
2. $z = x^2 - y^2$ at (3, –2, 5)
3. $z = 5 + (x-1)^2 + (y+2)^2$ at (2, 0, 10)
4. $z = \sin(x + y)$ at (1, –1, 0)
5. $z = \ln(2x + y)$ at (–1, 3, 0)
6. $z = e^x \cdot \ln(y)$ at (3, 1, 0)
7. $z = xy$ at (–1, 2, –2)
8. $z = \sqrt{x - y}$ at (5, 1, 2)

In problems 9 – 18, find the differential of the given function.

9. $z = x^2 y^3$
10. $z = x^4 - 5x^2 y + 6xy^3 + 10$
11. $z = \frac{1}{x^2 + y}^2$
12. $z = y \cdot e^{xy}$
13. $u = e^x \cdot \cos(xy)$
14. $v = \ln(2x - 3y)$
15. $w = x^2 y + y^2 z$
16. $w = x \sin(yz)$
17. $w = \ln(\sqrt{x^2 + y^2 + z^2})$
18. $w = \frac{x + y}{y + z}$
19. If $z = 5x^2 + y2$ and (x, y) changes from (1, 2) to (1.05, 2.1), compare the values of Δz and dz.
20. If $z = x^2 - xy + 3y^2$ and (x, y) changes from (3, –1) to (2.96, –0.95), compare the values of Δz and dz.

In problems 21 – 24, use differentials to approximate the value of f at the given point.

21. $f(x, y) = \sqrt{x^2 - y^2}$ at (5.01, 4.02)

22. $f(x, y) = \sqrt{20 - x^2 - 7y^2}$ at (1.95, 1.08)

23. $f(x, y) = \ln(x - 3y)$ at (6.9, 2.06)

24. $f(x, y) = y \cdot e^{xy}$ at (0.2, 1.96)

25. The length and width of a rectangle are measured as 30 cm and 24 cm, respectively, with an error in measurement of at most 0.1 cm in each. Use differentials to estimate the maximum error in the calculated area of the rectangle.

26. The dimensions of a closed rectangular box are measured as 80 cm, 60 cm, and 50 cm, respectively, with a possible error of 0.2 cm in each dimension. Use differentials to estimate the maximum error in calculating the surface area of the box.

27. Use differentials to estimate the amount of tin in a closed tin can with diameter 8 cm and height 12 cm if the tin is 0.04 cm thick.

28. Use differentials to estimate the amount of metal in a closed cylindrical can that is 10 cm high and 4 cm in diameter if the metal in the wall is 0.05 cm thick and the metal in the top and bottom is 0.1 cm thick.

29. A boundary stripe 3 in. wide is painted around a rectangle whose dimensions are 100 ft. by 200 ft. Use differentials to approximate the number of square feet of paint in the stripe.

30. The pressure, volume, and temperature of a mole of an ideal gas are related by the equation $PV = 8.31T$, where P is measured in kilopascals, V in liters, and T in ^oK (= $^oC + 273$). Use differentials to find the approximate change in pressure if the volume increases from 12 L to 12.3 L and the temperature decreases from 310^oK to 305^oK.

PRACTICE ANSWERS

Practice 1: $z = f(a,b) + f_x(a,b)(x - a) + f_y(a,b)(y - b)$

with $f(x,y) = 5xy^2 + 7y + \sin(xy) - 2$, $a = 0$, $b = 1$, and $f(0,1) = 5$.

$f_x(x,y) = 5y^2 + y \cdot \cos(xy)$ so $f_x(0,1) = 5 + 1 = 6$.

$f_y(x,y) = 10xy + 7 + x \cdot \cos(xy)$ so $f_y(0,1) = 7$.

Then the equation of the tangent plane to f at (0,1,5) is $z = 5 + 6(x - 0) + 7(y - 1) = 6x + 7y - 2$.

Practice 2: (a) $\mathbf{dz} = f_x(a, b)\,dx + f_y(a, b)\,dy = \{ x\cdot\cos(2xy)\cdot 2y + \sin(2xy)\}\,dx + \{ x\cdot\cos(2xy)\cdot 2x \}\,dy$

(b) at $(1, \pi/2)$,

$\mathbf{dz} = \{1\cdot\cos(2\cdot 1\cdot\pi/2)\cdot 2\cdot\pi/2 + \sin(2\cdot 1\cdot\pi/2)\}\,dx + \{1\cdot\cos(2\cdot 1\cdot\pi/2)\cdot 2\cdot 1\}\,dy$ so

$\mathbf{dz} = (-\pi)\,dx + (-2)\,dy$.

(c)&(d) For $f(1.3, \frac{\pi}{2} - 0.1)$, $dx = 0.3$ and $dy = -0.1$ so $dz = (-\pi)(0.3) + (-2)(-0.1) \approx -0.742$.

Then $f(1.3, \frac{\pi}{2} - 0.1) \approx f(1, \frac{\pi}{2}) + dz = 3 + (-0.742) = 2.258$.

Actually, $f(1.3, \frac{\pi}{2} - 0.1) = 2.18006691401$ so the approximation "error" is 0.078, an "error" of less than 4%.

For $f(0.99, \frac{\pi}{2} + 0.2)$, $dx = -0.01$ and $dy = 0.2$ so $dz = (-\pi)(-0.01) + (-2)(0.2) = -0.369$.

Then $f(0.99, \frac{\pi}{2} + 0.2) \approx f(1, \frac{\pi}{2}) + dz = 3 + (-0.369) = 2.631$.

Actually, $f(0.99, \frac{\pi}{2} + 0.2) = 2.64700487061$ so the approximation "error" is 0.016, an "error" of less than 1%.

13.5 DIRECTIONAL DERIVATIVES and the GRADIENT VECTOR

Directional Derivatives

In Section 13.3 the partial derivatives f_x and f_y were defined as

$$\mathbf{f_x(x,y)} = \lim_{h \to 0} \frac{f(x+h,y) - f(x,y)}{h} \quad \text{(if the limit exists and is finite)} \quad \text{and}$$

$$\mathbf{f_y(x,y)} = \lim_{h \to 0} \frac{f(x,y+h) - f(x,y)}{h} \quad \text{(if the limit exists and is finite)}\cdot$$

These partial derivative measured the instantaneous rate of change of $z = f(x,y)$ as we moved in the increasing x–direction (while holding y constant) and in the increasing y–direction (while holding x constant). Sometimes, however, we are interested in the rate of change of $z = f(x,y)$ as we move in some other direction, and that leads to the idea of a **directional derivative** to measure the instantaneous rate of change of $z = f(x,y)$ as we move in any direction.

Definition:
The **directional derivative** of $z = f(x,y)$ in the direction of a unit vector $\mathbf{u} = \langle a, b \rangle$ is

$$D_{\mathbf{u}}f(x,y) = \lim_{h \to 0} \frac{f(x+ah, y+bh) - f(x,y)}{h} \quad \text{(if the limit exists and is finite)}\cdot$$

Figures 1a and 1b illustrate the slope of the line tangent to the curve where the plane above the vector u intersects the surface $z = f(x,y)$.

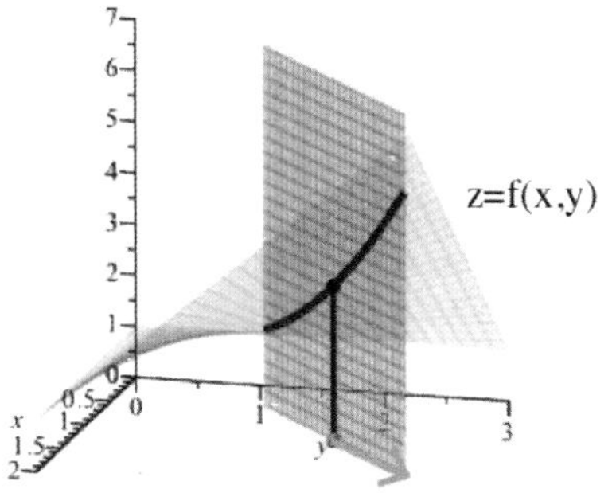

Fig. 1a

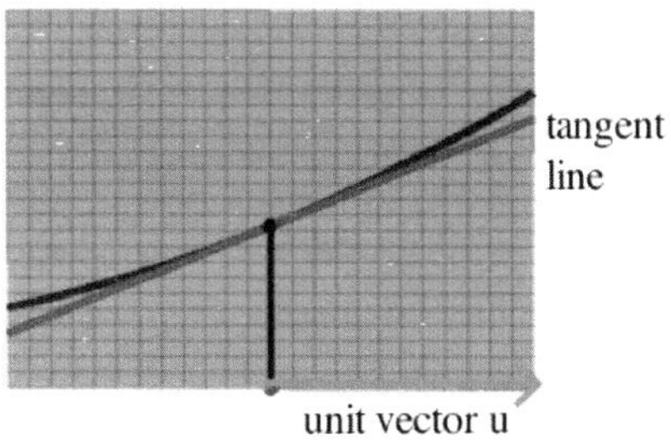

Fig. 1b

Using this definition can be tedious and algebraically messy, but it is worth doing once. Fortunately we will soon see a much more efficient method.

Example 1: Use this definition to calculate $D_{\mathbf{u}}f(1, 2)$ for $f(x,y) = 1 + xy$ and $\mathbf{u} = \langle 0.6, 0.8 \rangle$

Solution: $D_{\mathbf{u}}f(1, 2) = \lim_{h \to 0} \frac{f(1+0.6h, 2+0.8h) - f(1,2)}{h} = \lim_{h \to 0} \frac{[1+(1+0.6h)(2+0.8h)] - 3}{h}$

$$\lim_{h \to 0} \frac{[1+2++2(0.6)h+0.8h] - 3}{h} = 2(0.6)+0.8 = 2$$

Practice 1: Calculate $D_{\mathbf{u}}f(2, 1)$ for $f(x,y) = x^2 + 2x + 3y + 1$ and $\mathbf{u} = \left\langle \frac{5}{13}, \frac{12}{13} \right\rangle$.

For more complicated functions it becomes extremely difficult to use the definition to calculate directional derivatives. Fortunately there is a much easier way given by the next theorem.

> Theorem:
> If $f(x,y)$ is differentiable,
> then f has a directional derivative in the direction of every <u>unit vector</u> $\mathbf{u} = \langle a, b \rangle$, and
> $$D_{\mathbf{u}}f(x,y) = f_x(x,y)\, a + f_y(x,y)\, b .$$

A proof of this result uses a multivariable "Chain Rule" that we will discuss in Section 13.8.

Example 2: Verify that this theorem gives the same answer for $D_{\mathbf{u}}f(x,y)$ as our use of the directional derivative definition in Example 1.

Solution: $f_x(x,y)=y$ and $f_y(x,y)=x$ so $f_x(1, 2)= 2$ and $f_y(1, 2)= 1$. $\mathbf{u} = \langle 0.6, 0.8 \rangle$ so

$D_{\mathbf{u}}f(1, 2) = (2)(0.6) + (1)(0.8) = 2$, the same result as in Example 1.

Practice 2: Verify that this theorem gives the same answer for $D_{\mathbf{u}}f(x,y)$ as your use of the directional derivative definition in Practice 1.

Example 3: Calculate the directional derivatives of $z = f(x,y) = x + 5x^2y^3$ at the point (2,1) in the directions of the unit vectors (a) $\mathbf{u} = \langle 0.6, 0.8 \rangle$, (b) $\mathbf{u} = \langle -0.6, -0.8 \rangle$, (c) $\mathbf{u} = \langle 0.8, 0.6 \rangle$, (d) $\mathbf{u} = \mathbf{i} = \langle 1, 0 \rangle$, and (e) $\mathbf{u} = \mathbf{j} = \langle 0, 1 \rangle$.

Solution: $f_x(x,y) = 1 + 10xy^3$ and $f_y(x,y) = 15x^2y^2$ so $f_x(2,1) = 21$ and $f_y(2,1) = 60$. Then, by the previous Theorem,

(a) for $\mathbf{u} = \langle 0.6, 0.8 \rangle$, $D_{\mathbf{u}}f(2,1) = f_x(2,1)\, a + f_y(2,1)\, b = (21)(0.6) + (60)(0.8) = 60.6$.

(b) for $\mathbf{u} = \langle -0.6, -0.8 \rangle$, $D_{\mathbf{u}}f(2,1) = (21)(-0.6) + (60)(-0.8) = -60.6$.

(c) for $\mathbf{u} = \langle 0.8, 0.6 \rangle$, $D_\mathbf{u}f(2,1) = (21)(0.8) + (60)(0.6) = 52.8$.

(d) for $\mathbf{u} = \mathbf{i} = \langle 1, 0 \rangle$, $D_\mathbf{u}f(2,1) = (21)(1) + (60)(0) = 21$ $(= f_x(2,1))$.

(e) for $\mathbf{u} = \mathbf{j} = \langle 0, 1 \rangle$, $D_\mathbf{u}f(2,1) = (21)(0) + (60)(1) = 60$ $(= f_y(2,1))$.

The Gradient Vector

You might have noticed that the pattern $D_\mathbf{u}f(x,y) = f_x(x,y)\,a + f_y(x,y)\,b$ for calculating the directional derivative can be viewed as the dot product of the vector $\langle f_x(x,y), f_y(x,y) \rangle$ with the unit direction vector $\mathbf{u} = \langle a, b \rangle$. This vector $\langle f_x(x,y), f_y(x,y) \rangle$ shows up in a variety of contexts and is called the **gradient** of f.

> Definition of the **Gradient Vector:**
>
> The **gradient vector** of f(x,y) is $\nabla f(x,y) = \langle f_x(x,y), f_y(x,y) \rangle = \frac{\partial f}{\partial x}\mathbf{i} + \frac{\partial f}{\partial y}\mathbf{j}$.
>
> The symbol " ∇f " is read as "grad f" or "del f."

Example 4: Calculate $\nabla f(x,y)$ and $\nabla f(0,1)$ for (a) $f(x,y) = x + 5x^2y^3$ and (b) $f(x,y) = y^2 + \sin(x)$.

Solution: (a) For $f(x,y) = x + 5x^2y^3$, $f_x(x,y) = 1 + 10xy^3$ and $f_y(x,y) = 15x^2y^2$ so

$\nabla f(x,y) = \langle 1 + 10xy^3, 15x^2y^2 \rangle$ and $\nabla f(0,1) = \langle 1, 15 \rangle$.

(b) For $f(x,y) = y^2 + \sin(x)$, $f_x(x,y) = \cos(x)$ and $f_y(x,y) = 2y$ so

$\nabla f(x,y) = \langle \cos(x), 2y \rangle$ and $\nabla f(0,1) = \langle 1, 2 \rangle$.

Practice 3: Calculate $\nabla f(x,y)$ and $\nabla f(1,2)$ for (a) $f(x,y) = e^{xy} + 2x^3y + y^2$ and (b) $f(x,y) = \cos(2x+3y)$.

The directional derivative can now be written simply as the dot product of the gradient and the unit direction vector.

$$D_\mathbf{u}f(x,y) = \nabla f(x,y) \cdot \mathbf{u}$$

The gradient vector $\nabla f(x,y)$ is useful for much more than a compact notation for directional derivatives. $\nabla f(x,y)$ has a number of special features that make it useful for investigating the behavior of surfaces.

Three very important properties of the gradient vector $\nabla f(x,y)$:

(1) At a point (x,y), the maximum value of the directional derivative $D_{\mathbf{u}}f(x,y)$ is $|\nabla f(x,y)|$.

(2) At a point (x,y), the maximum value of $D_{\mathbf{u}}f(x,y)$ occurs when $\mathbf{u}$ has the same direction as the gradient vector $\nabla f(x,y)$. (At each point (x,y), the gradient vector $\nabla f(x,y)$ "points" in the direction of maximum increase for $f(x,y)$.)

(3) At a point (x,y), the gradient vector $\nabla f(x,y)$ is normal (perpendicular) to the level curve that goes through the point (x,y).

One of the beauties of mathematics is that sometimes a result like the powerful and non-obvious properties of the gradient can be proven in rather simple ways.

Proof of (1) and (2): $D_{\mathbf{u}}f(x,y) = \nabla f(x,y) \cdot \mathbf{u}$

$= |\nabla f(x,y)|\,|\mathbf{u}| \cos(\theta)$ θ is the angle between the vectors $\nabla f(x,y)$ and $\mathbf{u}$

$= |\nabla f(x,y)| \cos(\theta)$

The maximum value of $\cos(\theta)$ is 1 (when $\theta = 0$), so the maximum value of $D_{\mathbf{u}}f(x,y)$ is $|\nabla f(x,y)|$ and this maximum occurs when $\theta = 0$, when $\nabla f(x,y)$ and $\mathbf{u}$ have the same direction.

Proof of (3): f is constant along the level curve at the point (x,y) so $D_{\mathbf{u}}f(x,y) = 0$ when $\mathbf{u}$ is the direction of the level curve. But $D_{\mathbf{u}}f(x,y) = \nabla f(x,y) \cdot \mathbf{u}$ so $\nabla f(x,y)$ is perpendicular to $\mathbf{u}$, the direction of the level curve.

Example 5: Find (a) the maximum rate of change of $f(x,y) = xe^y$ at the point (2,0,) and (b) the direction in which this maximum rate of change occurs.

Solution: $f_x(x,y) = e^y$ and $f_y(x,y) = xe^y$ so $f_x(2,0) = e^0 = 1$ and $f_y(2,0) = 2e^0 = 2$.

(a) The maximum value of the rate of change of f is $D_{\mathbf{u}}f(x,y) = |\nabla f(1,2)| = |\langle 1, 2 \rangle| = \sqrt{5}$.

(b) This maximum value occurs when $\mathbf{u}$ is in the direction of $\nabla f(1,2)$: $\mathbf{u} = \langle 1/\sqrt{5}, 2/\sqrt{5} \rangle$.

Practice 4: Find (a) the maximum rate of change of $f(x,y) = \sqrt{2x+3y}$ at the point $(5, 2)$ and (b) the direction in which this maximum rate of change occurs.

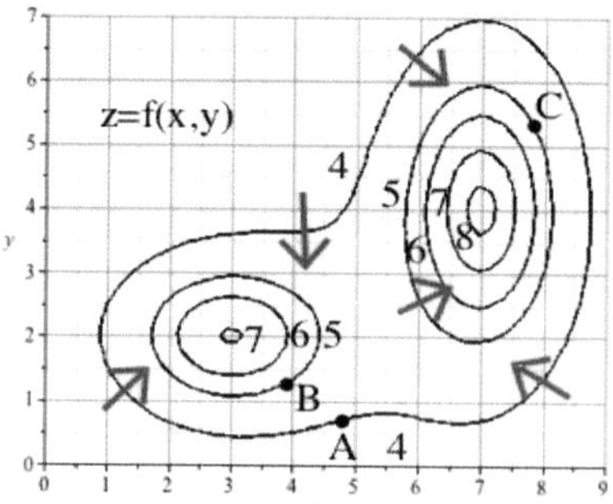

Fig. 2

Fig. 2 shows several level curves for a function $z = f(x,y)$ and the gradient vector at several locations. (Note: the lengths of these gradient vectors are exaggerated.)

Practice 5: Sketch the gradient vector $\nabla f(x,y)$ for the function f in Fig. 2 at A, B and C.

A ball placed at (x,y) will begin to roll in the direction $u = -\nabla f(x,y)$.

Climbing to a (local) maximum

Property (2) is the foundation for using the gradient vector $\nabla f(x,y)$ in iterative methods for finding local maximums of functions of several variables:

(i) At any point (x,y) we take a short "step" in the direction of $\nabla f(x,y)$ — this takes us "uphill" along the steepest route at that point.

(ii) Repeat step (i) until a (local) maximum is reached.

Property (3) provides an easy way to geometrically determine the direction of the gradient from the level curves of a surface.

Fig. 3 shows level curves for a function z = f(x,y) and the "uphill gradient" paths for several starting points.

Practice 6: Sketch the "uphill gradient" path for the function f in Fig. 3 at starting points A, B and C.

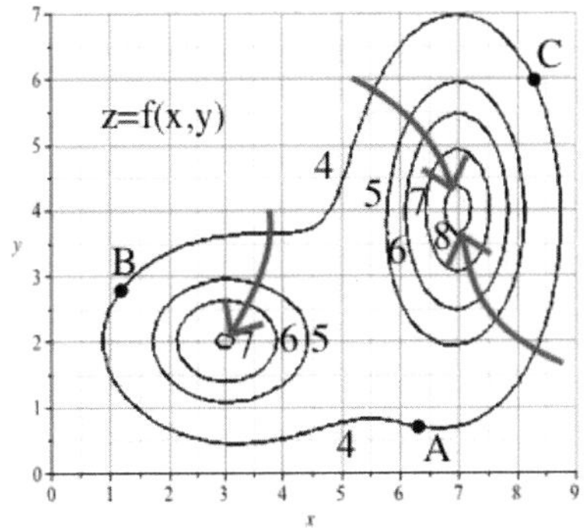

Fig. 3

Beyond z = f(x,y)

So far all of the examples have dealt with functions of two variables, z = f(x,y), but that was just for convenience. The definitions and ideas of gradient vectors and directional derivatives and their properties extend in a natural way to functions of three (or more) variables.

Extensions to w = f(x,y,z)

Definition: $\nabla f(x,y,z) = \langle f_x(x,y,z), f_y(x,y,z), f_z(x,y,z) \rangle = \frac{\partial f}{\partial x}\,\mathbf{i} + \frac{\partial f}{\partial y}\,\mathbf{j} + \frac{\partial f}{\partial z}\,\mathbf{k}$.

Theorem: For a differentiable function f(x,y,z) and a unit direction vector $\mathbf{u} = \langle a, b, c \rangle$,

$$D_{\mathbf{u}}f(x,y,z) = \nabla f(x,y,z) \cdot \mathbf{u}$$

Features: (1) The maximum value of the directional derivative $D_{\mathbf{u}}f(x,y,z)$ is $|\nabla f(x,y,z)|$.

(2) The maximum value of $D_{\mathbf{u}}f(x,y,z)$ occurs when **u** has the same direction as $\nabla f(x,y,z)$.

(2) The gradient vector $\nabla f(x,y,z)$ is normal (perpendicular) to the level surface through the point (x,y,z).

These same ideas also extend very naturally to functions of more than three variables.

For example, if x, y and z (all in meters) give the location in a room then w=f(x,y,z) could be the temperature (°C) at that location. Then instead of a level curve (in 2D), the points (x,y,z) where w=70 would be a level **surface** in 3D. $D_{\mathbf{u}}f(x,y,z)$ would be the instantaneous rate of change of temperature at location (x,y,z) in the direction **u** , and the units of $D_{\mathbf{u}}f(x,y,z)$ would be °C/m . The gradient vector $\nabla f(x,y,z)$ would still give the maximum value of the directional derivative and would point in the direction of maximum rate of temperature increase. A heat-seeking flying bug in the room would follow a path in the direction of the gradient at each point.

PROBLEMS

In problems 1 – 4, find the directional derivative of f at the given point in the direction indicated by the given angle θ. (Note: θ is the angle the direction vector **u** makes with the positive x–axis, so the components of **u** are $\langle \cos(\theta), \sin(\theta) \rangle$.)

1. $f(x,y) = x^2y^2 + 2x^4y$ at $(1,-2)$ with $\theta = \pi/3$
2. $f(x,y) = (x^2 - y)^3$ at $(3,1)$ with $\theta = 3\pi/4$
3. $f(x,y) = y^x$ at $(1,2)$ with $\theta = \pi/2$
4. $f(x,y) = \sin(x + 2y)$ at $(4,-2)$ with $\theta = -2\pi/3$

In problems 5 – 8, (a) find the gradient of f, (b) evaluate the gradient at the given point P, and (c) find the rate of change of f at P in the direction of the given vector **u** .

5. $f(x,y) = x^3 - 4x^2y + y^2$ at $P = (0,-1)$ with $\mathbf{u} = \langle 3/5, 4/5 \rangle$.
6. $f(x,y) = e^x \cdot \sin(y)$ at $P = (1,\pi/4)$ with $\mathbf{u} = \langle -1/\sqrt{5}, 2/\sqrt{5} \rangle$.
7. $f(x,y,z) = xy^2z^3$ at $P = (1,-2,1)$ with $\mathbf{u} = \langle 1/\sqrt{3}, -1/\sqrt{3}, 1/\sqrt{3} \rangle$.
8. $f(x,y,z) = xy + yz^2 + xz^3$ at $P = (2,0,3)$ with $\mathbf{u} = \langle -2/3, -1/3, 2/3 \rangle$.

In problems 9 – 14, find the directional derivative of the given function at the given point in the direction of the given vector **v** .

9. $f(x,y) = \sqrt{x-y}$ at $(5,1)$ with $\mathbf{v} = \langle 12, 5 \rangle$.
10. $f(x,y) = x/y$ at $(6,-2)$ with $\mathbf{v} = \langle -1, 3 \rangle$.
11. $g(x,y) = x \cdot e^{xy}$ at $(-3,0)$ with $\mathbf{v} = 2\mathbf{i} + 3\mathbf{j}$.
12. $g(x,y) = e^x \cos(y)$ at $(1,\pi/6)$ with $\mathbf{v} = \mathbf{i} - \mathbf{j}$.
13. $f(x,y,z) = \sqrt{xyz}$ at $(2,4,2)$ with $\mathbf{v} = \langle 4, 2, -4 \rangle$.
14. $f(x,y,z) = z^3 - x^2y$ at $(1,6,2)$ with $\mathbf{v} = \langle 3, 4, 12 \rangle$.

In problems 15 – 20, find the maximum rate of change of f at the given point and the direction in which it occurs.

15. $f(x, y) = x \cdot e^{-y} + 3y$ at the point $(1, 0)$

16. $f(x, y) = \ln(x^2 + y^2)$ at the point $(1, 2)$

17. $f(x, y) = \sqrt{x^2 + 2y}$ at the point $(4, 10)$

18. $f(x, y, z) = x + y/z$ at the point $(4, 3, -1)$

19. $f(x, y) = \cos(3x + 2y)$ at the point $(\pi/6, -\pi/8)$

20. $f(x, y, z) = \frac{x}{y} + \frac{y}{z}$ at the point $(4, 2, 1)$

21. At each dot in Fig. 4 sketch the gradient vector.

22. At each dot in Fig. 5 sketch the gradient vector.

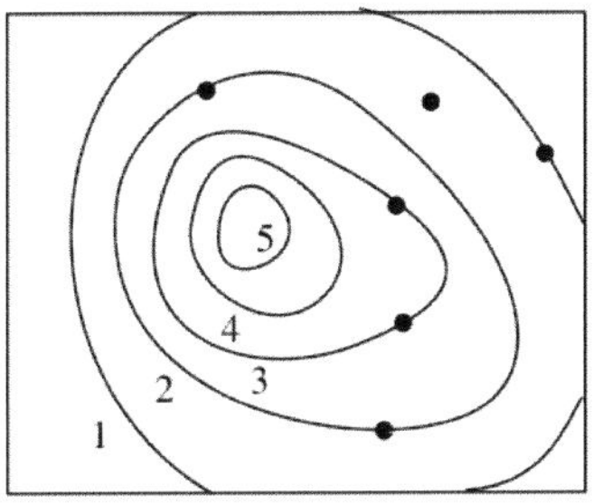

Fig. 4

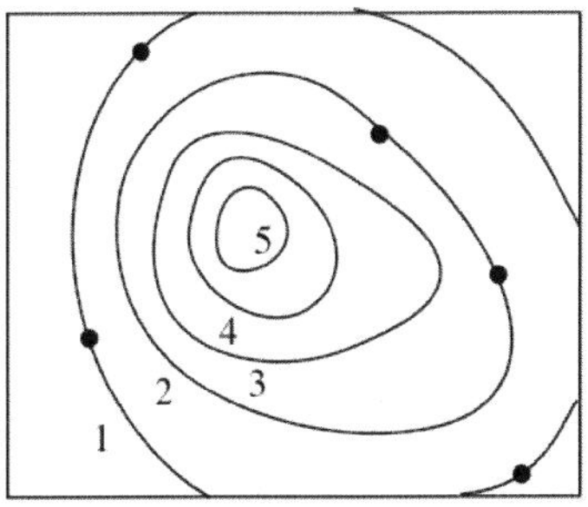

Fig. 5

23. At each dot in Fig. 6 sketch the "uphill gradient" path.

24. At each dot in Fig. 7 sketch the "uphill gradient" path.

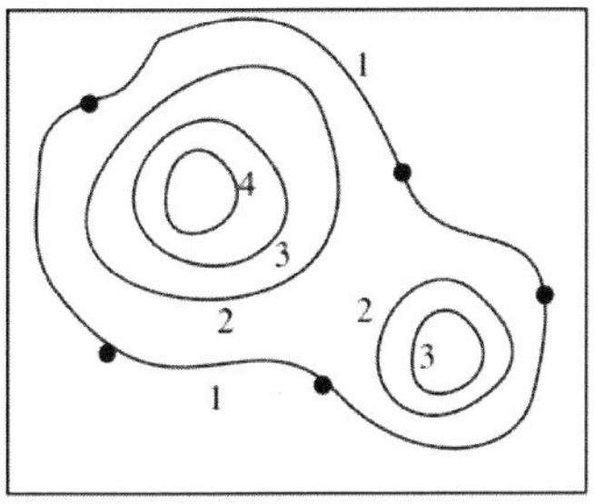

Fig. 6

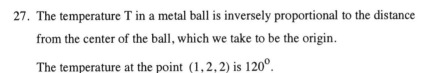

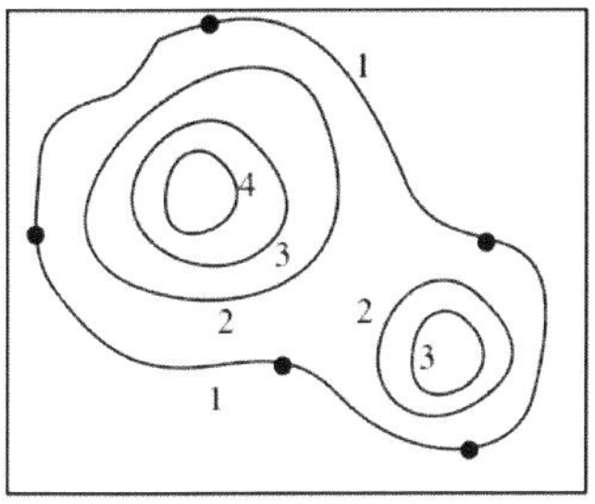

Fig. 7

25. Show that a differentiable function f decreases most rapidly at (x,y) in the direction opposite to the gradient vector, that is, in the direction $-\nabla f(x,y)$.

26. Use the result of Problem 23 to find the direction in which the function $f(x,y) = x^4y - x^2y^3$ decreases fastest at the point $(2, -3)$.

27. The temperature T in a metal ball is inversely proportional to the distance from the center of the ball, which we take to be the origin. The temperature at the point $(1, 2, 2)$ is 120°.
   (a) Find the rate of change of T at $(1, 2, 2)$ in the direction toward the point $(2, 1, 3)$.
   (b) Show that at any given point in the ball the direction of greatest increase in temperature is given by a vector that points toward the origin.

28. The temperature at a point (x, y, z) is given by $T(x,y,z) = 200 \cdot e^{(-x^2 - 3y^2 - 9z^2)}$ where T is measured in ^oC and x, y, and z in meters.
   (a) Find the rate of change of temperature at the point $P(2, -1, 2)$ in the direction toward the point $(3, -3, 3)$.
   (b) In which direction does the temperature increase fastest at P?
   (c) Find the maximum rate of increase at P.

29. Suppose that over a certain region of space the electrical potential V is given by $V(x,y,z) = 5x^2 - 3xy + xyz$.

(a) Find the rate of change of the potential at P(3, 4, 5) in the direction of the vector $\mathbf{v} = \mathbf{i} + \mathbf{j} - \mathbf{k}$.

(b) In which direction does V change most rapidly at P?

(c) What is the maximum rate of change at P?

30. Suppose that you are climbing a hill whose shape is given by the equation $z = 1000 - 0.01x^2 - 0.02y^2$ and you are standing at the point with coordinates (60, 100, 764).

(a) In which direction should you proceed initially in order to reach the top of the hill fastest?

(b) If you climb in that direction, at what angle above the horizontal will you be climbing initially?

31. Let F be a function of two variables that has continuous partial derivatives and consider the points A(1,3), B(3,3), C(1,7), and D(6,15). The directional derivative of A in the direction of the vector AB is 3, and the direction derivative at A in the direction of AC is 26. Find the direction derivative of f at A in the direction of the vector AD.

Practice Answers

Practice 1: f(x,y) = 1 + xy and $\mathbf{u} = \langle 0.6, 0.8 \rangle$

$$D_{\mathbf{u}}f(2,1) = \lim_{h\to 0} \frac{f\left(1+\frac{5}{13}h, 2+\frac{12}{13}h\right) - f(2,1)}{h} = \lim_{h\to 0} \frac{\left[\left(1+\frac{5}{13}h\right)^2 + 2\left(1+\frac{5}{13}h\right) + 3\left(2+\frac{12}{13}h\right) + 1\right] - 12}{h}$$

$$= \lim_{h\to 0} \frac{\left[\left(4+\frac{20}{13}h+\frac{25}{13}h^2\right) + \left(4+\frac{10}{13}h\right) + \left(2+\frac{36}{13}h\right) + 1\right] - 12}{h} = \lim_{h\to 0} \frac{\frac{66}{13}h + \frac{25}{13}h^2}{h} = \frac{66}{13}$$

Practice 2: $f_x(x,y) = 2x+2$ and $f_y(x,y) = 3$ so $f_x(2, 1) = 6$ and $f_y(2, 1) = 3$. $\mathbf{u} = \left\langle \frac{5}{13}, \frac{12}{13} \right\rangle$ so

$$D_{\mathbf{u}}f(2, 1) = (6)\left(\frac{5}{13}\right) + (3)\left(\frac{12}{13}\right) = \frac{66}{13}$$, the same result as in Example 2 but much easier.

Practice 3: (a) For $f(x,y) = e^{xy} + 2x^3y + y^2$, $f_x(x,y) = y\cdot e^{xy} + 6x^2y$ and $f_y(x,y) = x\cdot e^{xy} + 2x^3$ so $\nabla f(x,y) = \langle y\cdot e^{xy} + 6x^2y,\ x\cdot e^{xy} + 2x^3 \rangle$ and $\nabla f(1,2) = \langle 2e^2 + 12,\ e^2 + 2 \rangle$.

(b) For $f(x,y) = \cos(2x+3y)$, $f_x(x,y) = -2\sin(2x+3y)$ and $f_y(x,y) = -3\sin(2x+3y)$ so $\nabla f(1, 2) = \langle -2\sin(2x+3y), -3\sin(2x+3y) \rangle$ and $\nabla f(1, 2) = \langle -2\sin(8), -3\sin(8) \rangle$.

Practice 4: $f_x(x,y) = \dfrac{2}{\sqrt{2x+3y}}$ and $f_y(x,y) = \dfrac{3}{\sqrt{2x+3y}}$ so $f_x(5,2) = \dfrac{2}{16}$ and $f_y(5,2) = \dfrac{3}{16}$.

(a) The maximum value of the rate of change of f is $D_{\mathbf{u}}f(5,2) = |\nabla f(1,2)| = \left|\left\langle \frac{2}{16}, \frac{3}{16} \right\rangle\right| = \frac{5}{16}$.

(b) This maximum value occurs when **u** is in the direction of $\nabla f(5,2)$: $\mathbf{u} = \left\langle \frac{2}{5}, \frac{3}{5} \right\rangle$.

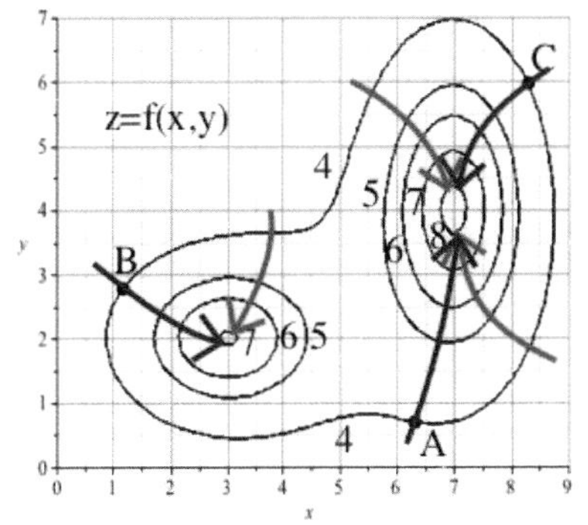

Fig. 8

Practice 5: See Fig. 8. Note that each gradient vector is perpendicular to the level curve and points uphill.

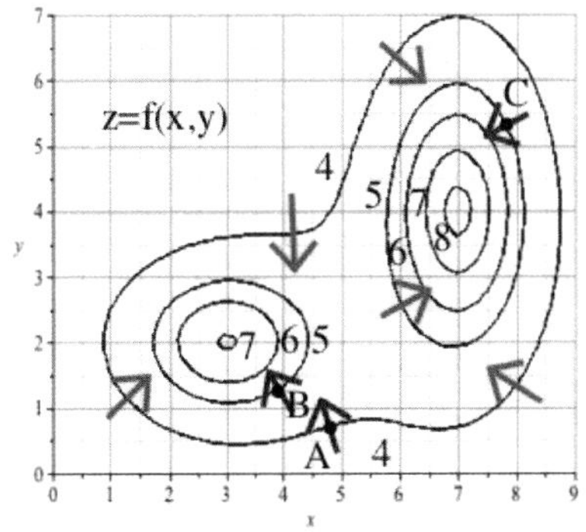

Fig. 9

Practice 6: See Fig. 9. Note that "uphill gradient" path is always perpendicular to the level curves.

13.6 MAXIMUMS AND MINIMUMS

One important use of derivatives in beginning calculus was to find maximums and minimums of functions of a single variable. Similarly, an important use of partial derivatives is to find maximums and minimums of functions of two (or more) variables.

We are going to consider three situations, and each situation will require a different method.

Three max/min situations: (a) domain = entire xy-plane,

(b) domain = bounded region of the xy-plane, and

(c) domain = a path in the xy-plane.

Situation (a) might ask for the highest elevation anywhere on earth (= at the summit of Mt. Everest = 29,029 ft), (b) might ask for the highest elevation in the state of Washington (= at the summit of Mt. Rainier = 14,4011 ft), and (c) might ask for the highest elevation we achieved during a hike on Mt. Rainier even if we did not reach the summit.

We consider situations (a) and (b) in this section and situation (c) in the next section.

(a) Domain of our max/min search of f is the ENTIRE xy-plane

Definition: A function of two variables f(x,y) has a **local maximum** at (a,b) if $f(x,y) \le f(a,b)$ for all points (x,y) in some disk with center (a,b). The value f(a,b) is called a **local maximum value** of f.

The next theorem tells us where we should look for maximums and minimums.

Theorem: If f is differentiable and has a local maximum or minimum at (a,b), then $f_x(a,b) = 0$ and $f_y(a,b) = 0$.

Note: It is **possible** that $f_x(a,b) = 0$ and $f_y(a,b) = 0$ and that f(a,b) is **not** a local maximum or minimum. (See Example 2 below.)

Proof: The proof is simply a process of elimination. If one (or both) of $f_x(a,b)$ or $f_y(a,b)$ is positive, then moving a small distance Δ in the direction of that variable will increase the value of f so either $f(a+\Delta,b) > f(a,b)$ or $f(a,b+\Delta) > f(a,b)$ and f(a,b) is not a local maximum. A similar argument also shows that f(a,b) can not be a local minimum.

Example 1: Find all local maximums and minimums of $f(x,y) = x^2 + y^2 - 2x - 6y + 7$.

Solution: $f_x(x,y) = 2x - 2$ so $f_x(x,y) = 0$ when $x = 1$. $f_y(x,y) = 2y - 6$ so $f_y(x,y) = 0$ when $y = 3$.

The only possible location of a maximum or minimum of $f(x,y)$ is at the point $(1,3)$, but we do not know if we have a maximum or minimum or neither at that point. (The graph of $z = f(x,y)$ in Fig. 1 indicates that $f(1,3)$ is a maximum.)

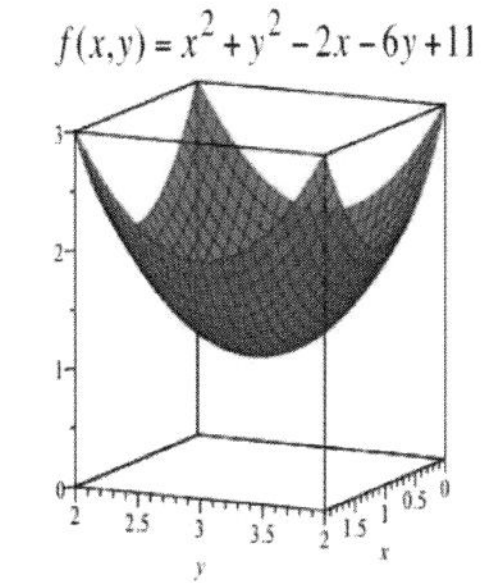

Fig. 1

Example 2: Find all local maximums and minimums of $f(x,y) = x^2 - y^2$.

Solution: $f_x(x,y) = 2x$ so $f_x(x,y) = 0$ when $x = 0$. $f_y(x,y) = -2y$ so $f_y(x,y) = 0$ when $y = 0$.

The only possible location of a maximum or minimum of $f(x,y)$ is at the point $(0,0)$, but $f(0,0) = 0$ is neither a local maximum nor a minimum of f: for any $a \neq 0, f(a,0) = a^2 > f(0,0)$ so $f(0,0)$ is not a maximum; for any $b \neq 0, f(0,b) = -b^2 < f(0,0)$ so $f(0,0)$ is not a minimum. (The surface $z = x^2 - y^2$ in Fig. 2 is called a "saddle.")

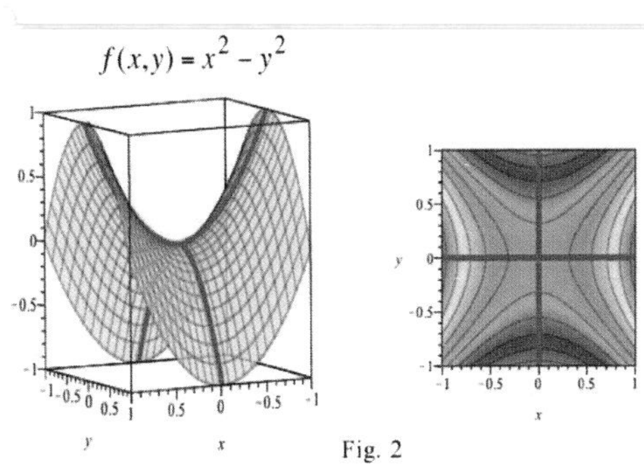

Fig. 2

Example 3: Find all critical points of $f(x,y) = x^2y - 2x^2 - y^3 + 3y + 7$.

Solution: $f_x(x,y) = 2xy - 4x$ and $f_y(x,y) = x^2 - 3y^2 + 3$ so we need to solve the system $\{2xy - 4x = 0$ and $x^2 - 3y^2 + 3 = 0\}$. In order for $0 = 2xy - 4x = 2x(y-2)$ we know that either x=0 or y=2.

x=0 case: Putting x=0 into f_y we have $f_y(0,y) = -3y^2 + 3 = 0$ so $y^2 = 1$ and $y = \pm 1$. This gives us two critical points: (0, 1) and (0, -1).

y=2 case: Putting y=02 into f_y we have $f_y(x,2) = x^2 - 12 + 3 = 0$ so $x^2 = 9$ and $x = \pm 3$. This gives us two new critical points: (3, 2) and (-3, 2).

This function has 4 critical points, and any local maximums or minimums can only occur at one of those 4 locations. Unfortunately it is not easy to decide whether each critical point gives us a local maximum, a local minimum or a saddle. For that we need a graph or the Second Derivative Test.

Practice 1: Find all critical points of $f(x,y) = 2x^3 + xy^2 + 5x^2 + y^2 + 3$.

In beginning calculus we had a Second Derivative Test to help determine whether a critical point was a local maximum or a local minimum. There is also a Second (Mixed Partial) Derivative Test to help us determine whether a critical point of a function of two variables is a local maximum or minimum or saddle.

Second Derivative Test for Maximums and Minimums

Suppose the second partial derivatives f_{xx}, f_{xy}, f_{yx}, and f_{yy} are continuous in a disk with center (a,b) and $f_x(a,b) = 0$ and $f_y(a,b) = 0$. Let $D = D(a,b) = f_{xx}(a,b)f_{yy}(a,b) - \{ f_{xy}(a,b) \}^2$.

(i) If $D > 0$ and $f_{xx}(a,b) > 0$, then f(a,b) is a local minimum.

(ii) If $D > 0$ and $f_{xx}(a,b) < 0$, then f(a,b) is a local maximum.

(iii) If $D < 0$, then f(a,b) is not a local minimum or a local maximum. (It is a saddle point.)

(iv) If $D = 0$, then the test is "indeterminate": f(a,b) could be a local maximum or a local minimum or neither.

The proof of this theorem is given in an appendix after the Practice solutions.

Example 4: In Example 3 we found 4 critical points of $f(x,y) = x^2y - 2x^2 - y^3 + 3y + 7$: (0, 1), (0, -1), (3, 2) and (-3, 2). Use the Second Derivative Test to determine whether each of these gives a local maximum of f, a local minimum of f, or is a saddle point.

Solution: $f_{xx}(x,y) = 2y - 4$, $f_{yy}(x,y) = -6y$, and $f_{xy}(x,y) = 2x$. Many people find it easiest (and safest) to arrange the numerical information in a table. Fig. 3 shows the surface and level curves for this $z = f(x,y)$.

point	(0, 1)	(0, -1)	(3, 2)	(3, -2)
f_{xx}	-2	-6	0	0
f_{yy}	-6	6	-12	-12
f_{xy}	0	0	6	-6
D	12	-36	-36	-36
result	max	saddle	saddle	saddle

Practice 2: Use the Second Derivative Test to determine whether each critical point of $f(x,y) = 2x^3 + xy^2 + 5x^2 + y^2 + 3$ (from Practice 1) gives a local maximum of f, a local minimum of f, or is a saddle point.

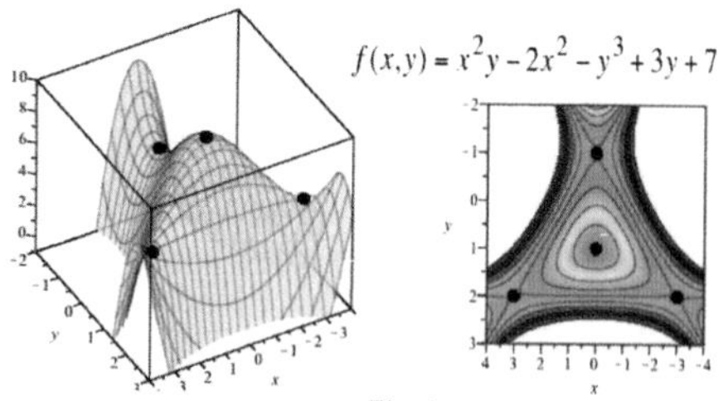

Fig. 3

Example 5: Use the ideas of this section to find the shortest distance from the point (1, 0 –2) to the plane x + 2y + z = 4.
(Suggestion: Minimize f(x,y) = the square of the distance)

Solution: The square of the distance of (x,y,z) to the point (1,0,–2) is $(x-1)^2 + (y-0)^2 + (z+2)^2$, is a function of three variables. However, we know that (x,y,z) is on the plane $x + 2y + z = 4$ so $z = 4 - x - 2y$. Replacing z with $4 - x - 2y$ in the distance (squared) formula, we want to minimize $f(x,y) = (x-1)^2 + (y-0)^2 + (6 - x - 2y)^2$.

$f_x = 2(x - 1) + 2(6 - x - 2y)(-1) = 4x + 4y - 14$ and $f_y = 2y + 2(6 - x - 2y)(-2) = 4x + 10y - 24$. We need to find the values of x and y that make f_x and f_y both equal to zero, and the only place that occurs is at the point (a,b) = (11/6, 5/3).

At the point (a,b) = (11/6, 5/3) we have $f_{xx}(a,b) = 4$, $f_{xy} = 4$, and $f_{yy} = 10$ so $D(a,b) = f_{xx}f_{yy} - (f_{xy})^2 = 24 > 0$ and $f_{xx} > 0$. Then by part (i) of the Second Derivative Test we can conclude that f(x,y) has a local minimum at (11/6, 5/3): the shortest distance is $\sqrt{f(11/6, 5/3)} = \sqrt{(5/6)^2 + (5/3)^2 + (5/6)^2} = \sqrt{\frac{5}{6}} = \frac{5\sqrt{6}}{6}$.

(Note: This problem probably would be easier using the ideas from section 11.6.)

(b) Domain of our max/min of f search is a BOUNDED REGION of the xy-plane

In beginning calculus we sometimes needed to find the maximum or minimum value of a function f(x) for x in an interval [a, b]. In that situation we found critical points in [a, b] where f '(x)= 0 or was undefined and then we also needed to check the values of f at the ENDPOINTS when x=a and x=b. When looking for max/mins on a bounded region R we still need to find critical points (x,y) in R, but we also need to consider values f on the BOUNDARY of R.

> Method for finding Maximums and Minimums on Bounded Domains:
>
> To find the maximum and minimum values of a differential function on a closed bounded region R:
>
> (1) Find the values of f at the critical points of f in R.
>
> (2) Find the extreme values of f on the boundary of R.
>
> (3) The largest value of f from steps (1) and (2) is the absolute maximum value of f on R; the smallest value of f is the absolute minimum value of f on R.
>
> Note: The Second Derivative Test is not used in this situation.

Example 6: Find the absolute maximum and minimum values of $f(x,y) = x^2 - 2xy + 2y + 3$ on the rectangle $R = \{ (x,y) : 0 \le x \le 3 \text{ and } 0 \le y \le 2 \}$.

Solution: Step (1): Find the critical points of f in R. These occur when $f_x(x,y) = 2x - 2y = 0$ and $f_y(x,y) = -2x + 2 = 0$, so (solving algebraically) we have $x = 1$ and $y = 1$. The only critical point from step (1) is (x,y) = (1,1) and f(1,1) = 4.

Step (2): Find the critical points and extreme values of f on the boundary of R.

The boundary of the rectangle R consists of four line segments L1, L2, L3, and L4 where L1 = segment from (0,0) to (3,0), L2 = segment from (3,0) to (3,2), L3 = segment from (3,2) to (0,2), and L4 = segment from (0,2) to (0,0).

On L1, $0 \le x \le 3$ and $y = 0$ so $f(x,y) = f(x,0) = x^2 + 3$ which has minimum value $f(0,0) = 3$ and maximum value $f(3,0) = 12$.

On L2, $x = 3$ and $0 \le y \le 2$ so $f(x,y) = f(3,y) = 9 - 6y + 2y + 3 = 12 - 4y$ which has minimum value $f(3,2) = 4$ and maximum value $f(3,0) = 12$.

On L3, $0 \le x \le 3$ and $y = 2$ so $f(x,y) = f(x,2) = x^2 - 4x + 4 + 3 = x^2 - 4x + 7$. Using the methods of Chapter 3 ($f' = 2x - 4 = 0$ when $x = 2$, $f'' = 2 > 0$) we know $f(2,2) = 3$ is a minimum and (checking endpoints $x = 0$ and $x = 3$) that $f(0,2) = 7$ is a maximum.

On L4, $x = 0$ and $0 \le y \le 2$ so $f(0,y) = 2y + 3$ which is a linear function with minimum $f(0,0) = 3$ and maximum $f(0,2) = 7$.

On the boundary (L1, L2, L3, and L4) the minimum value is $3 = f(0,0) = f(2,2)$ and the maximum value is $12 = f(3,0)$.

Comparing the minimum and maximum values from step (2) with $f(1,1) = 4$ from step (1) we have that the absolute minimum is $3 = f(0,0) = f(2,2)$ and the absolute maximum value is $12 = f(3,0)$. See Fig. 4.

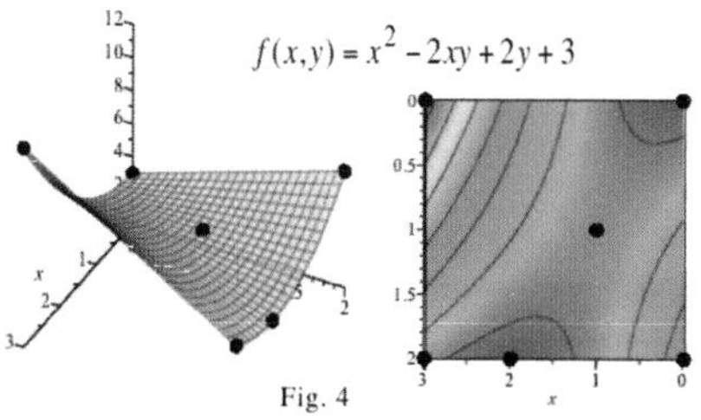

Fig. 4

To fond maximums and minimums on a bounded region, we do not use the Second Derivative test, we simply evaluate the function at each critical point and select the largest and smallest values of the function

Practice 3: Find the locations and maximum and minimum values of $f(x,y) = 16 - xy$ on the elliptical domain $2x^2 + y^2 \le 36$.

Example 7: Rewrite the function $f(x,y) = x^2 - 2xy + 2y + 3$ on the boundary lines of the set $D = \{ (x,y) : 0 \le x \le 3 \text{ and } 0 \le y \le 2x \}$

Solution: The boundary of D consists of the three line segments L1 = segment from (0,0) to (3,0), L2 = segment from (3,0) to (3,6), and L3 = the segment from (0,0) to (3,6) along the line y = 2x.

On L1, $0 \le x \le 3$ and $y = 0$ so $f(x,y) = f(x,0) = x^2 + 3$.

On L2, $x = 3$ and $0 \le y \le 6$ so $f(x,y) = f(3,y) = 9 - 6y + 2y + 3 = 12 - 4y$.

On L3, $0 \le x \le 3$ and $y = 2x$ so $f(x,y) = f(x,2x) = x^2 - 4x^2 + 4x + 3 = -3x^2 + 4x + 3$.

To actually maximize or minimize f on D, we would now need to apply steps (1) and (2).

PROBLEMS

In problems 1 – 20, find the local maximums, minimums, and saddle points of the given function.

1. $f(x, y) = x^2 + y^2 + 4x - 6y$
2. $f(x, y) = 4x^2 + y^2 - 4x + 2y$
3. $f(x, y) = 2x^2 + y^2 + 2xy + 2x + 2y$
4. $f(x, y) = 1 + 2xy - x^2 - y^2$
5. $g(x,y) = xy^2 - x^3 + 3x - 2y^2 + 5$
6. $g(x,y) = x^3 + 2y^2 - xy^2 - 3x + 7$
7. $f(x, y) = x^2 + y^2 + x^2y + 4$
8. $f(x, y) = 2x^3 + xy^2 + 5x^2 + y^2$
9. $f(x, y) = x^3 - 3xy + y^3$
10. $f(x, y) = y\sqrt{x} - y^2 - x + 6y$
11. $f(x, y) = xy - 2x - y$
12. $f(x, y) = xy(1 - x - y)$
13. $f(x, y) = \dfrac{x^2y^2 - 8x + y}{xy}$
14. $f(x, y) = x^2 + y^2 + \dfrac{1}{x^2y^2}$
15. $f(x, y) = e^x \cdot \cos(y)$
16. $f(x, y) = (2x - x^2)(2y - y^2)$
17. $f(x, y) = 3x^2y + y^3 - 3x^2 - 3y^2 + 2$
18. $f(x, y) = xy \cdot e^{(-x^2 - y^2)}$
19. $f(x, y) = x \cdot \sin(y)$
20. $f(x,y) = 2x^3 + y^2 - 6xy + 10$

In problems 21 – 32 find the maximum and minimum values of f on the set R.

21. $f(x,y) = 5 - 3x + 4y$, R is the closed triangular region with vertices (0,0), (4,0), and (4,5).

22. $f(x,y) = x^2 + 2xy + 3y^2$, R is the closed triangular region with vertices (–1,1), (2,1), and (–1,–2).

23. $f(x,y) = x^2 + y^2 + x^2y + 4$, $R = \{ (x,y) \mid |x| \le 1, |y| \le 1 \}$.

24. $f(x,y) = y\sqrt{x} - y^2 - x + 6y$, $R = (x,y) \mid 0 \le x \le 9, 0 \le y \le 5 \}$.

25. $f(x,y) = 1 + xy - x - y$, R is the region bounded by the parabola $y = x^2$ and the line $y = 4$.

26. $f(x,y) = y^2 - 2x^2 + 10$. R is the region bounded by the parabola $y = x^2$ and the line $y = 4$.

27. $f(x,y) = 2x^2 - y^2 + 30$. R is the region bounded by the parabola $y = x^2$ and the line $y = 4$.

28. $f(x,y) = x^2 + 3y + 7$. R is he region shown in Fig. 5 (include the boundary).

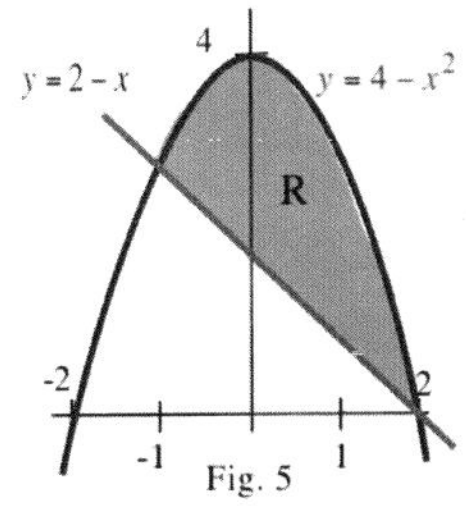

Fig. 5

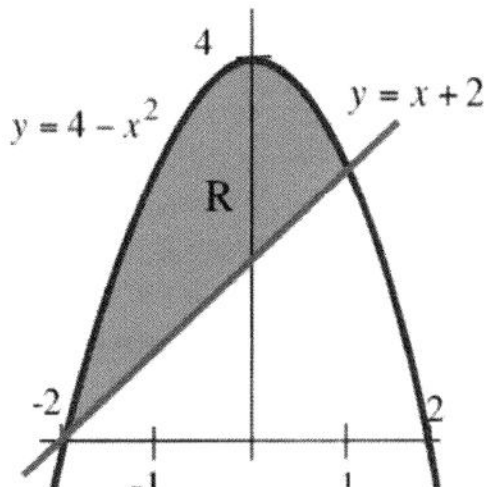

Fig. 6

29. $f(x,y) = x^2 + 3y + 7$. R is he region shown in Fig. 6 (include the boundary).

30. $f(x,y) = 2x^2 + x + y^2 - 2$, $R = \{ (x,y) \mid x^2 + y^2 \le 4 \}$.

31. $f(x,y) = 2x^3 + y^4$, $R = \{ (x,y) \mid x^2 + y^2 \le 1 \}$.

32. $f(x,y) = x^3 - 3x - y^3 + 12y$, R is the quadrilateral whose vertices are $(-2,3)$, $(2,3)$, $(2,2)$, and $(-2,2)$.

33. Find the point on the plane $x + 2y + 3z = 4$ that is closest to the origin.

34. Find the point on the plane $2x - y + z = 1$ that is closest to the point $(-4, 1, 3)$.

35. Find three consecutive numbers whose sum is 100 and whose product is a maximum.

36. Find three positive numbers x, y, and z whose sum is 100 such that $x^a y^b z^c$ is a maximum.

37. Find the volume of the largest rectangular box with edges parallel to the axes that can be inscribed in the ellipsoid $9x^2 + 36y^2 + 4z^2 = 36$.

38. Solve the problem in problem 31 for a general ellipsoid $x^2/a^2 + y^2/b^2 + z^2/c^2 = 1$.

39. Find the volume of the largest rectangular box in the first octant with three faces in the coordinate planes and one vertex in the plane $x + 2y + 3z = 6$.

40. Solve the problem in problem 33 for a general plane $x/a + y/b + z/c = 1$ where a, b, and c are positive numbers.

41. Find the dimensions of a rectangular box of maximum volume such that the sum of the lengths of its 12 edges is a constant c.

Practice Solutions

Practice 1: $f(x,y) = 2x^3 + xy^2 + 5x^2 + y^2 + 3$. $f_x(x,y) = 6x^2 + y^2 + 10x$ and $f_y(x,y) = 2xy + 2y$ so we need to solve the system $\{ 6x^2 + y^2 + 10x = 0$ and $2xy + 2y = 0 \}$. The second equation is easier: in order for $0 = 2xy+2y = 2y(x+1)$ we know that either y=0 or x=-1.

y=0 case: Putting y=0 into f_x we have $f_x(x, 0) = 6x^2 + 10x = 0$ so $2x(3x+5) = 0$ and x=0 or x= -5/3. This gives us two critical points: (0, 0) and (-5/3, 0).

x= -1 case: Putting x=-1 into f_x we have $f_x(-1, y) = 6 + y^2 - 10 = 0$ so $y^2 = 4$ and $y = \pm 2$. This gives us two new critical points: (-1, 2) and (-1, -2).

This function has 4 critical points: (0, 0), (-5/3, 0), (-1, 2) and (-1, -2).

Practice 2: $f(x,y) = 2x^3 + xy^2 + 5x^2 + y^2 + 3$, $f_{xx}(x,y) = 12x + 10$, $f_{yy}(x,y) = 2x + 2$, $f_{xy}(x,y) = 2y$. The information for the Second Derivative Test is organized in the table. f has a Local minimum at (0,0), saddle points at (-1,2) and (-1,-2), And a local maximum at (-5/3, 0).

point	(0, 0)	(-1, 2)	(-1,-2)	(-5/3,0)
f_{xx}	10	-2	-2	-20
f_{yy}	2	0	0	-4/3
f_{xy}	0	4	-4	0
D	20	-16	-16	80/3
result	min	saddle	saddle	max

Practice 3: $f(x,y) = 16 - xy$, $f_x = -y$, so the only interior point with $f_x = f_y = 0$ is (0,0).

Boundary: $2x^2 + y^2 \le 36$ so $y = \pm\sqrt{36 - 2x^2}$ with $-\sqrt{18} \le x \le \sqrt{18}$. Substituting this y into f we have $f(x,y) = 16 - x \cdot \sqrt{36 - 2x^2}$ which is a function of the single variable x. Then

$f\,'(x) = \dfrac{2x^2}{\sqrt{36-2x^2}} - \sqrt{36-2x^2}$. Setting f '(x)=0 and solving for x, we get $x = \pm 3$

so our critical points on the boundary are (3, $\sqrt{18}$), (3, $-\sqrt{18}$), (-3, $\sqrt{18}$), (-3, $-\sqrt{18}$), and the endpoints ($\sqrt{18}$, 0) and ($-\sqrt{18}$, 0) , Evaluating f at each of these critical points we see that the maximum value of f is $16 + 9\sqrt{2} \approx 28.73$ at (3, $\sqrt{18}$) and (-3, $-\sqrt{18}$). The minimum value of f is $16 - 9\sqrt{2} \approx 3.27$ at (3, $-\sqrt{18}$) and (-3, $\sqrt{18}$)

$f(x,y) = 16 - xy$ on $2x^2 + y^2 \le 36$

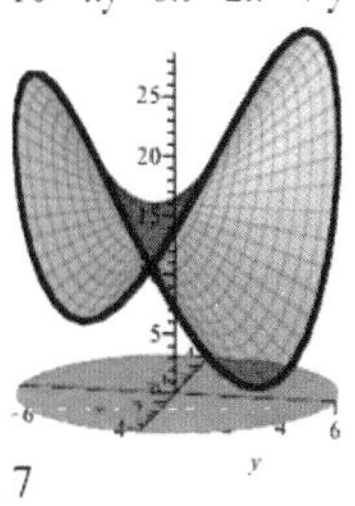

Fig. 7

Appendix: Proof of parts (i) and (ii) of the Second Derivative Test (and part (iv) too)

Suppose we have a critical point (a,b) so $\{f_x(a,b) = 0 \ \textit{and} \ f_y(a,b) = 0\ \}$. The proof involves calculating the second partial derivative in the direction $u = \langle h,k \rangle$ and then determining when that second partial derivative is positive and negative. The directional derivative in the direction $u = \langle h,k \rangle$ is $D_u f = \nabla f \bullet u = f_x h + f_y k$. Then the second derivative in the direction $u = \langle h,k \rangle$ is

$$D_u^2 f = D_u(\ D_u f\ ) = \frac{\partial}{\partial x}(D_u f)h + \frac{\partial}{\partial y}(D_u f)k$$

$$= \frac{\partial}{\partial x}(f_x h + f_y k)h + \frac{\partial}{\partial y}(f_x h + f_y k)k$$

$$= (f_{xx}h + f_{xy}k)h + (f_{xy}h + f_{yy}k)k$$

$$= f_{xx}h^2 + f_{xy}kh + f_{xy}hk + f_{yy}k^2$$

$$= f_{xx}h^2 + 2f_{xy}hk + f_{yy}k^2$$

$$= f_{xx}h^2 + 2f_{xy}hk + \frac{f_{xy}^2}{f_{xx}}k^2 + f_{yy}k^2 - \frac{f_{xy}^2}{f_{xx}}k^2$$

$$= f_{xx}\left(h^2 + 2\frac{f_{xy}}{f_{xx}}hk + \frac{f_{xy}^2}{f_{xx}^2}k^2\right) + k^2\left(f_{yy} - \frac{f_{xy}^2}{f_{xx}}\right)$$

$$= f_{xx}\left(h + \frac{f_{xy}}{f_{xx}}k\right)^2 + \frac{k^2}{f_{xx}}\left(f_{xx}f_{yy} - f_{xy}^2\right)$$

$$= f_{xx}\left\{\left(h + \frac{f_{xy}}{f_{xx}}k\right)^2 + \frac{k^2}{f_{xx}^2}\left(f_{xx}f_{yy} - f_{xy}^2\right)\right\}$$

If $D = f_{xx}f_{yy} - f_{xy}^2 > 0$ then the part in the curly brackets is positive, so the sign of $D_u^2 f$ is the same as the sign of f_{xx}:

(i) if $f_{xx} > 0$ then $D_u^2 f > 0$ and f is concave up in every direction u so our critical point gives a local minimum

(ii) if $f_{xx} < 0$ then $D_u^2 f < 0$ and f is concave down in every direction u so our critical point gives a local maximum.

Part (iv): $f(x,y) = x^2y^2$, $g(x,y) = -x^2y^2$, $h(x,y) = x^3y^3$ all have the critical point (0,0) and D = 0 at that critical point. But f has a local minimum at (0,0), g has a local maximum at (0,0) and h has neither a local min or max at (0,0). So D = 0 at a critical point does not tell us whether we have a local max or a local min or neither.

13.7 LAGRANGE MULTIPLIER METHOD

Suppose we go on a walk on a hillside, but we have to stay on a path. Where along this path are we at the highest elevation? That is the basic problem we consider in this section: how to find a maximum or minimum subject to a constraint (staying on a path). Our method, Lagrange Multipliers, is very algebraic, but it also has a geometric interpretation.

Example 1: Fig. 1 shows the level curves for a hill and several paths. The dots on path A and B are at the highest elevations along those two paths.

Practice 1: Mark the locations of maximum elevation along paths C and D.

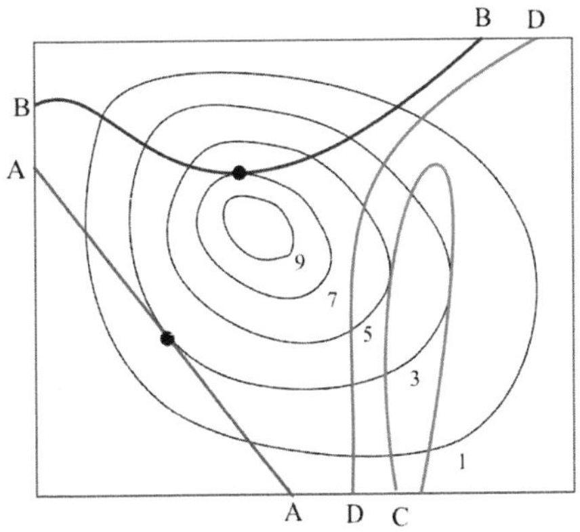

Fig. 1: A hillside and several paths

You night have noticed that at each maximum along a path the path was tangent to a level curve. This is the basis of the Lagrange Multiplier method – find the points along a path (constraint) where the path is tangent to the level curve of the function. But finding the tangent to level curves can be difficult so instead we use the fact that if two curves have parallel tangent vectors then they have parallel gradient vectors. And it is easy to calculate the gradient vector for a function.

Lagrange Multiplier Method to Maximize/Minimize f(x,y) along a path:

To find the maximum and minimum values of $f(x,y)$ subject to the constraint (condition) that $g(x,y) = k$

(a) Find all values of x, y, and λ such that $\nabla f(x,y) = \lambda \nabla g(x,y)$ and $g(x,y) = k$.

(b) Evaluate f at the points (x,y) found in step (a). The largest value of $f(x,y)$ at these points is the maximum value of f. The smallest value of $f(x,y)$ is the minimum.

Example 2: Find the maximum and minimum values of $f(x,y) = y^2 - x^2$ on the ellipse $x^2 + 4y^2 = 4$.

Solution: $f(x,y) = y^2 - x^2$ and $g(x,y) = x^2 + 4y^2 = 4$.

Then $\nabla f(x,y) = -2x\mathbf{i} + 2y\mathbf{j}$ and $\nabla g(x,y) = 2x\mathbf{i} + 8y\mathbf{j}$. The Lagrange condition that $\nabla f(x,y) = \lambda \nabla g(x,y)$ means $-2x\mathbf{i} + 2y\mathbf{j} = \lambda(2x\mathbf{i} + 8y\mathbf{j}) = 2\lambda x\mathbf{i} + 8\lambda y\mathbf{j}$ so

i: $-2x = 2\lambda x$

j: $2y = 8\lambda y$ (3 equations in 3 unknowns x, y, and λ)

constraint: $x^2 + 4y^2 = 4$

If $x = 0$, then $y^2 = 1$ so $y = +1$ or $y = -1$ (from constraint) and $\lambda = 1/4$ (from $\mathbf{j}$ condition).

Then we have the points $(x,y) = (0, 1)$ and $(0, -1)$

If $x \neq 0$, then $-2 = 2\lambda$ (from $\mathbf{i}$ condition) so $\lambda = -1$. Then $2y = -8y$ (from $\mathbf{j}$ condition) so $y = 0$ and $x^2 = 4$ so $x = +2$ or $x = -2$. Then we have the points $(x,y) = (2, 0)$ and $(-2, 0)$.

Finally, $f(0, 1) = 1$ $f(0, -1) = 1$ (maximum value of f is 1)

$f(2, 0) = -4$ $f(-2, 0) = -4$ (minimum value of f is -4).

Fig. 2 shows our constraint, the surface and the path along the surface.

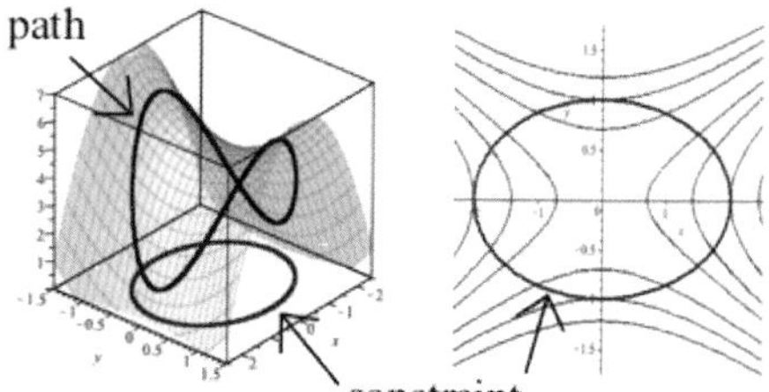

Fig. 2

The Lagrange Multiplier method allows us to trade a calculus problem for the algebra problem of solving a system of equations. But that algebra can be difficult.

Practice 2: Use the Lagrange Multiplier method to find the maximum and minimum values of $f(x,y) = 7x + 3y + 25$ on the circle $x^2 + y^2 = 9$.

Example 3: Find the maximum and minimum values of $f(x,y) = x^2 + y + 2$ on the circle $x^2 + y^2 = 1$.

Solution: $f(x,y) = x^2 + y + 2$ and $g(x,y) = x^2 + y^2 = 1$.

Then $\nabla f(x,y) = 2x\mathbf{i} + 1\mathbf{j}$ and $\nabla g(x,y) = 2x\mathbf{i} + 2y\mathbf{j}$. The Lagrange condition that $\nabla f(x,y) = \lambda \nabla g(x,y)$ means $2x\mathbf{i} + 1\mathbf{j} = \lambda(2x\mathbf{i} + 2y\mathbf{j}) = 2\lambda x\mathbf{i} + 2\lambda y\mathbf{j}$ so

i: $2x = 2\lambda x$

j: $1 = 2\lambda y$ (3 equations in 3 unknowns x, y, and λ)

constraint: $x^2 + y^2 = 1$

If $x = 0$, then $y^2 = 1$ so $y = +1$ or $y = -1$ (from the constraint $x^2 + y^2 = 1$).

If $y = +1$, then $\lambda = 1/2$ (from $\mathbf{j}$ condition) so one solution is $x = 0, y = 1$, and $\lambda = 1/2$.

If $y = -1$, then $\lambda = -1/2$ (from $\mathbf{j}$ condition) so one solution is $x = 0, y = -1$, and $\lambda = -1/2$.

Then we have the points $(x,y) = (0, 1)$ and $(0, -1)$

If $x \neq 0$, then $2x = 2\lambda x$ (from $\mathbf{i}$ condition) so $\lambda = 1$. Then $1 = 2y$ (from $\mathbf{j}$ condition) so $y = 1/2$ and $x^2 + (1/2)^2 = 1$ so $x = +\sqrt{3}/2$ or $x = -\sqrt{3}/2$. Then we have the points $(x,y) = (+\sqrt{3}/2, 1/2)$ and $(-\sqrt{3}/2, 1/2)$.

Finally, $f(0, 1) = 3$

$f(0, -1) = 1$ (minimum of f is 1)

$f(+\sqrt{3}/2, 1/2) = f(-\sqrt{3}/2, 1/2) = 13/4$

(maximum value of f is 13/4).

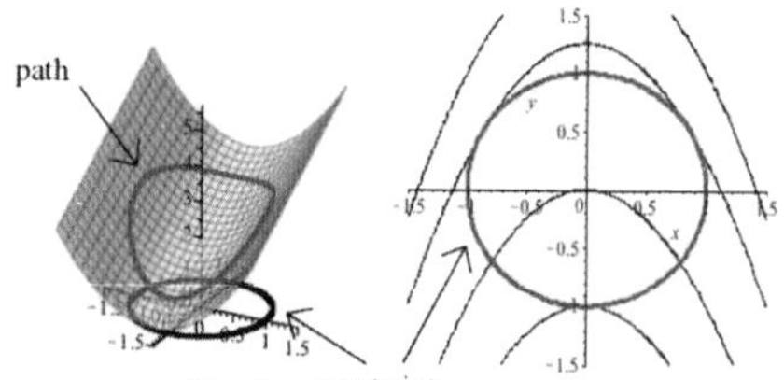

Fig. 3

Fig. 3 shows the constraint, the surface and the path.

The same idea also works for functions and constraints with three (or more) variables:

Find all values of x, y, z and λ such that $\nabla f(x,y,z) = \lambda \nabla g(x,y,z)$ and $g(x,y,z) = k$,

but now we need to solve four equations in four unknowns.

Example 4: Find the volume of the largest rectangular box with a divider but no top (see Fig. 4) that can be constructed from 288 square inches of material.

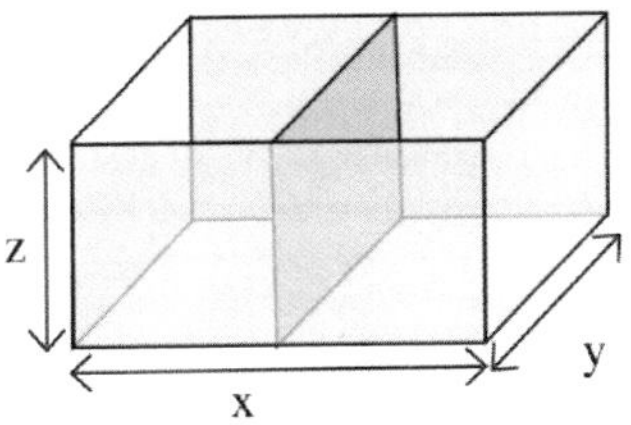

Fig. 4: Rectangular box with one divider and no top

Solution: $V(x,y,z) = xyz$ and $g(x,y,z) = xy + 2xz + 3yz = 288$.

Then $\nabla V(x,y,z) = yz\mathbf{i} + xz\mathbf{j} + xy\mathbf{k}$ and

$\nabla g(x,y,z) = (y + 2z)\mathbf{i} + (x + 3z)\mathbf{j} + (2x + 3y)\mathbf{k}$.

The Lagrange condition that $\nabla f(x,y) = \lambda \nabla g(x,y)$ means

$$yz\mathbf{i} + xz\mathbf{j} + xy\mathbf{k} = \lambda(y + 2z)\mathbf{i} + \lambda(x + 3z)\mathbf{j} + \lambda(2x + 3y)\mathbf{k} \quad \text{so}$$

i: $yz = \lambda y + \lambda 2z$

j: $xz = \lambda x + \lambda 3z$ (4 equations in 4 unknowns x, y, z and λ)

k: $xy = \lambda 2x + \lambda 3y$

constraint: $xy + 2xz + 3yz = 288$

There are a variety of ways to solve this system, but this algebraic way is relatively easy. Multiply the first equation by x, the second by y, and the third by z to get

i: $xyz = \lambda xy + \lambda 2xz$

j: $xyz = \lambda xy + \lambda 3yz$

k: $xyz = \lambda 2xz + \lambda 3yz$

Then all 3 equations are equal to xyz so $\lambda xy + \lambda 2xz = \lambda xy + \lambda 3yz = \lambda 2xz + \lambda 3yz$.
Since $\lambda xy + \lambda 2xz = \lambda xy + \lambda 3yz$ then $y = \frac{2}{3}x$. Since $\lambda xy + \lambda 2xz = \lambda 2xz + \lambda 3yz$ then $z = \frac{1}{3}x$.
Putting those values for y and z into the constraint we get

$$288 = x\left(\frac{2}{3}x\right) + 2x\left(\frac{1}{3}x\right) + 3\left(\frac{2}{3}x\right)\left(\frac{1}{3}x\right) = 2x^2$$ so x = 12 inches, y = 8 inches and z = 4 inches.

The maximum volume is $V(12, 8, 4) = (12)(8)(4) = 384$ cubic inches.

Practice 3: Find the dimensions of the largest volume rectangular box with two dividers but no top (see Fig. 5) that can be constructed from 384 square centimeters of material.

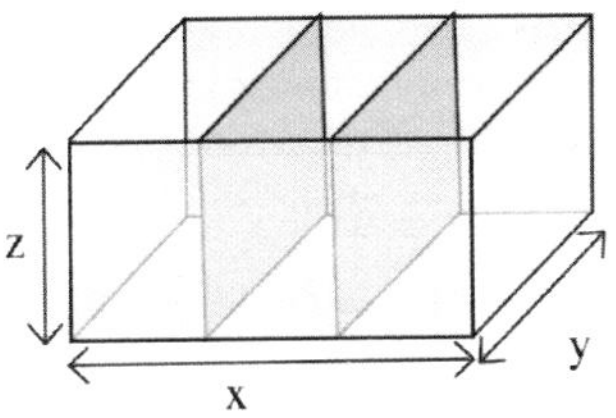

Fig. 5: Rectangular box with two dividers and no top

PROBLEMS

1. In Fig. 6 locate the maximum and minimum values of z along each path and estimate their values.

2. In Fig. 7 locate the maximum and minimum values of z along each path and estimate their values.

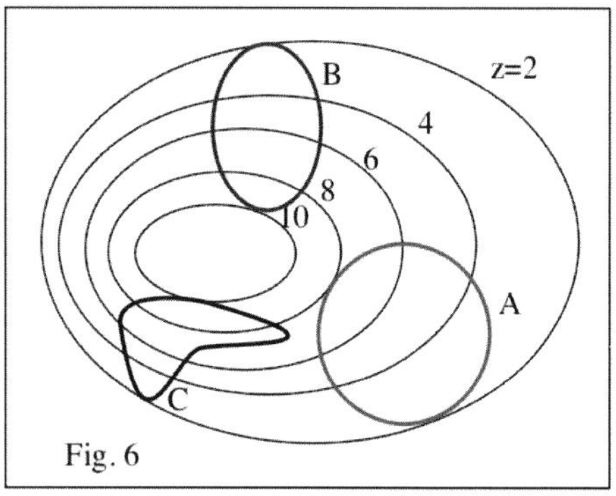

Fig. 6

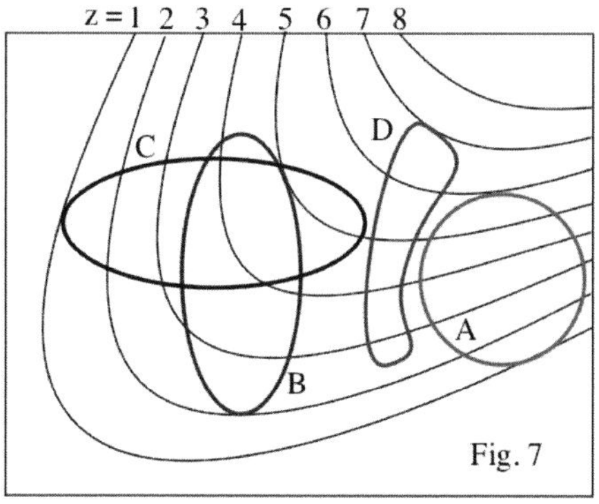

Fig. 7

In problems 3 – 11, use the Lagrange Multiplier method to find the maximum and minimum values of the given function subject to the given constraint.

3. $f(x,y) = x^2 - y^2$; $x^2 + y^2 = 1$

4. $f(x,y) = 2x + y$; $x^2 + 4y^2 = 1$

5. $f(x,y) = xy$; $9x^2 + y^2 = 4$

6. $f(x,y) = x^2 + y^2$; $x^4 + y^4 = 1$

7. $f(x,y,z) = x + 3y + 5z$; $x^2 + y^2 + z^2 = 1$

8. $f(x,y,z) = x - y + 3z$; $x^2 + y^2 + 4z^2 = 4$

9. $f(x,y,z) = xyz$; $x^2 + 2y^2 + 3z^2 = 6$

10. $f(x,y,z) = x^2y^2z^2$; $x^2 + y^2 + z^2 = 1$

11. $f(x,y,z) = x^2 + y^2 + z^2$; $x^4 + y^4 + z^4 = 1$

12. Find the maximum volume of a rectangular box with no top that has a surface area of 48 square inches.

13. Find the maximum volume of a rectangular box with no top that has a surface area of A square inches.

14.. Using your result from problem 11*, show that the area of the bottom is A/3, the total area of the front and back sides is A/3, and the total area of the two end sides is A/3.

15. Find the maximum volume of a rectangular box with no top that be built at a cost of \$15.00 if the bottom material costs \$0.50/in^2 and the materials for the sides costs \$0.01/in^2 .

16.. Find the maximum volume of a rectangular box with no top that be built at a cost of \$T if the bottom material costs \$B/in^2 and the materials for the 4 sides costs \$S/in^2 .

17. Find the maximum volume of a cylinder with no top that has a surface area of 48 square inches.

18. Find the maximum volume of a cylinder with a top that has a surface area of 48 square inches.

Practice Answers

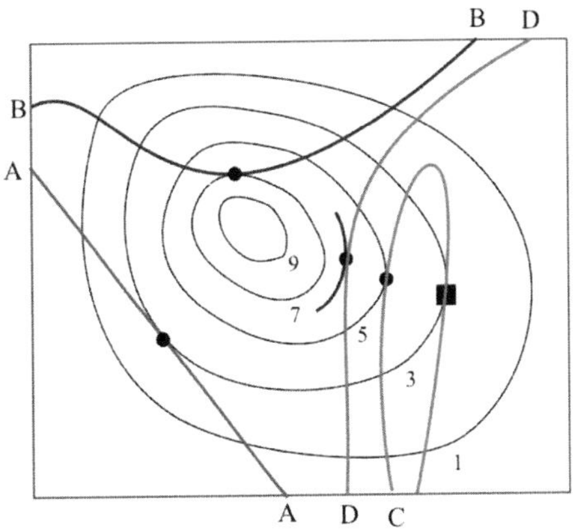

Fig. P1: A hillside and several paths

Practice 1: The dots in Fig. P1 shows the locations of the maximum elevations along paths C and D. The little square on path C is a local maximum along that path. The figure also includes part of the level curve that goes through the dot on path D.

Practice 2: f(x,y)= 7x + 3y + 25 and g(x,y) = $x^2 + y^2$ so $\nabla g = \langle 2x,\ 2y\rangle$ and $\nabla f = \langle\ 7,\ 3\rangle$. Putting these into the Lagrange equation $\nabla f = \lambda \cdot \nabla g$ we have the algebraic system

i: $7 = 2\lambda x$ (so $x \neq 0$)

i: $3 = 2\lambda y$ (so $y \neq 0$)

constraint: $x^2 + y^2 = 9$

Then (from **i**) $x = \dfrac{7}{2\lambda}$ and (from **j**) $y = \dfrac{3}{2\lambda}$. Putting these into the constraint $\left(\dfrac{7}{2\lambda}\right)^2 + \left(\dfrac{3}{2\lambda}\right)^2 = 9$

Then $49 + 9 = 36\lambda^2$ and $\lambda = \pm\dfrac{\sqrt{58}}{6} \approx \pm 1.269$. Putting these back into x and y equations, for $\lambda = \dfrac{\sqrt{58}}{6}$ we have $x = \dfrac{21}{\sqrt{58}} \approx 2.757$, $y = \dfrac{9}{\sqrt{58}} \approx 1.182$ and $f\left(\dfrac{21}{\sqrt{58}}, \dfrac{9}{\sqrt{58}}\right) = \dfrac{169}{\sqrt{58}} + 25 \approx 57.19$ the maximum value of f on the elliptical path. For $\lambda = -\dfrac{\sqrt{58}}{6}$, we have $x = \dfrac{-21}{\sqrt{58}}$, $y = \dfrac{-9}{\sqrt{58}}$ and

$$f\left(\frac{-21}{\sqrt{58}}, \frac{-9}{\sqrt{58}}\right) = \frac{-169}{\sqrt{58}} + 25 \approx 2.81 \cdot$$

Figure P2 shows the surface f, the contours for f, and the constraint g.

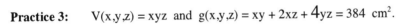

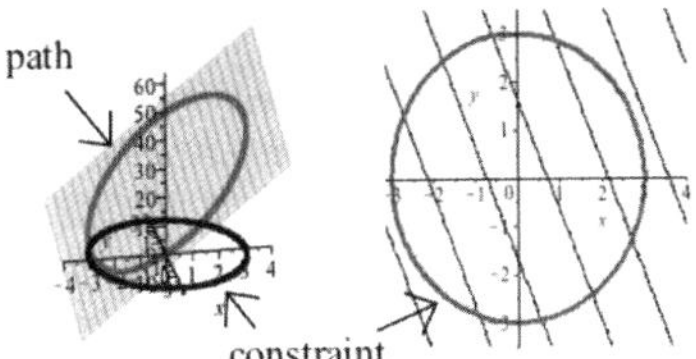

Fig. P2

Practice 3: V(x,y,z) = xyz and g(x,y,z) = xy + 2xz + 4yz = 384 cm^2.

Then ∇V(x,y,z) = yz**i** + xz**j** + xy**k** and

∇g(x,y,z) = (y + 2z)**i** + (x + 4z)**j** + (2x + 4y)**k** so

yz**i** + xz**j** + xy**k** = λ(y + 2z)**i** + λ(x + 4z)**j** + λ(2x + 4y)**k** so

i: $yz = \lambda y + \lambda 2z$

j: $xz = \lambda x + \lambda 4z$ (4 equations in 4 unknowns x, y, z and λ)

k: $xy = \lambda 2x + \lambda 4y$

constraint: $xy + 2xz + 4yz = 384$

Multiply the first equation by x, the second by y, and the third by z,

Then all 3 equations are equal to xyz so $\lambda xy + \lambda 2xz = \lambda xy + \lambda 4yz = \lambda 2xz + \lambda 4yz$. Then

$384 = x\left(\frac{1}{2}x\right) + 2x\left(\frac{1}{4}x\right) + 4\left(\frac{1}{2}x\right)\left(\frac{1}{4}x\right) = \frac{3}{2}x^2$ so x=16 cm, y=8 cm and z=2 cm

and he maximum volume is 256 cm^3 .

13.8 The Chain Rule for Functions of Several Variables

In Section 2.4 we saw the Chain Rule for a function of one variable.

> **Chain Rule (Leibniz notation form)**
>
> If y is a differentiable function of u, and u is a differentiable function of x,
>
> then y is a differentiable function of x and $\frac{dy}{dx} = \frac{dy}{du} \cdot \frac{du}{dx}$.

One interpretation of this Chain Rule is that if x is a signal that is amplified by a factor of 3 by u (du/dx=3) and the signal u gets amplified by a factor of 2 by y (dy/du=3) then the total amplification of x by the combination of u followed by y is by a factor of 6:

$$\frac{dy}{dx} = \frac{dy}{du} \cdot \frac{du}{dx} = (2)(3) = 6.$$

We can also represent this pattern graphically as in Fig. 1.

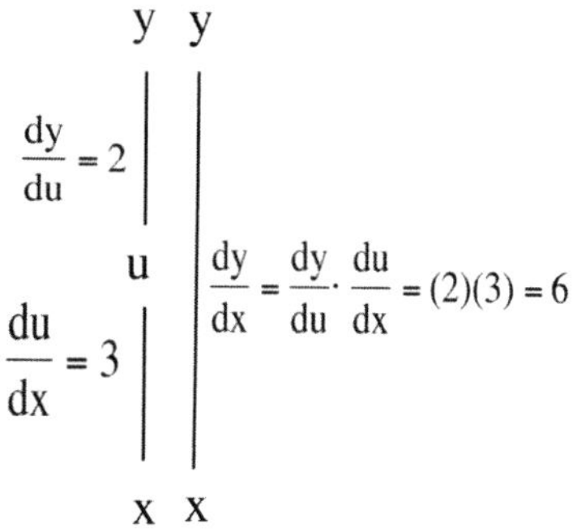

Fig. 1

f is a function of x and y, and each of x and y is a function of t

But suppose that the original signal at t is 1 db (decibel) and that there are two intermediate amplifiers x and y that feed into our final amplifier z as in Fig. 2. If x amplifies t by a factor of 3 (dx/dt=3), z amplifies x by a factor of 2 (dz/dx=2), y amplifies t by a factor of 4 (dy/dt=4), and z amplifies y by a factor of 5 (dz/dy=5), then we can ask what is the total amplification of the original signal 1 db signal at t to the final output z. The original 1 db signal at t becomes 6 db (along the txz path) and 20 db (along the tyz path) for a total output of 26 db.

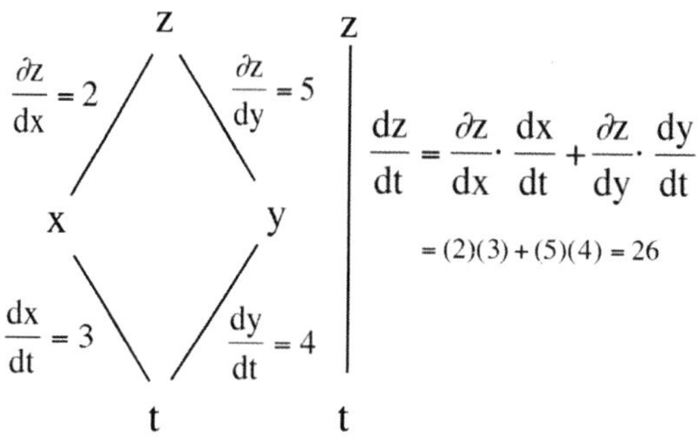

Fig. 2

This is essentially the Chain Rule for a function of two variables: we multiply the rates of change along each path and then add the results to get the total rate of change.

> **The Chain Rule for a Function of Two Dependent Variables**
>
> If $z = f(x, y)$ is a differentiable function of x and y, and x(t) and y(t) are differentiable functions of t, then z is a differentiable function of t, and
>
> $$\frac{df}{dt} = \frac{dz}{dt} = \frac{\partial z}{\partial x}\frac{dx}{dt} + \frac{\partial z}{\partial y}\frac{dy}{dt}.$$

A tree diagram (Fig. 3) is a visual way to organize information that may help you remember "which derivatives go where" in the Chain Rule formula. Write z, the name of the first function at the top of the diagram. Draw branches to x and y, the names of the independent variables of z. Then draw more branches to t, the independent variable of x and y. Then add in the derivative notations as shown in the diagram below.

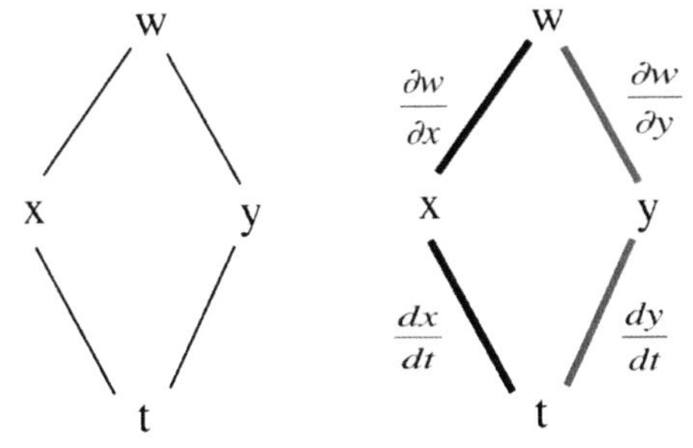

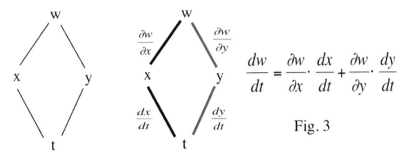

$$\frac{dw}{dt} = \frac{\partial w}{\partial x}\cdot\frac{dx}{dt} + \frac{\partial w}{\partial y}\cdot\frac{dy}{dt}$$

Fig. 3

The two paths from z to t indicate the Chain Rule formula will have two terms. The number of pieces in a path from z to t indicates the number of factors in the corresponding term. The two paths from z to t and two pieces along each path correspond to the two terms of two factors in the chain rule formula.

Example 1: (a) Use the Chain Rule to find the rate of change of $f(x,y) = xy$ with respect to t along the path $x = \cos t$, $y = \sin t$ when $t = \frac{\pi}{3}$.

(b) If the units of t are seconds, the units of x and y are meters and the units of f are °C, then what are the units of $\frac{df}{dt}$?

Solution: (a) At $t = \frac{\pi}{3}$ we have $\frac{\partial f}{dx} = y(t) = \sin(\pi/3) = \frac{\sqrt{3}}{2}$, $\frac{\partial f}{dy} = x(t) = \cos(\pi/3) = \frac{1}{2}$,

$\frac{dx}{dt} = -\sin(t) = -\sin(\pi/3) = -\frac{\sqrt{3}}{2}$ and $\frac{dy}{dt} = \cos(t) = \cos(\pi/3) = \frac{1}{2}$ so

$$\frac{df}{dt} = \frac{\partial f}{dx}\cdot\frac{dx}{dt} + \frac{\partial f}{dy}\cdot\frac{dy}{dt} = \left(\frac{\sqrt{3}}{2}\right)\left(-\frac{\sqrt{3}}{2}\right) + \left(\frac{1}{2}\right)\left(\frac{1}{2}\right) = -\frac{1}{2}.$$

(b) Clearly the units of $\frac{df}{dt} = \frac{\text{units of f}}{\text{units of t}} = \frac{{}^oC}{\text{sec}}$. Following the pieces of the Chain Rule we have $\frac{df}{dt} = \frac{\partial f}{dx}\cdot\frac{dx}{dt} + \frac{\partial f}{dy}\cdot\frac{dy}{dt} = \left(\frac{{}^oC}{m}\right)\left(\frac{m}{s}\right) + \left(\frac{{}^oC}{m}\right)\left(\frac{m}{s}\right) = \frac{{}^oC}{s}$ the same result.

Note: In this example we could have simply replaced x and y with the appropriate functions of t so $f(t) = \cos(t)\cdot\sin(t)$ and then differentiated, but such a replacement is not always easy.

Practice 1: Use the Chain Rule to calculate the value of $\frac{df}{dt}$ when t = 2 for the functions

$$f(x,y) = x^4y^3 + 3x^2y,\ x(t) = t^2 + t - 5,\ y(t) = t^3 - 2t^2 - t + 4.$$

The Chain Rule for Functions of Three Dependent Variables adds one term to our previous pattern.

> If $w = f(x,y,z)$ is a differentiable function of x, y and z, and x, y, and z are differentiable functions of t, then w is a differentiable function of t, and
>
> $$\frac{df}{dt} = \frac{dw}{dt} = \frac{\partial w}{\partial x}\frac{dx}{dt} + \frac{\partial w}{\partial y}\frac{dy}{dt} + \frac{\partial w}{\partial z}\frac{dz}{dt}.$$

Many students find that using a tree diagram (Fig/ 4) for this situation makes it easy to keep track of the pattern: multiply the derivatives along each of the three paths and then add those results together.

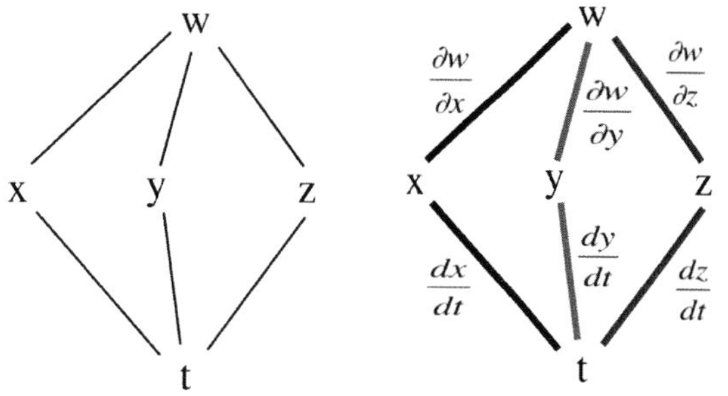

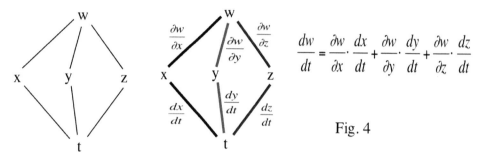

$$\frac{dw}{dt} = \frac{\partial w}{\partial x}\cdot\frac{dx}{dt} + \frac{\partial w}{\partial y}\cdot\frac{dy}{dt} + \frac{\partial w}{\partial z}\cdot\frac{dz}{dt}$$

Fig. 4

<u>**Example 2**</u>: Use the Chain Rule to find the value of $\frac{df}{dt}$ for $f(x,y,z) = xy + z$ along the helix $x(t) = \cos(t), y(t) = \sin(t), z(t) = t$. What is the derivative's value at $t = 0$? (This is the instantaneous rate of change as our point moves along a helix.)

Solution: If $t = 0$ then x=1, y=0, z=0, $\frac{dx}{dt} = -\sin(t) = 0$, $\frac{dy}{dt} = \cos(t) = 1$, and $\frac{dz}{dt} = 1$. At (1, 0, 0) $\frac{\partial f}{\partial x} = y = 0$, $\frac{\partial f}{\partial y} = x = 1$ and $\frac{\partial f}{\partial z} = 1$. Finally,

$$\frac{df}{dt} = \frac{\partial f}{\partial x}\cdot\frac{dx}{dt} + \frac{\partial f}{\partial y}\cdot\frac{dy}{dt} + \frac{\partial f}{\partial z}\cdot\frac{dz}{dt} = (0)(0) + (1)(1) + (1)(1) = 2.$$

Practice 2: Use the Chain Rule to find the value of $\frac{df}{dt}$ for the Example 2 functions when $t = \pi/3$?

In General

There are many ways in which functions of several variables can be combined. Rather than stating or memorizing a Chain Rule pattern for each new situation, just keep in mind the general pattern:

> **Build a tree dependency diagram,**
> **multiply along each path and add these path results.**

Example 3: Suppose $w = f(x,y,z)$, $x = g(r,s)$, $y = h(r,s)$, and $z = k(r,s)$ and that all four functions are differentiable,

(a) Build a tree dependency diagram for these functions,

(b) Create a Chain Role for $\frac{\partial w}{\partial r}$ and $\frac{\partial w}{\partial s}$.

Solution:

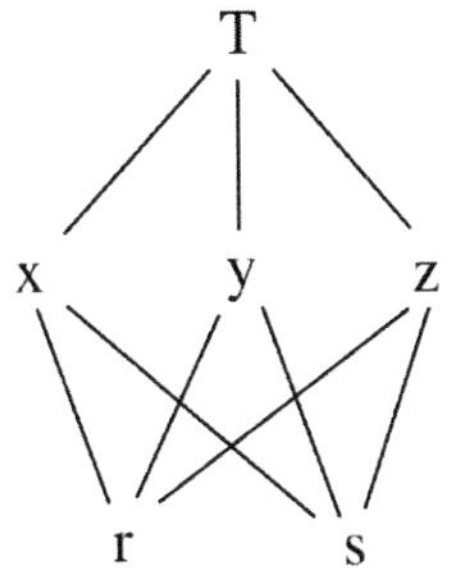

dependency tree

Fig. 5

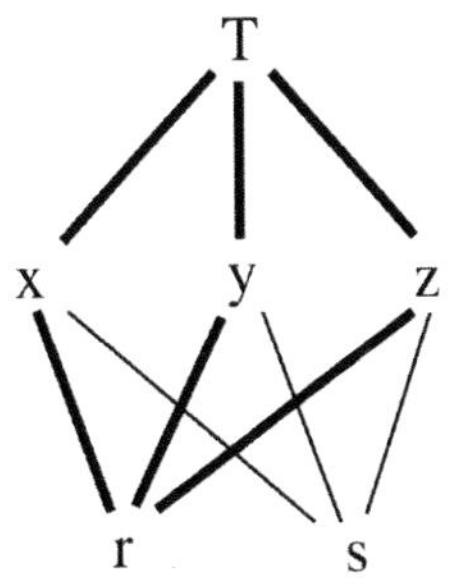

Three paths from p to T

$$\frac{\partial T}{\partial p} = \frac{\partial T}{\partial x}\cdot\frac{\partial x}{\partial p} + \frac{\partial T}{\partial y}\cdot\frac{\partial y}{\partial p} + \frac{\partial T}{\partial z}\cdot\frac{\partial z}{\partial p}$$

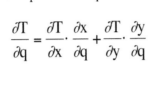

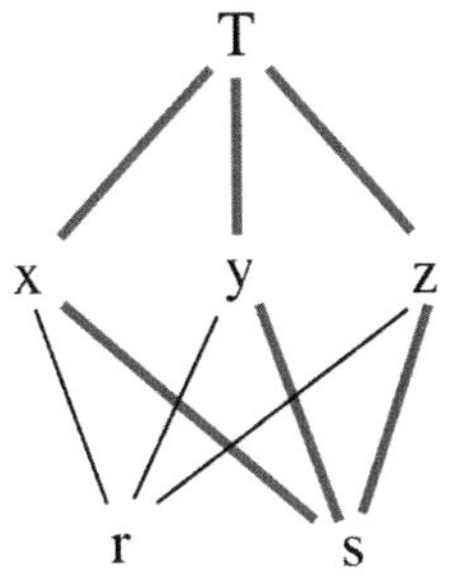

Two paths from q to T

$$\frac{\partial T}{\partial q} = \frac{\partial T}{\partial x}\cdot\frac{\partial x}{\partial q} + \frac{\partial T}{\partial y}\cdot\frac{\partial y}{\partial q}$$

Practice 3: Suppose $T = f(x, y, z)$, $x = g(p, q, r)$, $y = h(p, q)$ and $z = k(p, r)$ and that all of these functions are differentiable.

(a) Build a dependency tree diagram for these functions.

(b) Create Chain Rules for $\frac{\partial T}{\partial p}$ and $\frac{\partial T}{\partial q}$.

Example 4: (a) The voltage V in a circuit satisfies the law V = IR. Write the Chain Rule for $\frac{dV}{dt}$.

(b) If the voltage is dropping because the battery is wearing out and the resistance s increasing because the circuit is heating up, then how fast is the current I changing when R = 500 ohms, I = 0.04 amps, dR/dt = 0.5 ohms/sec, and dV/dt = -0.01 volt/sec?

Solution: (a) $\frac{dV}{dt} = \frac{\partial V}{\partial I}\cdot\frac{\partial I}{\partial t} + \frac{\partial V}{\partial R}\cdot\frac{\partial R}{\partial t}$

(b) $\frac{\partial V}{\partial I} = \frac{\partial (IR)}{\partial I} = R = 500\ ohms$ and $\frac{\partial V}{\partial R} = \frac{\partial (IR)}{\partial R} = I = 0.04\ amps$.

Putting all of this information into the equation in part (a) we have

$$\left(-0.01\frac{amp\cdot ohms}{\text{sec}}\right) = (500\ ohms)\cdot\left(\frac{\partial I}{\partial t}\frac{amps}{\text{sec}}\right) + (0.04\ amps)\cdot\left(0.5\frac{ohms}{\text{sec}}\right) \text{ so}$$

$\frac{\partial I}{\partial t}$ = -0.00006 amps/sec.

PROBLEMS

In problems 1 and 2, use the information in Table 1.

1. Calculate $\frac{df}{dt}$ when t = 1 and t = 3

2. Calculate $\frac{df}{dt}$ when t = 2 and t = 4

$\frac{\partial f}{\partial x} =$

x\y	0	1	2	3
0	5	4	6	2
1	1	9	7	10
2	3	5	4	11
3	7	8	5	13

$\frac{\partial f}{\partial y} =$

x\y	0	1	2	3
0	3	7	2	4
1	6	1	8	3
2	1	4	5	6
3	5	2	9	7

t	1	2	3	4
x	2	0	1	3
y	3	1	0	2

t	1	2	3	4
$\frac{dx}{dt}$	-1	5	-2	6
$\frac{dy}{dt}$	-3	7	-1	8

Table 1

In exercises 37, express $\frac{df}{dt}$ as a function of t by using the Chain Rule. Then evaluate $\frac{df}{dt}$ at the given value of t.

3. $f(x,y) = x^2 + y^2, \quad x = \cos t, \quad y = \sin t, \quad t = \pi$

4. $f(x,y) = x^2y^2 + 3x + 4y + 1, \quad x = 3 + t^2, \quad y = 1 + 2t, \; t = 2$

5. $f(x,y,z) = x^2y + yz + xz, \quad x = 3 + 2t, \quad y = t^2, \; z = 5t, \; t = 2$

6. $f(x,y,z) = \frac{x}{y} + \frac{y}{z} + \frac{z}{x}, \quad x = 1 + 2t, \quad y = 2 + 3t, \; z = 3 + 4t, \; t = 1$

7. $f(x,y,z) = 2ye^x - \ln z, \quad x = \ln(t^2 + 1), \quad y = \tan^{-1} t, \quad z = e^t, \quad t = 1$

8. $f(x,y,z) = xyz, \quad x = 2\cos(t), \quad y = \sin(t), \; z = 3t, \; t = \pi$

9. Express $\frac{\partial w}{\partial u}$ and $\frac{\partial w}{\partial v}$ as functions of u and v by using the Chain Rule.
Then evaluate $\frac{\partial w}{\partial u}$ and $\frac{\partial w}{\partial v}$ at the given point (u, v).
$w = xy + yz + xz, \quad x = u + v, \quad y = u - v, \quad z = uv, \quad (u, v) = (-2, 0)$

10. Find $\frac{\partial w}{\partial r}$ when r = 1, s = −1, if $w = (x + y + z)^2, \quad x = r - s, \quad y = \cos(r + s), \quad z = \sin(r + s)$.

11. Find $\frac{\partial z}{\partial u}$ when u = 0, v = 0, if $z = \cos(xy) + x \cdot \sin(y), \quad x = u + v, \quad y = uv$.

12. The lengths a, b, and c of the edges of a rectangular box are changing with time. At the instant in question, a = 1 meter, b = 2 meters, c = 3 meters, $\frac{da}{dt} = \frac{db}{dt} = 1$ m/sec, and $\frac{dc}{dt} = -3$ m/sec. At what rates are the box's volume V and surface area S changing at that instant? Are the box's interior diagonals increasing or decreasing in length,?

13. In an ideal gas the pressure P (in kilopascals kPa), the volume V (in liters L), and the temperature T (in kelvin K) satisfy the equation PV = 8.31T. How fast is the pressure changing, dP/dt , when the temperature is 310 K and is decreasing at a rate of 0.2 K/s, and the volume is 80 L and is increasing at a rate of 0.1 L/s?

14. Given: w is a function of x, y, and z; x is function of r; y is a function of r and s; z is a function of s and t; and s is a function of t. Make a tree diagram, and then write a chain rule formula for $\frac{\partial w}{\partial t}$.

Practice Answers

Practice 1: $f(x,y) = x^4y^3 + 3x^2y$, $x(t) = t^2 + t - 5$, $y(t) = t^3 - 2t^2 - t + 4$.

If t = 2, then x = 1, y = 2, $\frac{dx}{dt} = 2t + 1 = 5$ and $\frac{dy}{dt} = 3t^2 - 4t - 1 = 3$. At x=1, y=2

$\frac{\partial f}{\partial x} = 4x^3y^3 + 6xy = 46$ and $\frac{\partial f}{\partial y} = 3x^4y^2 + 3x^2 = 15$ so $\frac{df}{dt} = \frac{\partial f}{dx}\cdot\frac{dx}{dt} + \frac{\partial f}{dy}\cdot\frac{dy}{dt} = (46)(5) + (15)(3) = 275.$

Each Chain Rule problem has a lot of pieces so you need to be organized.

Note: Substituting x(t) and y(t) into f gives

$f(t) = (t^2 + t - 5)^4(t^3 - 2t^2 - t + 4)^3 + 3(t^2 + t - 5)^2(t^3 - 2t^2 - t + 4)$ and that would be a messy derivative.

Practice 2: $f(x,y,z) = xy + z$, $x(t) = \cos(t)$, $y(t) = \sin(t)$, $z(t) = t$.

If $t = \pi/3$ then $x=\cos(\pi/3)= \frac{1}{2}$, $y=\sin(\pi/3)= \frac{\sqrt{3}}{2}$, $z= \frac{\pi}{3}$, $\frac{dx}{dt} = -\sin(t) = -\sin(\pi/3) = -\frac{\sqrt{3}}{2}$,

$\frac{dy}{dt} = \cos(t) = \cos(\pi/3) = \frac{1}{2}$, and $\frac{dz}{dt} = 1$. At $\left(\frac{1}{2}, \frac{\sqrt{3}}{2}, \frac{\pi}{3}\right)$, $\frac{\partial f}{\partial x} = y = \frac{\sqrt{3}}{2}$, $\frac{\partial f}{\partial y} = x = \frac{1}{2}$, and $\frac{\partial f}{\partial z} = 1$.

Putting this all together $\frac{df}{dt} = \frac{\partial f}{\partial x}\cdot\frac{dx}{dt} + \frac{\partial f}{\partial y}\cdot\frac{dy}{dt} + \frac{\partial f}{\partial z}\cdot\frac{dz}{dt} = \left(\frac{\sqrt{3}}{2}\right)\left(-\frac{\sqrt{3}}{2}\right) + \left(\frac{1}{2}\right)\left(\frac{1}{2}\right) + (1)(1) = \frac{1}{2}.$

Practice 3:

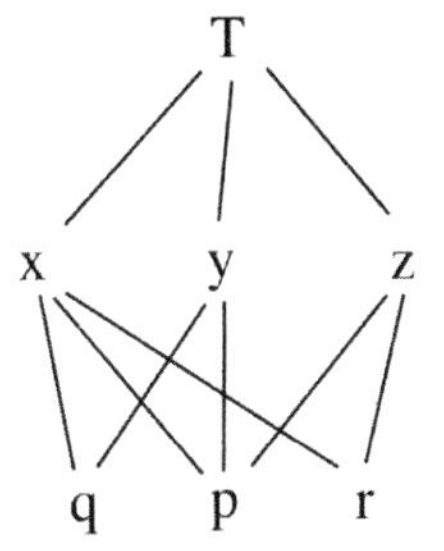

dependency tree

(there is no line from y to r or from z to q)

Fig. 6

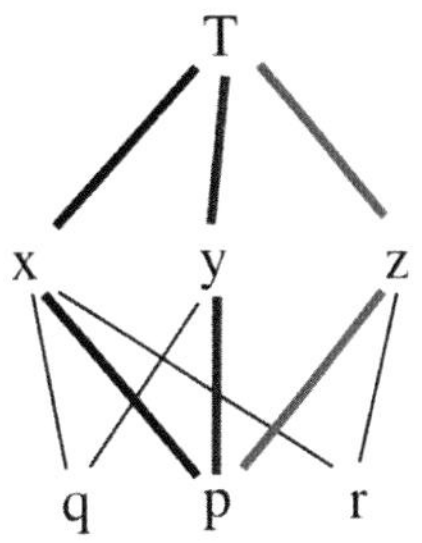

Three paths from p to T

$$\frac{\partial T}{\partial p} = \frac{\partial T}{\partial x}\cdot\frac{\partial x}{\partial p} + \frac{\partial T}{\partial y}\cdot\frac{\partial y}{\partial p} + \frac{\partial T}{\partial z}\cdot\frac{\partial z}{\partial p}$$

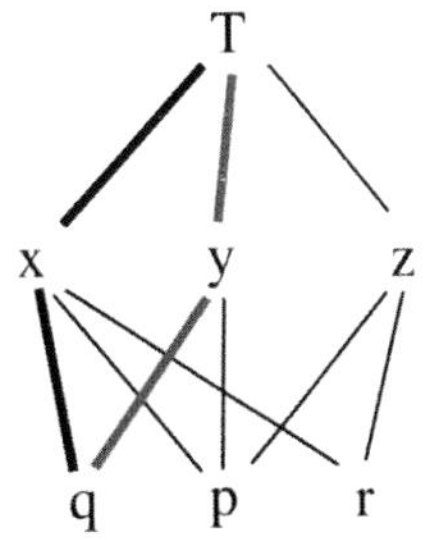

Two paths from q to T

$$\frac{\partial T}{\partial q} = \frac{\partial T}{\partial x}\cdot\frac{\partial x}{\partial q} + \frac{\partial T}{\partial y}\cdot\frac{\partial y}{\partial q}$$

13.1 Selected Answers

1. (a) minimum = 2, maximum = 30 (b) down 2, down 2 more, up 9 and up 5 more
   (c) up 5, up 5, up 5 up 5, then no change

3. (a) maximum = 10 m (b) depth increases by 3, decreases by 4, increase 1, decreases 1
   (c) depth decreases 2, decreases 1, decreases 1

5. (el ires

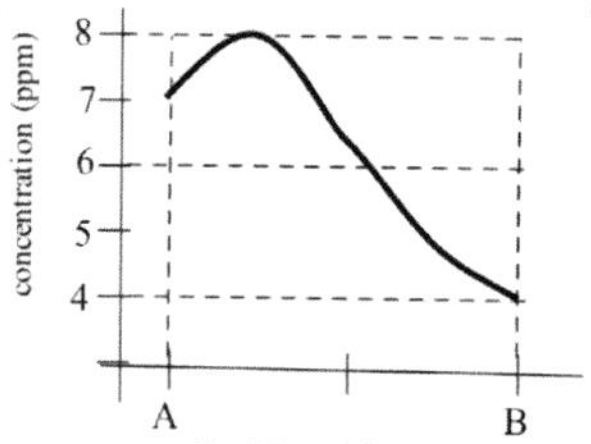

Problem 5(b)

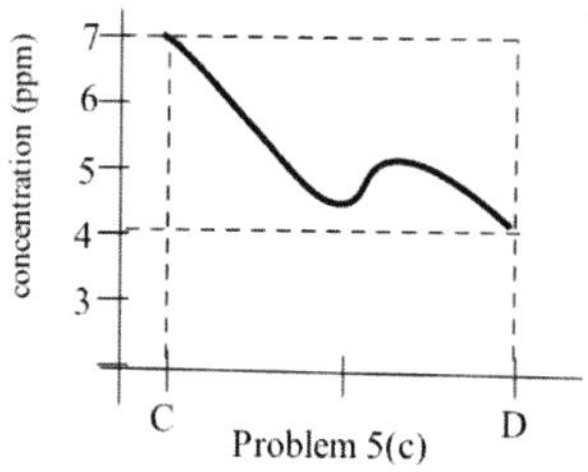

Problem 5(c)

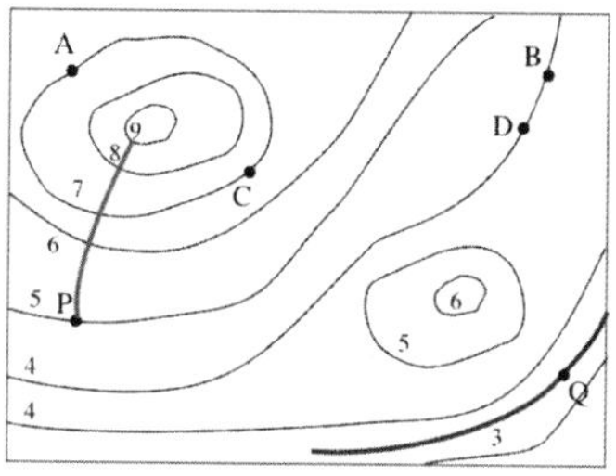

Problem 5(d) and (e)

7 to 15. See figures below

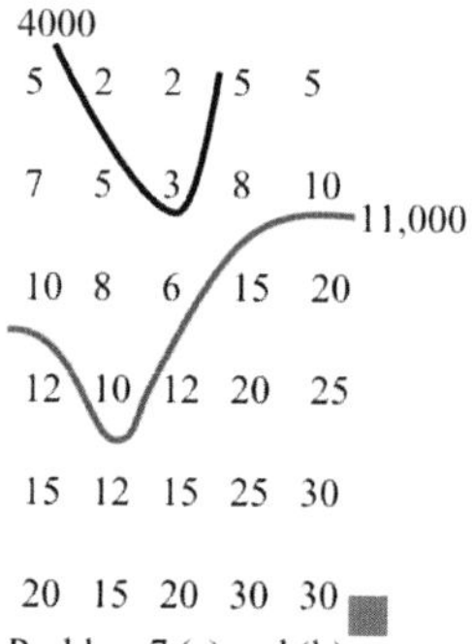

Problem 7 (a) and (b)

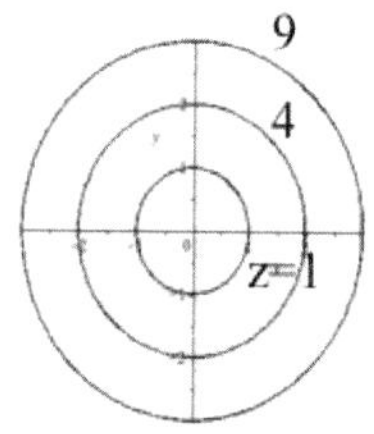

Problem 9

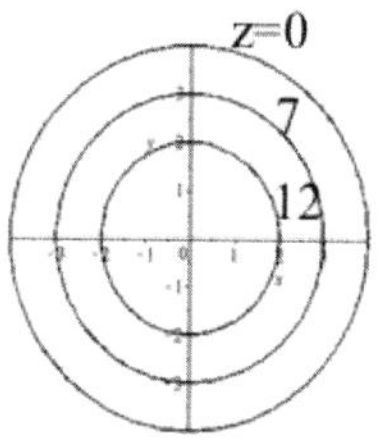

Problem 11

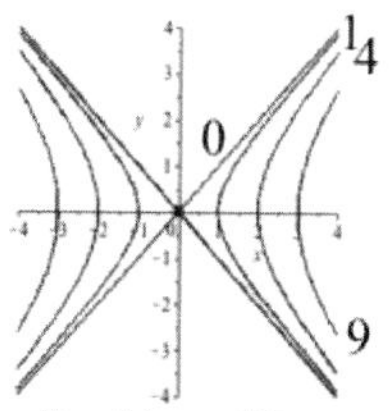

Problem 13

17. level curves D
18. level curves C
19. level curves A
20. level curves G
21. level curves B
22. level curves F
23. level curves E

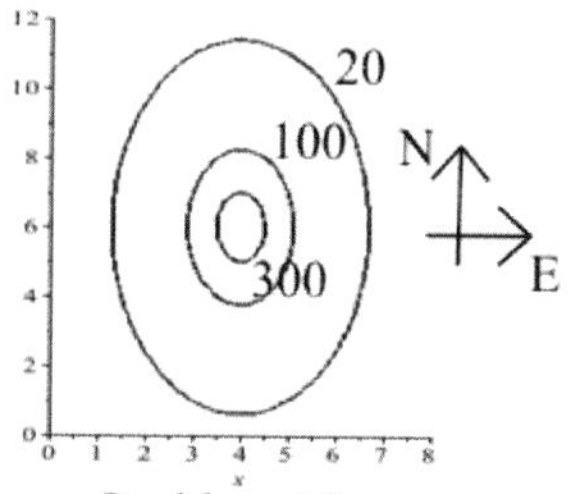

Problem 15

13.2 Selected Answers

1. (a) 3 (b) 2 (c) 4 (d) about 2.4

3. (a) 3 (b) 2 (c) dne (d) dne

5. -921

7. -5/2

9. π

11. dne If y=0 then limit = 1. If y= -x then the function is undefined.

13. dne Paths y=0 and y=x give different limit values.

15. dne Paths y=0 and y=x give different limit values.

17. dne Paths y=0 and y=x give different limit values.

19. dne Paths y=0 and y=x give different limit values.

21. 1

23. dne Paths y=0 and y= x^2 give different limit values.

25. 2 Try rationalizing the denominator.

27. 0 The function equals $\frac{x(y-1)}{(x-1)^2+(y-1)^2}$ which is easier to analyze.

29. -3/5

31. dne If x=y=0 and z->0 then limit = -1. If y=z=0 and x->0 then limit = 1.

33. dne If y=z=0 and x->0 then limit = 0. If z=0 and yx=x ->0 then limit = ½..

35. Define f(2,1) = 4

37. No value of f(1,2) or f(3,2) will make f continuous at those points.
f(1,2) = 3 makes f continuous at (1,2).

39. Not continuous (not defined) on the circle where $x^2 + y^2 = 1$.

41. Not continuous (not defined) if $2x+3y \le 0$ or $y \le -2x/3$.

43. In order for T to be continuous at (x,y) we need both $x+y \ge 0$ (so $y \ge -x$) and $x-y \ge 0$ (so $x \ge y$). That requires $x \ge 0$ and $-x \le y \le x$.

45. F is continuous at (x,y,z) if yz >0.

13.3 Selected Answers

1. $f_x(1,2) = -8$, $f_y(1,2) = -4$

2. $f_x(1,0) = -1/\sqrt{3}$, $f_y(1,0) = 0$

3. $f_x(3,-1) = -27$

4. 2

5. $\frac{\partial z}{\partial x} = (x^4 + 3x^2y^2 - 2xy^3)/(x^2+y^2)^2$, $\frac{\partial z}{\partial y} = (y^4 + 3x^2y^2 - 2x^3y)/(x^2+y^2)^2$

6. $(1/y) - (y/x^2)$

7. $(y-z)/(x-y)$, $(x+z)/(x-y)$

8. $(x-y-z)/(x+z)$, $(y-x)/(x+z)$

9. $\frac{\partial z}{\partial x} = \frac{2xz}{2yz - x^2}$ $\frac{\partial z}{\partial y} = \frac{2y + z^2}{x^2 - 2yz}$

11. $y \sec(xy) + xy^2 \sec(xy)\tan(xy)$

12. 0

13. y + z, x + z, x + y

14. $f_x(x,y) = 3x^2y^5 - 4xy + 1$, $f_y(x,y) = 5x^3y^4 - 2x^2$

15. $f_x(x,y) = 4x^3 + 2xy^2$, $f_y(x,y) = 2x^2y + 4y^3$

16. $f_x(x,y) = 2y/(x+y)^2$, $f_y(x,y) = -2x/(x+y)^2$

17. $f_x = e^x \{ \tan(x-y) + \sec^2(x-y) \}$, $f_y = -e^x \sec^2(x-y)$

18. $f_s = -3s/\sqrt{2 - 3s^2 - 5t^2}$, $f_t = -5t/\sqrt{2 - 3s^2 - 5t^2}$

19. $f_u = v/(u^2 + v^2)$, $f_v = -u/(u^2 + v^2)$

20. $g_x = 2xy^4 \sec^2(x^2y^3)$, $g_y = \tan(x^2y^3) + 3x^2y^3 \sec^2(x^2y^3)$

21. $\frac{\partial z}{\partial x} = 1/\sqrt{x^2+y^2}$, $\frac{\partial z}{\partial y} = y/(x^2+y^2+x\sqrt{x^2+y^2})$

22. $\frac{\partial z}{\partial x} = \frac{3}{2}\cosh(\sqrt{3x+4y})/\sqrt{3x+4y}$, $\frac{\partial z}{\partial y} = 2\cosh(\sqrt{3x+4y})/\sqrt{3x+4y}$

23. $f_x = 2xyz^3 + y$, $f_y = x^2z^3 + x$, $f_z = 3x^2yz^2 - 1$

24. $f_x = yz\,x^{yz-1}$, $f_y = z\,x^{yz}\ln(x)$, $f_z = y\,x^{yz}\ln(x)$

25. $u_x = -yz\cos(y/(x+z))/(x+z)^2$, $u_y = z\cos(y/(x+z))/(x+z)$,
$u_z = \sin(y/(x+z)) - yz\cos(y/(x+z))/(x+z)^2$

26. $f_x = 1/(z-t)$, $f_y = -1/(z-t)$, $f_z = -(x-y)/(z-t)^2$, $f_t = (x-y)/(z-t)^2$

31. $\frac{\partial z}{\partial x} = f'(x)$, $\frac{\partial z}{\partial y} = g'(y)$

32. $\frac{\partial z}{\partial x} = f'(x+y)$, $\frac{\partial z}{\partial y} = f'(x+y)$

33. $\frac{\partial z}{\partial x} = f'(x/y)(1/y)$, $\frac{\partial z}{\partial y} = f'(x/y)(-x/y^2)$

34. $f_{xx} = 2y$, $f_{xy} = 2x + 1/(2\sqrt{y}) = f_{yx}$, $f_{yy} = -x/(4y\sqrt{y})$

35. $z_{xx} = 3(2x^2+y^2)/\sqrt{x^2+y^2}$, $z_{xy} = 3xy/\sqrt{x^2+y^2} = z_{yx}$, $z_{yy} = 3(x^2+2y^2)/\sqrt{x^2+y^2}$

13.4 Selected Answers

1. $4x + 8y - z = 8$

3. $2x + 4y - z + 6 = 0$

5. $2x + y - z = 1$

7. $2x - y - z + 2 = 0$

9. $dz = 2xy^3\,dx + 3x^2y^2\,dy$

11. $dz = -2x(x^2+y^2)^{-2}\,dx - 2y(x^2+y^2)^{-2}\,dy$

13. $du = e^x \cdot (\cos(xy) - y \cdot \sin(xy))\,dx - x \cdot e^x \cdot \sin(xy)\,dy$

15. $dw = 2xy\,dx + (x^2 + 2yz)\,dy + y^2\,dz$

17. $dw = (x^2+y^2+z^2)^{-1}(x\,dx + y\,dy + z\,dz)$

19. $\Delta z = 0.9225$ and $dz = 0.9$

21. 2.9923. −0.28

25. 5.4 cm^2

27. 16 cm^3

29. 150

13.5 Selected Answers

1. $-4-\sqrt{3}$ 3. 1

5. (a) $\nabla f(x,y) = \langle 3x^2 - 8xy, -4x^2 + 2y \rangle$ (b) $\langle 0, -2 \rangle$ (c) $-8/5$

7. (a) $\nabla f(x,y,z) = \langle y^2z^3, 2xyz^3, 3xy^2z^2 \rangle$ (b) $\langle 4, -4, 12 \rangle$ (c) $20/\sqrt{3}$

9. 7/52 11. $29/\sqrt{13}$

13. 1/6 15. $\sqrt{5}$, $\langle 1/\sqrt{5}, 2/\sqrt{5} \rangle$

17. $\sqrt{17}/6$, $\langle 4/\sqrt{17}, 1/\sqrt{17} \rangle$

19. $\sqrt{(13/2)}$, $\langle -3/\sqrt{13}, -2/\sqrt{13} \rangle$

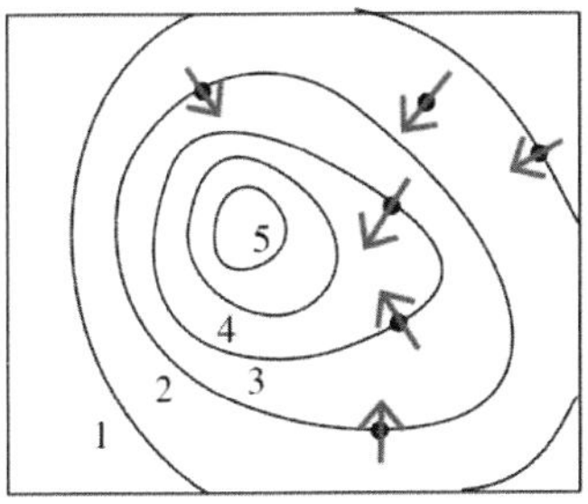

Fig. 10

21, See Fig. 10. Note that each gradient vector is perpendicular to the level curve and points uphill.

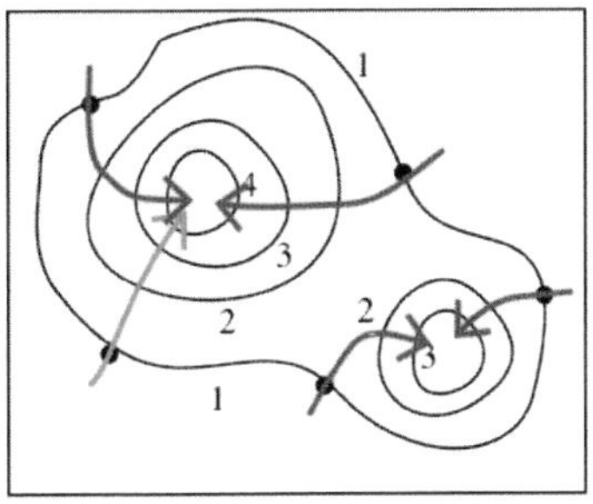

Fig. 11

23. See Fig. 11. Note that "uphill gradient" path is always perpendicular to the level curves.

27. (a) $-40/(3\sqrt{3})$

29. (a) $32/\sqrt{3}$ (b) $u = \langle 38, 6, 12 \rangle / (2\sqrt{406})$ (c) $2\sqrt{406}$

31. 327/13

13.6 Selected Answers

1. Minimum $f(-2, 3) = -13$ 3. Minimum $f(0, -1) = -1$

5. Local maximum: f(1,0)=12,. Saddle points: f(-1,0) = f(2,3) = f(2,-3) = 1

7. Local minimum: $f(0, 0) = 4$. Saddle points: $(\pm\sqrt{2}, -1)$

9. Local minimum: $f(1, 1) = -1$. Saddle point $f(0, 0) = 0$

11. Saddle point $f(1, 2) = -2$

13. Local maximum $f(-1/2, 4) = -6$ 15. None

17. Local maximum $f(0, 0) = 2$, local minimum $f(0, 2) = -2$, saddle points $(\pm 1, 1)$

19. Saddle points $(0, n\pi)$, n and integer 20. Minimum $f(4, 0) = -7$, maximum $f(4, 5) = 13$

21. Maximum $f(\pm 1, 1) = 7$, minimum $f(0, 0) = 4$

23. Maximum $f(2, 4) = 3$, minimum $f(-2, 4) = -9$

25. Critical points: f1,1)=0, f(-2,4)=- -9 minimum, f(2,4)= 3 maximum, f(-1/3, 1/9) = 32/27

27. Critical points: f(0,0) = 30 maximum, f(-2,4)=22, f(2,4)=22, f(0.4) = 14 minimum

29. Critical points: f(-1,3)=17, f(2,0)=11, f(0,4) = 19 maximum, f(3/2,1/2) = 43/4 minimum

31. Critical points: f(0,0)=0, f(1/2, 3/4)=145/256, f(1,0) = 2 maximum, f(-1,0)= -2 minimum

33. (2/7, 4/7, 6/7)

35. $\frac{100}{3}$, $\frac{100}{3}$, $\frac{100}{3}$

37. $16/\sqrt{3}$

39. 4/3

41. Cube, edge length c/12

13.7 Selected Answers

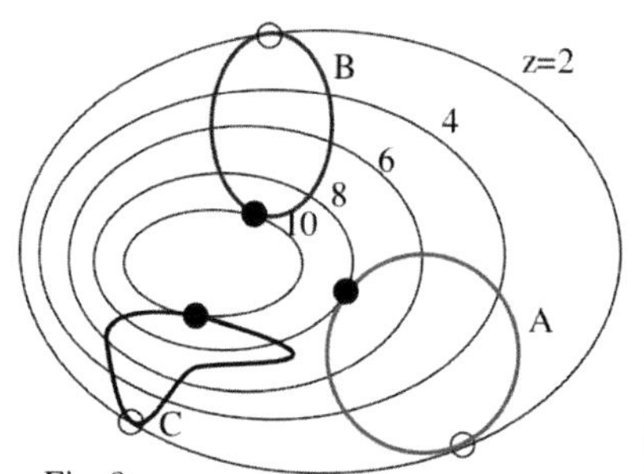

Fig. 8

1. See Fig. 8. The solid circles marks the locations of the maximum z values along each path, and the open circles mark the locations of the minimums, A: max z=8, min z=2. B: max z=10, min z=2, C: max z=10, min z=2.

3. maximum f(±1,0) = 1, minimum f(0,±1) = – 1

4. maximum $f(\sqrt{2}/3, \sqrt{2}) = f(-\sqrt{2}/3, -\sqrt{2}) = 2/3$, minimum $f(\sqrt{2}/3, -\sqrt{2}) = f(-\sqrt{2}/3, \sqrt{2}) = -2/3$

7. maximum $f(1/\sqrt{35}, 3/\sqrt{35}, 5/\sqrt{35}) = \sqrt{35}$, minimum $f(-1/\sqrt{35}, -3/\sqrt{35}, -5/\sqrt{35}) = -\sqrt{35}$

9. $x = \pm\sqrt{2}$, $y = \pm 1$, $z = \pm\sqrt{\frac{2}{3}}$. maximum f is $2/\sqrt{3}$ (when all are positive or one is positive and two are negative), minimum f is $-2/\sqrt{3}$.

11. maximum is $\sqrt{3} = f\left(\pm\frac{1}{\sqrt{3}}, \pm\frac{1}{\sqrt{3}}, \pm\frac{1}{\sqrt{3}}\right)$, minimum is $1 = f(\pm1,0,0) = f(0,\pm1,0) = f(0,0,\pm1)$

13. V=xyz with xy+2xz+2yz=A. Maximum V is $\frac{1}{2}\left(\frac{A}{3}\right)^{3/2}$ and that occurs when $x = y = \sqrt{\frac{A}{3}}$ and $z = \frac{1}{2}\sqrt{\frac{A}{3}}$.

15. V=xyz with 5xy + 1(2xz + 2yz) = 1500 (working in cents). Maximum volume is 2500 in^3 when x = y = 10 inches and z = 25 inches. Note that the cost of the bottom is \$5.00, the total cost of the two ends is \$5.00, and the total costs of the other two sides is \$5.00 .

16. V = xyz with Bxy + S2xz + 2Syz = T, $x = y = \sqrt{\frac{T}{3B}}$, $z = \frac{1}{2S}\sqrt{\frac{BT}{3}}$ and maximum volume is $V = \frac{T}{3B} \cdot \frac{1}{2S} \cdot \sqrt{\frac{BT}{3}}$. The cost of the bottom is T/3.

17. maximum volume is $64/\sqrt{\pi}$ when $r = 4/\sqrt{\pi}$ and $h = 4/\sqrt{\pi}$.

13.8 Selected Answers

1. (a) When t = 1, x=2 and y=3 so $\frac{\partial f}{\partial x} = 11$ and $\frac{\partial f}{\partial y} = 6$. Also $\frac{dx}{dt} = -1$ and $\frac{dy}{dt} = -3$.

Then $\frac{df}{dt} = \frac{\partial f}{\partial x} \cdot \frac{dx}{dt} + \frac{\partial f}{\partial y} \cdot \frac{dy}{dt} = (11) \cdot (-1) + (6) \cdot (-3) = -29$.

(b) When t = 3, x=1 and y=0 so $\frac{\partial f}{\partial x} = 1$ and $\frac{\partial f}{\partial y} = 6$. Also $\frac{dx}{dt} = -2$ and $\frac{dy}{dt} = -1$.

Then $\frac{df}{dt} = \frac{\partial f}{\partial x} \cdot \frac{dx}{dt} + \frac{\partial f}{\partial y} \cdot \frac{dy}{dt} = (1) \cdot (-2) + (6) \cdot (-1) = -8$.

3. When t = π, then x = cos(π) = -1, y = sin(π) = 0, $\frac{dx}{dt} = -\sin(t) = -\sin(\pi) = 0$,

$\frac{dy}{dt} = \cos(t) = \cos(\pi) = -1$, $\frac{\partial f}{\partial x} = 2x = 0$, and $\frac{\partial f}{\partial y} = 2y = -2$ so

$$\frac{df}{dt} = \frac{\partial f}{\partial x} \cdot \frac{dx}{dt} + \frac{\partial f}{\partial y} \cdot \frac{dy}{dt} = (0) \cdot (0) + (-2) \cdot (-2) = 2$$

5. When t = 2, x = 5, y = 4, z = 10, $\frac{dx}{dt} = 2$, $\frac{dy}{dt} = t^2 = 4$, $\frac{dz}{dt} = 5$,

$\frac{\partial f}{\partial x} = 2xy + z = 50$, $\frac{\partial f}{\partial y} = x^2 + z = 35$ and $\frac{\partial f}{\partial z} = y + x = 9$.

$$\frac{df}{dt} = \frac{\partial f}{\partial x} \cdot \frac{dx}{dt} + \frac{\partial f}{\partial y} \cdot \frac{dy}{dt} + \frac{\partial f}{\partial z} \cdot \frac{dz}{dt}$$

$$= (50) \cdot (2) + (35) \cdot (4) + (9) \cdot (5) = 285$$

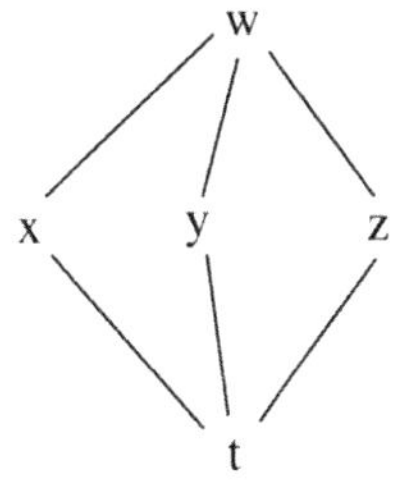

Fig. Problems 5 and 7

7. When t = 1, then $x = \ln(2)$, $y = \tan^{-1}(1) = \frac{\pi}{4}$, $z = e$, $\frac{dx}{dt} = \frac{2t}{1+t^2} = 1$, $\frac{dy}{dt} = \frac{1}{1+t^2} = \frac{1}{2}$,

$\frac{dz}{dt} = -\frac{1}{z} = -\frac{1}{e}$, $\frac{\partial f}{\partial x} = 2ye^x = 2 \cdot \frac{\pi}{4} \cdot e^{\ln(2)} = \frac{\pi}{4}$, $\frac{\partial f}{\partial y} = 2e^x = 2e^{\ln(2)} = 4$ and $\frac{\partial f}{\partial z} = -\frac{1}{e}$.

$$\frac{df}{dt} = \frac{\partial f}{\partial x} \cdot \frac{dx}{dt} + \frac{\partial f}{\partial y} \cdot \frac{dy}{dt} + \frac{\partial f}{\partial z} \cdot \frac{dz}{dt} = \left(\frac{\pi}{4}\right) \cdot (1) + (4) \cdot \left(\frac{1}{2}\right) + \left(-\frac{1}{e}\right) \cdot (e) = 1 + \frac{\pi}{4}$$

9. $w = xy + yz + xz$. When $(u, v) = (-2, 0)$ then $x = 2, y = -2, z = 0$,

$$\frac{dx}{du} = 1,\ \frac{dx}{dv} = 1,\ \frac{dy}{du} = 1,\ \frac{dy}{dv} = -1,\ \frac{dz}{du} = v = 0,\ \frac{dz}{dv} = u = -2$$

$$\frac{\partial w}{\partial x} = y + z = -2,\ \frac{\partial w}{\partial y} = x + z = 2,\ \frac{\partial w}{\partial z} = y + x = 0$$

$$\frac{dw}{du} = \frac{\partial w}{\partial x}\cdot\frac{dx}{du} + \frac{\partial w}{\partial y}\cdot\frac{dy}{du} + \frac{\partial w}{\partial z}\cdot\frac{dz}{du} = (2)(1) + (2)(1) + (0)(0) = 4$$

$$\frac{dw}{dv} = \frac{\partial w}{\partial x}\cdot\frac{dx}{dv} + \frac{\partial w}{\partial y}\cdot\frac{dy}{dv} + \frac{\partial w}{\partial z}\cdot\frac{dz}{dv} = (2)(1) + (2)(-1) + (0)(-2) = 0$$

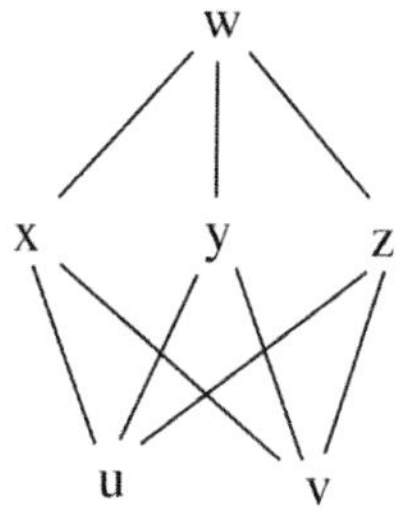

Fig. Problem 9

11. $z = \cos(xy) + x\cdot \sin(y),\ x = u + v + 2,\ y = uv$.

When $u = 0$ and $v = 0$ then $x = 2, y = 0$ and $z = 1$.

$$\frac{dx}{du} = 1,\ \frac{dx}{dv} = 1,\ \frac{dy}{du} = v = 0,\ \frac{dy}{dv} = u = 0$$

$$\frac{\partial z}{\partial x} = -y\cdot \sin(xy) + \sin(y) = 0,\ \frac{\partial z}{\partial y} = -x\cdot \sin(xy) + x\cdot \cos(y) = 2$$

$$\frac{dz}{du} = \frac{\partial z}{\partial x}\cdot\frac{dx}{du} + \frac{\partial z}{\partial y}\cdot\frac{dy}{du} = (0)(1) + (2)(0) = 0$$

$$\frac{dz}{dv} = \frac{\partial z}{\partial x}\cdot\frac{dx}{dv} + \frac{\partial z}{\partial y}\cdot\frac{dy}{dv} = (0)(1) + (2)(0) = 0$$

(That was a lot of work just to get a couple 0s.)

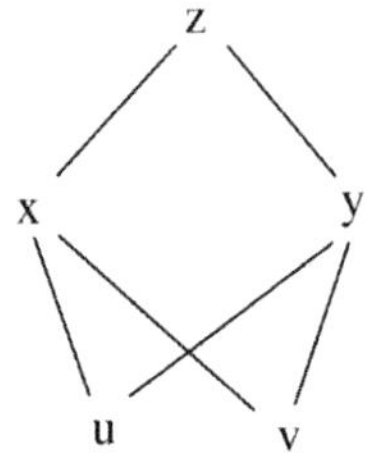

Fig. Problem 11

13. We know that $T = 310$ K, $\frac{dT}{dt} = -0.2\ \frac{K}{s}$, $V = 80$ L and $\frac{dV}{dt} = 0.1\ \frac{L}{s}$.

$P = 8.31\frac{T}{V}$ so $\frac{\partial P}{\partial T} = \frac{8.31}{V}$ and $\frac{\partial P}{\partial V} = -8.31\frac{T}{V^2}$. By the Chain Rule

$$\frac{dP}{dt} = \frac{\partial P}{\partial T}\cdot\frac{\partial T}{\partial t} + \frac{\partial P}{\partial V}\cdot\frac{\partial V}{\partial t} = \left(\frac{8.31}{V}\right)\cdot\left(\frac{\partial T}{\partial t}\right) + \left(-\frac{8.31T}{V^2}\right)\cdot\left(\frac{\partial V}{\partial t}\right)$$

$$= \left(\frac{8.31}{80}\right)\cdot(-0.2) + \left(-\frac{8.31\cdot 310}{80^2}\right)\cdot(0.1) = -0.061\ \frac{kPa}{sec}$$

14.1 DOUBLE INTEGRALS OVER RECTANGLES

Theorem:

If $f(x,y) \geq 0$ and f is integrable over the rectangle R,

then the volume V of the solid that lies above R and under the surface $z = f(x,y)$ is

$$V = \iint_R f(x,y)\, dA .$$

Properties of Double Integrals:

(1) $\iint_R f(x,y) + g(x,y)\, dA = \iint_R f(x,y)\, dA + \iint_R g(x,y)\, dA$

(2) $\iint_R K f(x,y)\, dA = K \iint_R f(x,y)\, dA .$

(3) If $f(x,y) \geq g(x,y)$ for all (x,y) in R, then $\iint_R f(x,y)\, dA \geq \iint_R g(x,y)\, dA .$

Example 1: Evaluate $\int_0^3 \int_1^2 x^2 y\, dy\, dx$

Solution: $\int_0^3 \int_1^2 x^2 y\, dy\, dx$ means $\int_0^3 \left\{ \int_1^2 x^2 y\, dy \right\} dx$ so first we evaluate the inside integral $\int_1^2 x^2 y\, dy$ **treating x as a constant**:

$$\int_1^2 x^2 y\, dy = \frac{1}{2} x^2 y^2 \Big|_{y=1}^{2} = \frac{1}{2} x^2 (2)^2 - \frac{1}{2} x^2 (1)^2 = \frac{3}{2} x^2 .$$

Then $\int_0^3 \left\{ \int_1^2 x^2 y\, dy \right\} dx = \int_0^3 \left\{ \frac{3}{2} x^2 \right\} dx = \frac{1}{2} x^3 \Big|_{x=0}^{3} = \mathbf{\frac{27}{2}} .$

Example 2: Evaluate $\int_1^2 \int_0^3 x^2 y \, dx \, dy$

Solution: $\int_1^2 \int_0^3 x^2 y \, dx \, dy$ means $\int_1^2 \left\{ \int_0^3 x^2 y \, dx \right\} dy$ so first we evaluate the inside integral $\int_0^3 x^2 y \, dx$ **treating y as a constant**:

$$\int_0^3 x^2 y \, dx = \frac{1}{3} x^3 y \Big|_{x=0}^{3} = \frac{1}{3}(3)^3 y - \frac{1}{3}(0)^3 y = 9y .$$

Then $\int_1^2 \left\{ \int_0^3 x^2 y \, dx \right\} dy = \int_1^2 \left\{ 9y \right\} dy = \frac{9}{2} y^2 \Big|_{y=1}^{2} = \frac{9}{2}(2)^2 - \frac{9}{2}(1)^2 = \mathbf{\frac{27}{2}}$.

A few important points:

- Always work from the inside out: first evaluate the inside integral.
- For $\int f(x,y) \, dx$ integrate with respect to x and **treat y as a constant.**
- For $\int f(x,y) \, dy$ integrate with respect to y and **treat x as a constant.**

It was not an accident that the answers to Example 1 and Example 2 were the same.

> Fubini's Theorem:
>
> If f is integrable over the rectangle $R = \{ (x,y) : a \le x \le b \text{ and } c \le y \le d \} = [a, b] \times [c, d]$
>
> then $\iint_R f(x,y) \, dA = \int_a^b \int_c^d f(x,y) \, dy \, dx = \int_c^d \int_a^b f(x,y) \, dx \, dy$

Fubini's Theorem says that we can integrate in either order and still get the same result — sometimes one order of integration is much easier than the other order.

Example 3: Evaluate $\iint_R y \sin(xy)\, dA$ where $R = [1,2] \times [0,\pi]$.

Solution: The notation $R = [1,2] \times [0,\pi]$ means the rectangle $1 \le x \le 2$ and $0 \le y \le \pi$.

By Fubini's Theorem we have a choice of evaluating $\int_0^{\pi}\int_1^2 y \sin(xy)\, dx\, dy$ or $\int_1^2\int_0^{\pi} y \sin(xy)\, dy\, dx$.

(a) $$\int_0^{\pi}\int_1^2 y \sin(xy)\, dx\, dy = \int_0^{\pi} \left\{ \int_1^2 y \sin(xy)\, dx \right\} dy$$

$$= \int_0^{\pi} \left\{ -\cos(xy) \Big|_{x=1}^{2} \right\} dy = \int_0^{\pi} \left\{ -\cos(2y) + \cos(1y) \right\} dy$$

$$= -\frac{1}{2}\sin(2y) + \sin(y) \Big|_{y=0}^{\pi}$$

$$= \left\{ -\frac{1}{2}\sin(2\pi) + \sin(\pi) \right\} - \left\{ -\frac{1}{2}\sin(2\cdot 0) + \sin(0) \right\} = 0$$

(b) $$\int_1^2\int_0^{\pi} y \sin(xy)\, dy\, dx = \int_1^2 \left\{ \int_0^{\pi} y \sin(xy)\, dy \right\} dx$$ so first we need to evaluate

$\int_0^{\pi} y \sin(xy)\, dy$ and that requires Integration by Parts, a more difficult situation than in part (a).

Example 4: Find the volume of the solid S that is bounded by the elliptic paraboloid $x^2 + 2y^2 + z = 16$, the planes $x = 2$ and $y = 2$, and the three coordinate planes (xy, xz, and yz–planes).

Solution: S lies under the surface $z = 16 - x^2 - 2y^2$ and above the square $0 \le x \le 2$, $0 \le y \le 2$. Then

$$V = \int_0^2\int_0^2 16 - x^2 - 2y^2\, dx\, dy = \int_0^2 \left\{ \int_0^2 16 - x^2 - 2y^2\, dx \right\} dy$$

$$= \int_0^2 \left\{ 16x - \frac{1}{3}x^3 - 2xy^2 \Big|_{x=0}^{2} \right\} dy = \int_0^2 \left\{ \frac{88}{3} - 4y^2 \right\} dy = \frac{88}{3}y - \frac{4}{3}y^3 \Big|_{y=0}^{2}$$

$$= \frac{88}{3}(2) - \frac{4}{3}(2)^3 = \frac{144}{3} = 48.$$

PROBLEMS — try all of these

For problems 1 – 4, find $\int_0^2 f(x,y)\,dx$ and $\int_0^1 f(x,y)\,dy$

1. $f(x,y) = x^2y^3$
2. $f(x,y) = 2xy - 3x^2$
3. $f(x,y) = x \cdot e^{x+y}$
4. $f(x,y) = \frac{x}{y^2 + 1}$

In problems 5 – 13, evaluate the double integrals.

5. $\int_0^4 \int_0^2 x\sqrt{y}\,dx\,dy$
6. $\int_{-1}^1 \int_0^1 (x^3y^3 + 3xy^2)\,dy\,dx$
7. $\int_0^3 \int_0^1 \sqrt{x+y}\,dx\,dy$
8. $\int_0^{\pi/4} \int_0^3 \sin(x)\,dy\,dx$
9. $\int_0^{\ln(2)} \int_0^{\ln(5)} e^{2x-y}\,dx\,dy$
10. $\iint_R (2y^2 - 3xy^3)\,dA$ where $R = \{ (x,y) : 1 \le x \le 2, 0 \le y \le 3 \}$.
11. $\iint_R x\sin(y)\,dA$ where $R = \{ (x,y) : 1 \le x \le 4, 0 \le y \le \pi/6 \}$.
12. $\iint_R x\sin(x+y)\,dA$ where $R = [0, \pi/6] \times [0, \pi/6]$.
13. $\iint_R \frac{1}{x+y}\,dA$ where $R = [1,2] \times [0,1]$.

ANSWERS

1. $\frac{8}{3}y^3, \frac{1}{4}x^2$
3. $e^{2+y} + e^y = e^y(e^2 + 1)$, $xe^{x+1} - xe^x = xe^x(e - 1)$
5. 32/3
6. 0
7. $\frac{4}{15}(31 - 9\sqrt{3})$
8. $3\left(1 - \frac{1}{\sqrt{2}}\right)$
9. 6
10. $-\frac{585}{8}$
11. $\frac{15}{4}(2 - \sqrt{3})$
12. $\frac{\sqrt{3} - 1}{2} - \frac{\pi}{12}$
13. ln(27/16)

14.2 DOUBLE INTEGRALS OVER GENERAL REGIONS

Sometimes we need the double integral of a function $z = f(x,y)$ over a domain D (in the xy–plane) that is not a rectangle, and in those cases we must chose the endpoints of the integrals very carefully.

A plane region **D** uses Vertical Slices if it lies between the graphs of two continuous functions of x:

$$\mathbf{D} = \{ (x,y) : a \le x \le b ,\ g_1(x) \le y \le g_2(x) \}$$ as in the figure.

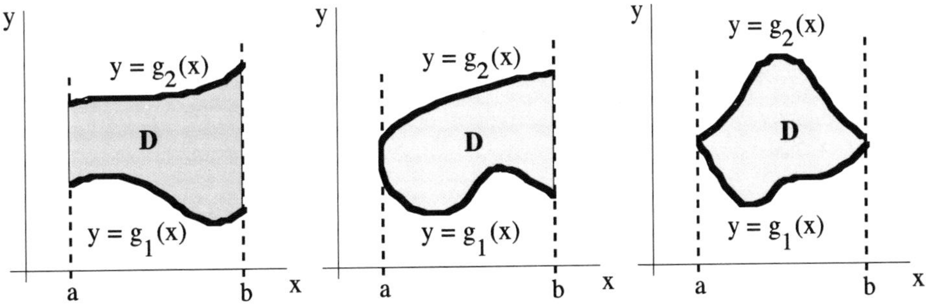

Some **Vertical Slice** Regions

In the Vertical Slice situation, it is usually easier to evaluate the double integral over **D** as

$$\iint_{\mathbf{D}} f(x,y)\ dA = \int_{x=a}^{b}\int_{y=g_1(x)}^{g_2(x)} f(x,y)\ dy\ dx$$

Example 1: Evaluate $\iint_D (x + 2y)\ dA$ where D is the region bounded by $2x^2 \le y \le 1 + x^2$ for $-1 \le x \le 1$.

Solution: $$\iint_D (x + 2y)\ dA = \int_{-1}^{1}\int_{2x^2}^{1+x^2} (x + 2y)\ dy\ dx = \int_{-1}^{1} (xy + y^2)\Big|_{y=2x^2}^{1+x2} dx$$

$$= \int_{-1}^{1} \{ (x(1+x^2) + (1+x^2)^2 \} - \{ (x(2x^2) + (2x^2)^2 \} \quad dx$$

$$= \int_{-1}^{1} \{ -3x^4 - x^3 + 2x^2 + x + 1 \} \quad dx = \frac{32}{15} .$$

Maple command: int(int(x+2*y, y=2*x^2..1+x^2) , x=–1..1); gives the result $\frac{32}{15}$.

Sometimes it is more useful to use Horizontal Slices and to integrate first with respect to x and then with respect to y (see the Horizontal Slice figure below).

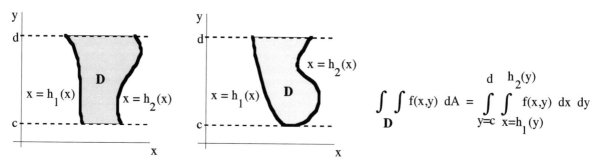

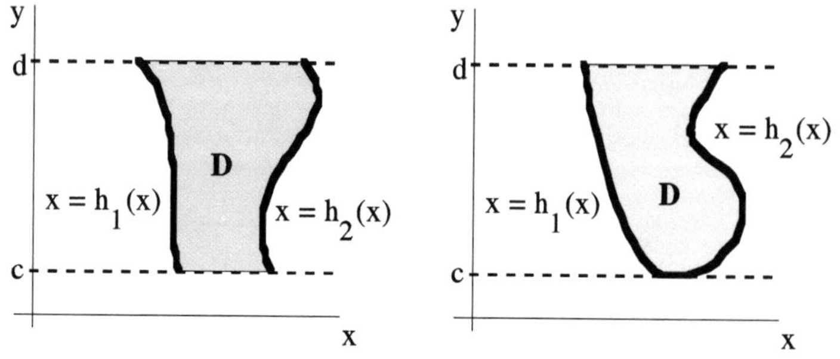

$$\iint_D f(x,y)\, dA = \int_{y=c}^{d} \int_{x=h_1(y)}^{h_2(y)} f(x,y)\, dx\, dy$$

Some **Horizontal Slice** Regions

Example 2: Evaluate $\iint_D (x^2 + y^2)\, dA$ where D is the region bounded by $0 \le y \le 4$ and $y/2 \le x \le \sqrt{y}$.

Solution: Sketch the region D.

$$\iint_D (x^2+y^2)\, dA = \int_0^4 \int_{y/2}^{\sqrt{y}} (x^2+y^2)\, dx\, dy = \int_0^4 \left(\tfrac{1}{3}x^3 + xy^2 \right) \Big|_{x=y/2}^{x=\sqrt{y}} dy$$

$$= \int_0^4 \left\{ \tfrac{1}{3}(\sqrt{y})^3 + (\sqrt{y})y^2 \right\} - \left\{ \tfrac{1}{3}(y/2)^3 + (y/2)y^2 \right\} dy$$

$$= \int_0^4 \left\{ \frac{y^{3/2}}{3} + y5/2 - \frac{y^3}{24} - \frac{y^3}{2} \right\} dy = \frac{2}{15}y^{5/2} + \frac{2}{7}y^{7/2} - \frac{13}{96}y^4 \Big|_0^4 = \frac{216}{35} .$$

Maple command: int(int(x^2 + y^2, x=y/2..sqrt(y)), y=0..4); gives $\frac{216}{35}$ after just a couple seconds.

The double integral over the region in Example 2 can also be expressed as a double integral in which we integrate first with respect to y and then with respect to x as

$$\iint_D (x^2+y^2)\, dA = \int_0^2 \int_{x^2}^{2x} (x^2+y^2)\, dy\, dx .$$

This double integral also evaluates to the value 216/35.

The next two figures illustrate two more regions on which we can use either Vertical Slices or Horizontal Slices.

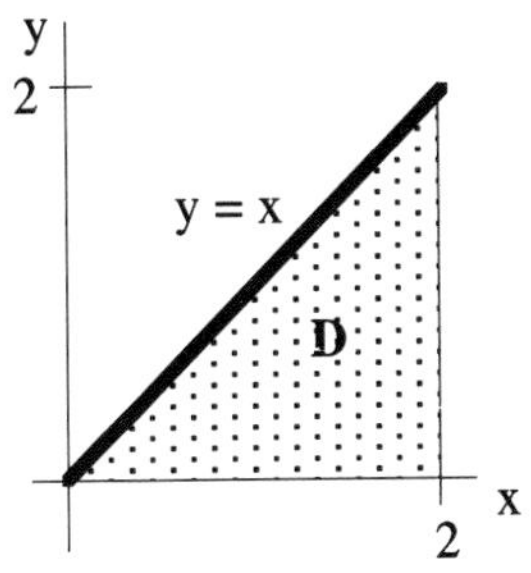

$$\int_{x=0}^{2}\int_{y=0}^{x} f(x,y)\ dy\ dx \qquad \text{vertical slices}$$

or

$$\int_{y=0}^{2}\int_{x=y}^{2} f(x,y)\ dx\ dy \qquad \text{horizontal slices}$$

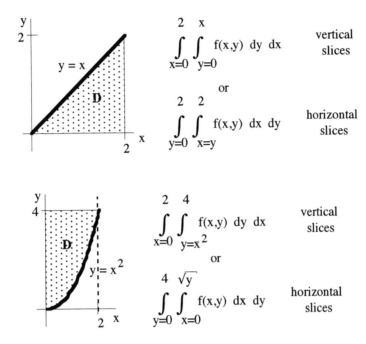

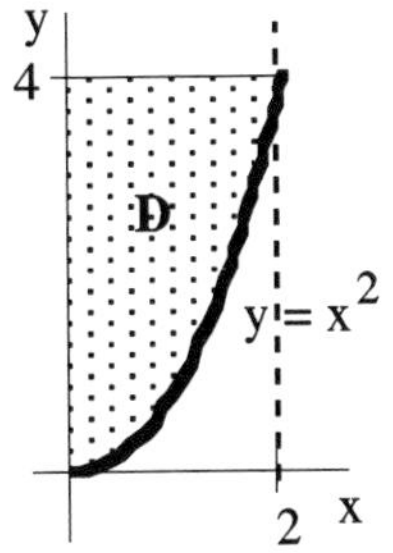

$$\int_{x=0}^{2}\int_{y=x^2}^{4} f(x,y)\ dy\ dx \qquad \text{vertical slices}$$

or

$$\int_{y=0}^{4}\int_{x=0}^{\sqrt{y}} f(x,y)\ dx\ dy \qquad \text{horizontal slices}$$

Two Properties of Double Integrals

(1) Using a double integral to find area: $\displaystyle\iint_D 1\ dA = \text{area of } D$.

(2) If $m \le f(x,y) \le M$ for all (x,y) in D, then $\displaystyle m\cdot(\text{area of } D) \le \iint_D f(x,y)\ dA \le M\cdot(\text{area of } D)$

Another Use of Double Integrals — Surface Area

If $z = f(x,y)$ for (x,y) in the plane region D and f_x and f_y are continuous,

then the area of the surface $z = f(x,y)$ that lies above the region D is

$$\text{surface area} = \iint_D \sqrt{1 + \{ f_x(x,y) \}^2 + \{ f_y(x,y) \}^2}\ dA\ .$$

PROBLEMS — try all of these

For problems 1 – 10, evaluate the integrals.

1. $\int_0^1 \int_0^y x \, dx \, dy$ 2. $\int_0^2 \int_{\sqrt{x}}^3 (x^2 + y) \, dy \, dx$ 3. $\int_0^1 \int_0^x \sin(x^2) \, dy \, dx$

4. $\iint_D xy \, dA$ where $D = \{ (x,y) : 0 \le x \le 1, x^2 \le y \le \sqrt{x} \}$.

5. $\iint_D (3x + y) \, dA$ where $D = \{ (x,y) : \pi/6 \le x \le \pi/4, \sin(x) \le y \le \cos(x) \}$.

6. $\iint_D (y - xy^2) \, dA$ where $D = \{ (x,y) : -y \le x \le 1+y, 0 \le y \le 1 \}$.

7. $\iint_D (e^{x/y}) \, dA$ where $D = \{ (x,y) : y \le x \le y^3, 1 \le y \le 2 \}$.

8. $\iint_D x \cdot \cos(y) \, dA$ where D is bounded by $y = 0$, $y = x^2$, and $x = 1$.

9. $\iint_D (x^2 + y) \, dA$ where D is bounded by $y = x^2$ and $y = 2 - x^2$.

10. $\iint_D 4y^3 \, dA$ where D is bounded by $y = x - 6$ and $y^2 = x$.

In problems 11 – 13, change the order of integration. (It helps to sketch the region.)

11. $\int_0^1 \int_0^x f(x,y) \, dy \, dx$ 12. $\int_1^2 \int_0^{\ln(x)} f(x,y) \, dy \, dx$ 13. $\int_0^4 \int_{y/2}^2 f(x,y) \, dx \, dy$

ANSWERS

1. $1/6$ 2. $16(1 - \frac{\sqrt{2}}{7})$ 3. $(1 - \cos(1))/2$ 4. $1/12$

5. $\pi(3\sqrt{2} - 1 - \sqrt{3})/4 + (14 - 13\sqrt{3})/8$ 6. $3/4$ 7. $\frac{1}{2} e^4 - 2e$

8. $(1 - \cos(1))/2$ 9. $16/5$ 10. $500/3$

11. $\int_0^1 \int_y^1 f(x,y) \, dx \, dy$ 12. $\int_0^{\ln(2)} \int_{e^y}^2 f(x,y) \, dx \, dy$ 13. $\int_0^2 \int_0^{2x} f(x,y) \, dy \, dx$

14.2 Additional Problems

For each shaded region, fill in the endpoints of the double integration for a function f(x,y) over that region.

1.

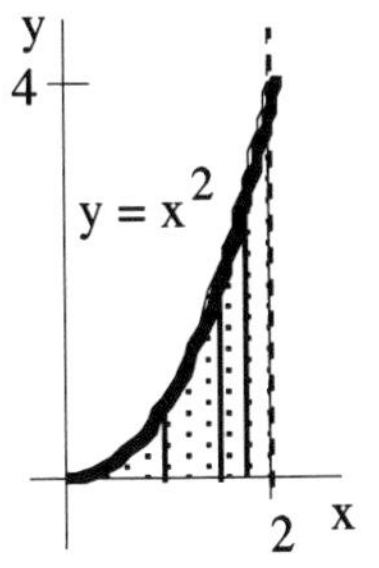

$\int\int f(x,y)\ dy\ dx$

vertical horizontal?

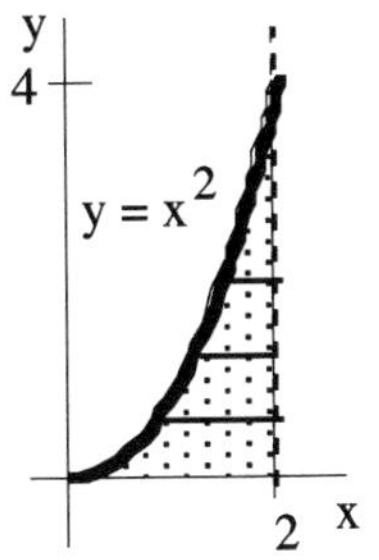

$\int\int f(x,y)\ dx\ dy$

vertical horizontal?

2.

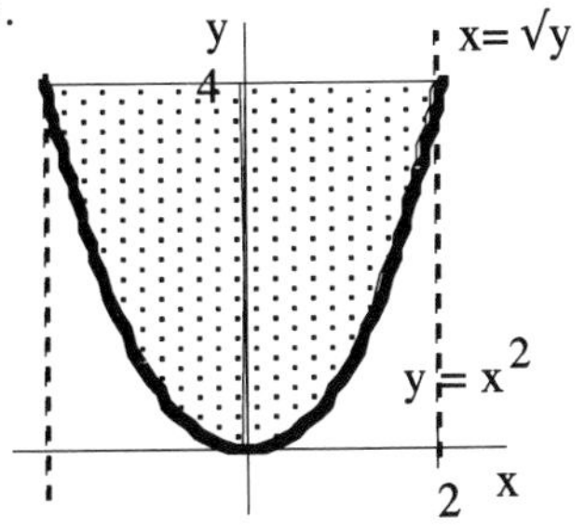

$\int\int f(x,y)\ dy\ dx$

$\int\int f(x,y)\ dx\ dy$

3.

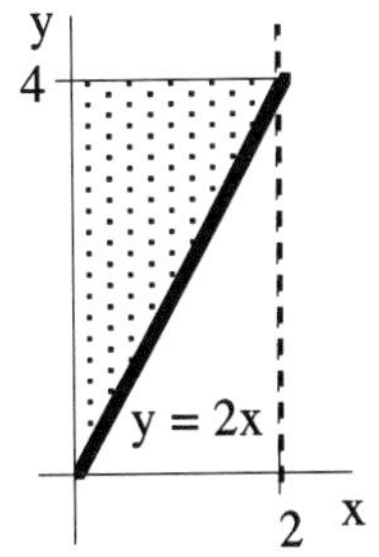

$\int\int f(x,y)\ dy\ dx$

$\int\int f(x,y)\ dx\ dy$

4.

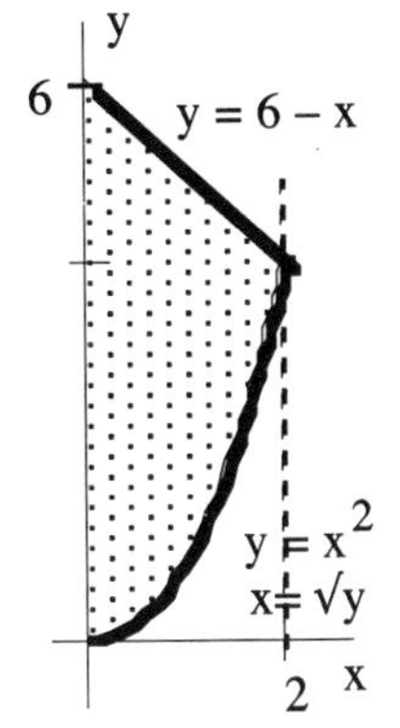

$\int\int f(x,y)\ dy\ dx$

$\int\int f(x,y)\ dx\ dy$

5.

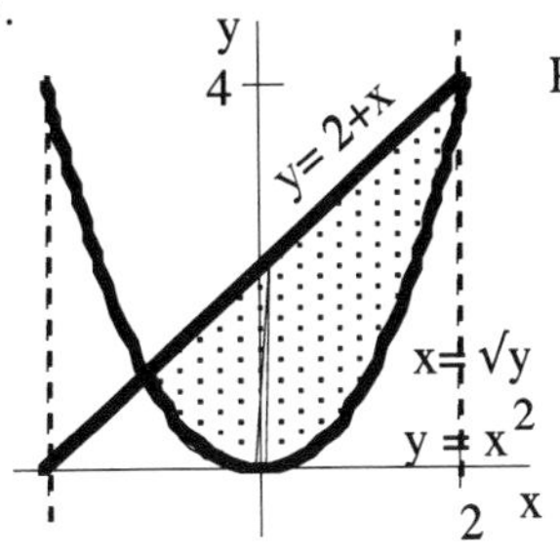

Pick the easier order of $\int$

$\int\int f(x,y)$

6. Sketch the domain of region of the integral, then change the order of integration.

$$\int_1^3\int_0^{x^3-1} f(x,y)\ dy\ dx = \int\int f(x,y)\ dx\ dy$$

7. Sketch the domain of region of the integral, then change the order of integration.

$$\int_0^2\int_1^{e^y} f(x,y)\ dx\ dy = \int\int f(x,y)\ dy\ dx$$